LASER SCANNER TECHNOLOGY

Edited by **J. Apolinar Munoz Rodriguez**

Laser Scanner Technology
Edited by J. Apolinar Munoz Rodriguez

Published by InTech

Janeza Trdine 9, 51000 Rijeka, Croatia

Publishing Process Manager Ivona Lovric

Technical Editor Teodora Smiljanic

Cover Designer InTech Design Team

Additional hard copies can be obtained from orders@intechopen.com

Laser Scanner Technology, Edited by J. Apolinar Munoz Rodriguez
p. cm.
ISBN 978-953-51-0280-9

For EU product safety concerns:
IN TECH d.o.o., Prolaz Marije Krucifikse Kozulić 3, 51000 Rijeka, Croatia,
info@intechopen.com or visit our website at intechopen.com.

Meet the editor

J. Apolinar Muñoz-Rodriguez received his BS degree in Computation and Systems Engineering from the Universidad del Bajio, his MS degree in Electrical Engineering from the Universidad de Guanajuato, and his PhD degree in Optics from the Universidad de Guanajuato - Centro de Investigaciones en Optica, A.C. in Leon Gto, Mexico. He worked in post-doctoral research as an associate researcher at Nanyang Technological University and now works as researcher in the Centro de Investigaciones en Optica, A.C. His current research interests include optical metrology, digital image processing, computer vision, artificial intelligence, mathematic morphology and visual cryptography. He has published more than forty scientific papers, six book chapters and edited one book. He has served as a reviewer in sixteen international journals and is a member of the National Researchers System (SNI) in Mexico.

Contents

Preface

Laser scanning technology has emerged as a research field in the science and engineering arena. Laser scanning performs a procedure to detect a surface using laser technology. A scanning device analyzes a real-world object to retrieve data on its shape and its appearance. The collected data can then be used to construct a digital three-dimensional model. The aim of the scanning is usually to create a point cloud on an object surface. Several technologies can be applied to perform the scanning of a scene. Among them are multi-faceted assessment, laser triangulation, 3D survey, range finder, tracking, LiDAR and so on. These methods are used extensively in applications such as industrial inspection, prototyping, quality control, reverse engineering, security systems and documentation of cultural buildings. Multiple scanning is sometimes performed via multiple cameras to obtain all slides of the scene under study. Common references are used to perform the alignment and the surface connection. Typically, optical tests are used to elucidate the power of laser scanning technology in the modern industry and in the research laboratories.

This book reports on the recent contributions of laser scanning technology in different areas around the world. The main topics of laser scanning described in this volume include full body scanning, traffic management, 3D survey process, bridge monitoring, tracking of scanning, human sensing, three-dimensional modelling, glacier monitoring and digitizing heritage monuments.

Body scanning aims to develop a device capable of producing sound waves which retract when a surface is hit. Thus, areas of high density could be shown by means of an oscilloscope. Also, it can be capable of identifying dangerous tools, whilst protecting personal data. This technique includes the current legislative matters pertaining to safety and security levels in the aviation sector.

Traffic management is performed by laser scanning, which provides a set of sensor communications able to detect and classify vehicles in real-time. Here, the reflected light by the vehicle is measured to determine vehicle shape. The profilometric method is based on the time elapsed between the emission and reception. The laser works in the infrared range, which is outside the visible range. Thus, possible distraction to road users is avoided. Laser scanner is configured by means of commands called telegrams, which contain the control commands.

The 3D survey process analyzes the 3D scanning inside industrial designs. Here, the role of the digital acquisition and modeling approach are discussed. The technique is based on the progressive refinements of sketches, in which the manual creativity of the designer is especially exploited. In this case, the modeling allows the recreation of a digital mould of the physical models. Then, reverse modeling updates the original object with the modifications.

Bridge monitoring is performed by a laser range finder and a scanning laser vibrometer. The laser range finder is based on the transmission and receipt of pulsed lights. Here, the point cloud is determined based on heterodyne laser beam phase shifts. The surface vibration is detected based on the measurement of Doppler effects of a returning continuous laser beam from a moving target. Thus, the damage within a bridge is performed based on the natural modal behaviors.

The section on tracking technique aims to give an overview of the multiple hypothesis tracking approach. The tracking algorithm takes as input a list of clustered lasers scanner measurements. This algorithm forms a track whenever there is "enough" evidence that a sequence of measurements represents a real target. Additionally, by using the appropriate filtering techniques the tracking algorithm estimates the kinematic state of the formed track. The hypothesis oriented presents an exhaustive method of enumerating all possible assignment track to measurement combinations. When a new measurement set is received, observations that fall within a track's validation region set a possible measurement to track assignment.

The human sensing section of the book is focused on human tracking and gait analysis. This method is well-suited to privacy protection because it does not use images but only range data. The experiment is performed in a crowded railway station in Tokyo in order to evaluate its effectiveness. This sensor measures distance by using the time of flight of laser light, and can also perform wide-area measurements. Here, a flat area off the floor is scanned with the sensors set on the floor. As a result, range data for ankles, including both static objects and moving objects are obtained. This procedure is performed by using multiple laser scanners in order to minimize occlusions in the wide-area sensing.

The section on three-dimensional modelling provides a general description of laser scanner methodology, dealing with the data capture stages and information processing. This technique deals with 3D laser scanner modelling in a typical survey and the improvement of laser scanner modelling. Here, the device carries out sweeps of the survey zone by laser beam to generate a 3D point cloud. The x, y and z coordinates of each point are stored in a reference system together with the reflected beam intensity and these provide us with a massive automated 3D metric data acquisition system.

The glacier monitoring dealt with in this book was performed by terrestrial laser scanning to retrieve surface data at the Belvedere Glacier, in the Italian Alps. The monitoring consisted of measurements of the central/lower part of the glacial body, of the glacier snout and measurement of a landslide affecting the terminal moraine of the Locce Glacier. The measurements were carried out by using GPS which provided only punctual evaluations of the phenomena. Here, the laser scanner worked as a range finder, projecting thousands of laser pulses towards the survey target and through the reflected beams, capable of reconstructing its geometry. This experiment performs the comprehension of glacier dynamics in time. Here, the laser scanner works without contact with the surveyed objects, so allowing to measure dangerous or inaccessible places too.

The digitizing monument was performed by the LiDAR system. The device provides a carrier of 3D laser scanner with lift-able and stable character. It was fixed on the ground to avoid vibration. Furthermore, it could be adjusted in height to provide the best survey. The device with a platform on the top can fix a scanner on the lift-pole. Each of the multi-supported legs implements a dynamic device for lifting up the truck so as to provide stable survey implementation without the shaking caused by rubber tires. For the work flows of 3D data acquisition of cultural heritage, the quantity and density of point cloud control the detail of 3D modelling by reverse engineering. In the registration of laser scanning, both software and hardware influence the representation and accuracy of point cloud. The purpose of 3D modelling is building up the same size of physical heritage in digital form that can export as visualization and virtual reality on the Internet.

Dr J. Apolinar Muñoz Rodriguez
Research Centre in Optics, A. C., Leon Guanajuato,
México

1

Foot Sole Scanning by Laser Metrology and Computer Algorithms

J. Apolinar Muñoz Rodriguez and Francisco Cháves Gutierrez
Centro de Investigaciones en Optica
México

1. Introduction

In recent years, the foot sole anatomy has been used in applications such as footwear manufacturing, person identification, foot deformation and health care (Xiong S., et. al 2010). Typically, the foot sole shape is used to make profitable shoes (Lee Au E. Y., et. al 2007). Nowadays, several researches have been developed in computer vision to detect the foot sole. Methods such as multiple camera, photogrametry, laser trinagulation, fringe projection habe been used to determine the foot sole. The ligthings methods, compute the surface of foot sole based on the calibration via perspective projection. Thus, the calibration provides the vision parameters to compute the object surface. These lighting methods provide a smoothed surface of the foot sole and some details are not profiled (Chen M. J. L. et. al, 2003). Furthermore, these systems do not provide the fitting of the shoe sole mould to the foot sole. The laser line technique can be used to improve these matters. In this technique, the three-dimensional surface is computed based on laser triangulation and the calibration via perspective projection model (Zhou F., et. al, 2005). The calibration includes the measurement of the distances of the setup geometry, focal distance, camera orientation, pixel scale and image centre. Typically, the calibration is performed by an external procedure to the vision system. This procedure is carried out by the recognition of the calibrated references. Typically, the setup geometry is modified online. In this case, a re-calibration is necessary to compute the three-dimensional surface (Wang G. et. al, 2004). But, the calibration references do not exist online and the calibration should be repeated. In this case, the perspective projection does not provide online re-calibration.

In the proposed chapter, a review of our methods for three-vision are applied to determine the foot sole. The proposed vision system moved in x-axis to perform the scanning. Here, an automatic re-calibration is performed to overcome limitations caused by online geometric modifications and physical measurements. This technique, retrieves the foot sole by laser line metrology and Bezier networks. Also, the modifications on the setup geometry are re-calibrated based on the data provided by the network. Thus, the accuracy and the performance are improved. The vision system is implemented based on laser line to scann the foot sole. Here, the three-dimensional model is performed by approximation networks based on behavior of the line position (Muñoz Rodríguez J.A, et. al, 2007). Also, the extrinsic and intrinsic parameters are computed based on the data provided by the network. Thus, setup modifications are re-calibrated during the visión task and external procedures are avoided. In this manner, limitations caused by online modification are avoided. Thus, a contribution is achieved in the field calibration of vision parameters for foot sole scanning. This technique is evaluated respect to the perspective projection model, which is used in laser triangulation for foot sole profiling. The error analysis is based on the root mean squared of error using references of a contact method. This technique is applied to fitting the shoe last to the foot sole. This procedure is carried out by Bezier curves using the foot sole data, which are outside of the contour of the shoe last. Also, foot length, foot width, heel width, mid foot width, arch length and arch angle are measured to achieve the fitting of the shoe last. In this manner, a profitable shoe last is achieved for footwear manufacturing. Thus, a contribution is achieved by the vision system on this field. Also, the time processing is short to obtain the foot sole topography.

2. Basic description of vision system

The proposed vision system includes an electromechanical device, a glass window, a CCD camera, a laser line projector and a computer. Fig. 1 shows the experimental setup to perform the foot sole scanning. In this setup, the laser line is projected perpendicularly to the glass window. Here, the laser line pass by the glass window and then it is projected on the foot sole. Then, the CCD camera captures the projected line on the foot sole. The camera and the laser diode are mounted on a linear slide, which can be moved in x-axis, y-axis and z-axis. To perform the scanning, the laser and the camera are moved in x-axis. From this scanning, a Bezier network computes the foot sole topography based on the position of laser line. Also, the three-dimensional modeling and the calibration are performed by means of the network. This modeling is performed based on the setup geometry shown in Fig.2. In this geometry, the x-axis and y-axis are located on the glass window, which correspond to the reference plane. The z-axis is located perpendicularly to the reference plane. The focal length f is the distance between the lens and the image plane. The center of image plane is indicated by x_c in the x-axis. The distance between the laser line and the optical axis is indicated by ℓ_a. In the z-axis, the foot sole depth is indicated by $h(x, y)$. The distance between the lens and the foot surface is defined by $z_i = D+h(x,y)$. In the proposed modeling, the surface depth is computed based on the position of laser line (Muñoz Rodríguez J.A, et. al, 2003). When the laser line is projected on the point A of the reference plane, the line position in the image plane is x_A. When the laser line is projected on the point B of the foot sole, the line position is shifted to the position x_B. This line shifting is described by next expression

$$s_i = x_A - x_B \tag{1}$$

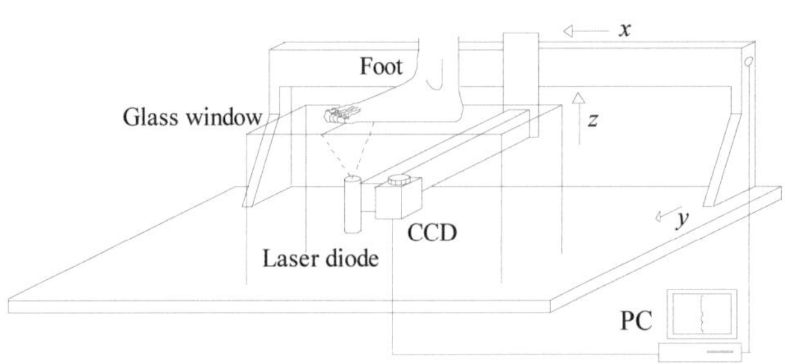

Fig. 1. Experimental setup for foot sole scanning.

Based on the camera position, the shifting s_i is directly proportional to the foot sole depth $h(x, y)$. This relationship is performed by a Bezier network to compute the surface depth. The network to compute the foot sole depth is described in section 3.

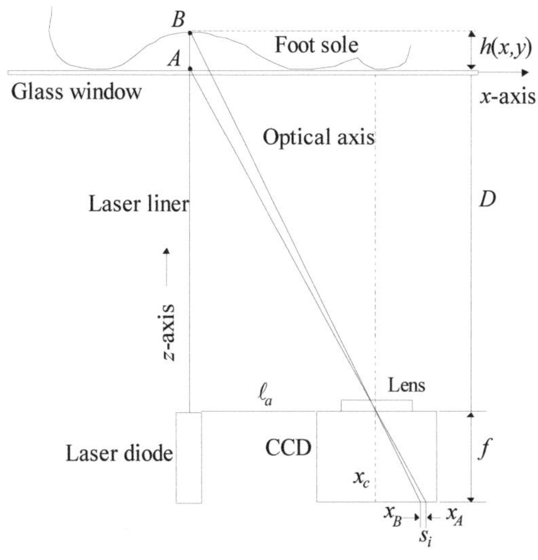

Fig. 2. Geometry of the experimental setup to compute the surface depth.

To compute the shifting, the line position x_B and x_A are detected in the image. This procedure is performed by measuring the intensity maximum in each row of the image. To carry it out, the pixels are approximated to a continuous function. Then, first and second derivative are computed to obtain the maximum. In this case, the pixels are represented by (x_0, I_0), (x_1, I_1),......, (x_n, I_n), where x_i is the pixel position, I_i is the pixel intensity and n is the pixel number. Bezier curves are used to fit a continuous function from the pixel intensity. The Bezier curves (Mortenson M. E., 1997) are is described by

$$P(u) = \sum_{i=0}^{n} \binom{n}{i} (1-u)^{n-i} u^i p_i, \quad \binom{n}{i} = \frac{n!}{i!(n-i)!}, \quad 0 \leq u \leq 1. \tag{2}$$

By applying the definition of Eq.(2), two equations are obtained, one for x and one for I:

$$x(u) = \binom{n}{0}(1-u)^n u^0 x_0 + \binom{n}{1}(1-u)^{n-1} u\, x_1 + \ldots\ldots + \binom{n}{n}(1-u)^0 u^n x_n, \quad 0 \leq u \leq 1. \tag{3}$$

$$I(u) = \binom{n}{0}(1-u)^n u^0 I_0 + \binom{n}{1}(1-u)^{n-1} u\, I_1 + \ldots\ldots + \binom{n}{n}(1-u)^0 u^n I_n, \quad 0 \leq u \leq 1. \tag{4}$$

Eq.(3) represents the pixel position and Eq.(4) represents the pixel intensity. Based on these equations, a continuous function is fitted from the pixels a laser line image. To carry it out, the pixel position $x_0, x_1, x_2,\ldots\ldots,x_n$, are substituted into Eq. (3) and the pixel intensity $I_0, I_1, I_2,\ldots,I_n$, are substituted into Eq.(4). These two equations are evaluated in the interval $0 \leq u \leq 1$ and the result of this curve is a concave function. Therefore, the second derivative $I''(u)$ is positive and the peak is a maximum global. In this manner, the maximum of $I(u)$ is computed based on the first derivative $I'(u)$=0 (Frederick H., et. al, 1982). To find the derivative $I'(u)$ = 0, the bisection method is applied. The Bezier function is defined in the interval $0 \le u \le 1$. Therefore, the initial values are u_i=0 and u_f=1. Then, the middle point is computed by $u^*=(u_i+u_f)/2$ to find u that converges to the $I'(u)$=0. Then, the first derivative $I'(u)$ is evaluated in middle point u^*. If the derivative $I'(u = u^*)$ is positive, then $u_i = u^*$. If the derivative $I'(u = u^*)$ is negative, then $u_f = u^*$. The next middle point u^* is obtained from the last pair of values u_i and u_f. These steps are repeated until to find $I'(u)$=0 based on a tolerance value. The value u=u^* where $I'(u)$ = 0 is substituted into Eq.(3) to determine the position of the intensity maximum $x(u)$.

3. Bezier network of a laser line

The network to perform the three-dimensional modeling includes an input vector, two parametric inputs, a hidden layer and an output layer. Each layer of this network is performed as follow. The input includes: the surface depth h_0, h_1, h_2,...,h_n, the line shifting s_i, and parametric values (u ,v). The value u is proportional to the shifting s_i according to the camera position in x-axis. The line shifting s_i is obtained by the procedure described in section 2. This shifting is represented by a value u, which is obtained by the next expresion.

$$u = a_0 + a_1 s, \tag{5}$$

where a_0 and a_1 are constant to be determined. By means of two values s and its respective value u, Eq.(5) is determined. The Bezier curves are defined in the interval $0 \le u \le 1$. Therefore, u=0 for the first line shifting and u=1 for the last line shifting s_n. Substituting these values in Eq.(5), two equation with two unknown constants are obtained. Solving these equations, a_0 and a_1 are determined. Thus, for each shifting s_i, a value u is computed via Eq.(5). The value v is proportional to the coordinate y in y-axis. This relationship is described by the next expression

$$v = b_0 + b_1 y \tag{6}$$

where b_0 and b_1 are the unknown constants. Using two values y and its respective v from Bezier curves, Eq.(6) is determined. Bezier curves are defined in the interval $0 \le v \le 1$. Therefore, v=0 for the first y and v=1 for the last y. Substituting these two values y and its respective v in Eq.(6), two equation with two unknown constants are obtained. Solving these equations, b_0 and b_1 are determined. Thus, for each coordinate y, a value v is computed via Eq.(6). The hidden layer is built by a Bezier basis function, which is described by

$$\mathrm{B}_{ij} = B_i(u) B_j(v), \tag{7}$$

where

$$B_i(u) = \binom{n}{i} u^i (1-u)^{n-i}, \; B_j(v) = \binom{m}{j} u^j (1-v)^{m-j}, \; \binom{n}{i} = \frac{n!}{i!(n-i)!}, \; \binom{m}{j} = \frac{m!}{j!(m-j)!}$$

The output layer is obtained by the summation of the neurons, which are multiplied by a weight. Thus, the output response is the surface dimension given by next expression

$$h(u,v) = \sum_{i=0}^{n} \sum_{j=0}^{m} w_{ij} h_i \, B_i(u) B_j(v), \quad 0 \le u \le 1, 0 \le v \le 1, \tag{8}$$

where w_{ij} are the weights, h_i is the known object surface, $B_i(u)$ and $B_j(v)$ are the Bezier basis function represented by Eq.(7). To construct the complete network Eq.(8), the suitable weights w_{ij} should be determined. To carry it out, the network is being forced to produce the correct surface depth $h(x,y)$ by means of an adjustment mechanism. This procedure is performed based on the shifting s_i produced by the known depth h_i. The coordinate y is the pixel number in the image of the vertical laser line. Then, the shifting s_i is converted to a value u via Eq.(5) and the coordinate y is converted to a value v via Eq.(6). Then, the depth h_i and its coordinate (u, v) are substituted in Eq.(8). Thus, next output $H(u, v)$ is generated

$$\mathrm{H}(u,v) = w_{00} h_0 B_0(u) B_0(v) + w_{01} h_1 B_0(u) B_1(v) + ,\ldots, + w_{0n} h_n B_0(u) B_n(v) +,\ldots, + \tag{9}$$
$$w_{m0} h_0 B_m(u) B_0(v) + ,\ldots, + w_{mn} h_n B_m(u) B_n(v)$$

Based on this output, a linear system is obtained and Eq.(9) is solved. In this manner, the weights w_{ij} are determined. Thus, the Bezier network H (u, v) has been completed. The result of this network is a model that computes the foot sole depth $h(x,y)$ based on a line shifting s_i. Fig. 3(a) shows a projected laser line on a foot sole mould. Here, the network is applied to obtain a profile of the mould of foot sole. To carry it out, the shifting s_i is detected in each image row, which corresponds to the coordinate y. Then, the shifting s_i is converted to a value u and the coordinate y is converted to the value v. Then, these values are substituted in the network Eq.(9) to compute the surface depth shown in Fig. 3(b). To know the accuracy, the data provided by the network are compared with the data provided by a contact method. The accuracy is computed based on a root means squared error (*rms*) (Masters T., 1993) by

$$rms = \sqrt{\frac{1}{n} \sum_{i=1}^{n} (ho_i - hc_i)^2} \tag{10}$$

where ho_i is the data provided by a coordinate measure machine (CMM), hc_i is the calculated data by the network and n is the number of data. For the data shown in Fig. 3(b), the error is a rms=0.118 *mm*.

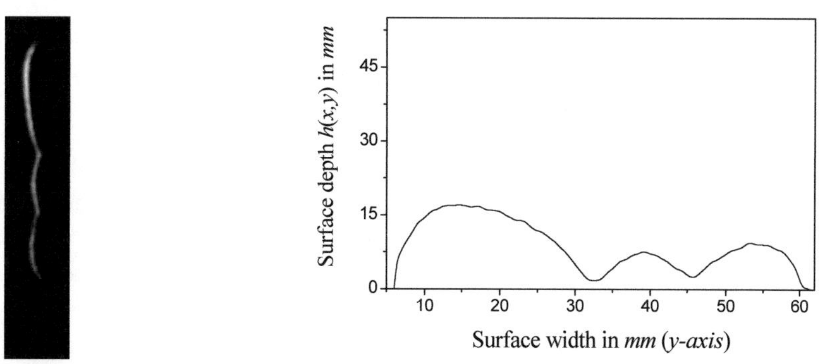

(a) (b)

Fig. 3. (a) Laser line projected on the foot sole mould. (b) Surface depth computed by the network form the laser line Fig.3 (a).

The depth resolution is provided by the network based on the minimum line shifting. In this setup, the magnitude of s_i is based on the camera position. Therefore, the depth resolution is defined as the minimum s_i based on the distance ℓ_a. In this case, the network is built using the minimum s_i and the maximum s_i at distance ℓ_a of the camera position. For this configuration, a shifting s_i=0.38 pixels is detected from the reference plane. To compute the surface depth, this s_i is substituted in the Eq.(8) and the result is h =0.28 *mm*.

4. Calibration of vision parameters

In the lighting methods, the object surface is computed based on the calibration of the vision parameters. This procedure is performed based on the perspective projection model (Mclvor A. M., 2002). Thus, the distances of the setup geometry, focal distance, camera orientation, pixel scale and image centre are determined. The calibration is carried out by means of a transformation matrix and calibrated references. This calibration is performed by the transformation $P_w=(x_w, y_w, z_w)$ to the camera coordinates $P_c=(x_c, y_c, z_c)$ by $P_c= R\cdot Pw+t$. Where R is a rotation matrix and t is a translation vector. Here, the transformation of P_c to the image coordinates (X_u,Y_u) is given by $X_u=fx_c/z_c$ and $Y_u=fy_c/z_c$. Considering radial distortion, the image coordinates are represented by $X_d+D_x=X_u$ and $Y_d+D_y=Y_u$. Where $D_x=X_d(\delta_1 r^2+\delta_2 r^4+\ldots)$, $D_y=Y_d(\delta_1 r^2+\delta_2 r^4+\ldots)$ and $r=(X_d^2+Y_d^2)^{1/2}$. In these expressions, X_d and Y_d are the distorted image coordinates. Also, these image coordinates in pixels should be converted to real coordinates by means a scale factor η. Thus, the parameters to be calibrated are the matrix R, the vector t, the focal distance f, the distortion coefficient δ_i, the image center (c_x, c_y) and the scale factor η. This procedure is carried out by using calibrated references such as rectangles or circles (Zhou F., et. al, 2005). In this step, coordinates of the calibrated references are detected. Then, the vision parameters are passed to the vision system to compute the object surface. The setup modifications play an important role to get good sensibility. Due to these modifications, a re-calibration is necessary to compute the three-dimensional shape (Wang G., et. al, 2004). But, the calibration references do not exist for online measurements. Therefore, an online re-calibration is not possible. In this case, the calibration is repeated again via calibrated references. Therefore, an automatic re-calibration is necessary to overcome limitations caused by online geometric modifications.

In the proposed setup, camera is placed around the central axis of the foot to perform the scanning. Here, the electromechanical device provides distance modification between the surface and the optical devices. When the vision system is moved in x-axis or y-axis, the setup geometry is not modified. But, when the vision system is moved in z-axis, the setup geometry is modified. Here, a re-calibration is performed based on the data provided by the network (Muñoz Rodríguez J. A., 2011). The camera orientation is described based on the geometry shown in Fig. 4. In this geometry, the shifting $s_1, s_2,\ldots, s_n$ correspond to the depth $h_1, h_2,\ldots h_n$, respectively. The distance between the lens and the surface depth is indicated by $z_i=D+h_i$. In this case, the optical axis is perpendicular to the reference plane. Thus, the depth h_i has a projection k_i in the reference plane. In this manner, the relation $(h_i/k_i)= [(D+h_i)/\ell_a]= [f/(\eta x_B-\eta x_c)]$ is obtained. In this expression, η is the pixel size. Also, the displacement can be defined by $s_i=(x_A-x_c)-(x_B-x_c)$. Thus, the projection k_i in the reference plane is deduced by

$$k_i=\frac{\eta(x_B-x_c)h_i}{f}=\frac{\eta[(x_A-x_c)-s_i]h_i}{f} \tag{11}$$

Considering radial distortion, $X_A = x_A + \delta_A$ and $X_B = x_B + \delta_B$, where x is the distorted image coordinate and δ is the distortion. Thus, the displacement is defined by $S_i=(X_A - x_c)-(X_B - x_c)$. Therefore, the projection k_i Eq.(11) is rewritten by

$$ku_i = \frac{\eta(X_B - x_c)h_i}{f} = \frac{\eta[(X_A - x_c) - S_i]h_i}{f} \tag{12}$$

From Eq.(12) f, η, x_c, X_A are constant and ku_i is the projection undistorted. In this case, a linear S_i produces a linear ku_i and the derivative ku_i respect to S_i dku/dS is a constant.

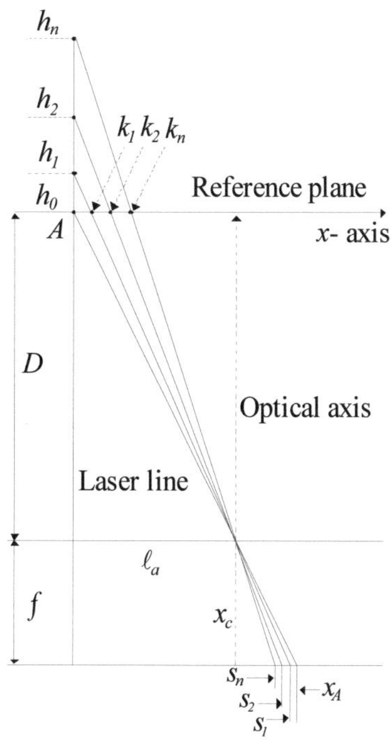

Fig. 4. Geometry of the relationship between the line shifting and the surface depth.

The camera orientation in y-axis is performed based on the geometry of Fig. 5. In this case, a surface pattern in the laser line is moved in steps y_i according to the depth h_i in the y-axis. Thus, the shifting pattern is computed by $t_i=\eta(y_A-y_i)=\eta(y_A-y_c)-\eta(y_i-y_c)$. Based on this geometry, the next expression is obtained by triangulation

$$q_i = \frac{\eta(y_i - y_c)h_i}{f} = \frac{\eta[(y_A - y_c) - t_i]h_i}{f}, \tag{13}$$

Considering radial distortion, $Y_A = y_A + \delta_A$ and $Y_i = y_i + \delta_i$, where y is the distorted image coordinate and δ is the distortion. In this case, $T_i = \eta(Y_A - y_c) - \eta(Y_i - y_c)$. Thus, the projection q_i in the reference plane is deduced by

$$qu_i = \frac{\eta(Y_i - y_c)h_i}{f} = \frac{\eta[(Y_A - y_c) - T_i]h_i}{f} \tag{14}$$

From Eq.(14) f, η, y_c, Y_A are constant. In this case, a linear T_i produces a linear qu_i and the derivative dqu/dT is a constant. Therefore, the camera orientation is defined by the dku/dS =constant and dqu/dT = constant.

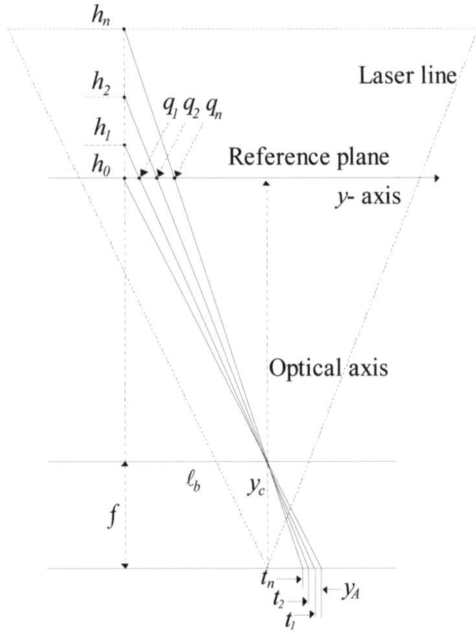

Fig. 5. Geometry of a perpendicular optical axis to the reference plane in y-axis.

For the camera orientation in x-axis, k_i is computed via Eq.(11) using the surface h_i provided by the network based on s_i. Due to the distortion, the derivative dk/ds is not exactly a constant. But, this derivative is the more similar to a constant. For the camera orientation in y-axis, k_i is computed via Eq.(13). Also, the derivative dq/dt is not exactly a constant. But, this derivative is the more similar to a constant. Therefore, the camera orientation perpendicular to reference plane is defined when the dk/ds and dq/dt are very similar to constant according to tolerance. To compute dku/ds=constant, the derivative of first projection k_1 is used. Thus, dku_i/ds = dk_1/ds = constant for i=1,2,...n and ku_i=i*k_1. The first derivative of k_i is selected because the distortion is observed in the last pixels of the image. Thus, using the terms from Eq.(11) and Eq.(12), the next expression is obtained

$$h_i = \frac{f\, ku_i}{\eta(x_B + \delta_B - x_c)} = \frac{fk_i}{\eta(x_B - x_c)}, \tag{15}$$

From Eq.(15), the distortion δ_B can be computed by the next expression

$$\delta_B = \frac{(x_B - x_c)(k_i - ku_i)}{k_i}, \tag{16}$$

From the terms s_i=(x_A−x_c) and S_i=(X_A −x_c)-(X_B −x_c)=(x_A−x_c)+(δ_A-δ_B), the expression S_i = s_i + (δ_A-δ_B) is obtained. Thus, using the terms from Eq.(11) and Eq.(12), the next the distortion δ_B can be computed by the next expression

$$h_i = \frac{f\,ku_i}{\eta[(X_A - x_c) - S_i]} = \frac{fk_i}{\eta[(x_B - x_c) - s_i]},\ \delta_B = \frac{ku_i}{k_i}(x_A - x_c - s_i) - (x_A - x_c) + s_i \qquad (17)$$

In this manner, the orientation and distortion have been deduced. Based on a perpendicular optical axis, the vision parameters are deduced. This procedure is carried out based on $S_i = s_i + (\delta_A - \delta_B)$ and the setup geometry Fig.4, which is described by

$$\frac{D + h_i}{\ell_a} = \frac{f}{\eta(X_B - x_c)} = \frac{D + h_i + f}{\eta(X_B - x_c) + \ell_a} = \frac{D + h_i + f}{\eta(X_A - x_c - S_i) + \ell_a}, \qquad (18)$$

From this equation η is the scale factor and D is the distance from the lens to the reference plane. The constant D, ℓ_a, f, η, x_c are determined by rewritten Eq.(18) in the next equation system

$$\begin{aligned}
h_0 &= \frac{f\,\ell_a}{\eta(X_A - x_c - S_0)} - D \\
h_1 &= \frac{f\,\ell_a}{\eta(X_A - x_c - S_1)} - D \\
h_2 &= \frac{f\,\ell_a}{\eta(X_A - x_c - S_2)} - D \\
h_3 &= \frac{f\,\ell_a}{\eta(X_A - x_c - S_3)} - D \\
h_4 &= \frac{f\,\ell_a}{\eta(X_A - x_c - S_4)} - D \\
h_5 &= \frac{f\,\ell_a}{\eta(X_A - x_c - S_5)} - D
\end{aligned} \qquad (19)$$

The values h_0, h_1,….,h_5, are computed by the network based on s_0, s_1,….,s_5 at the camera position x_c. Base on s_i, δ_A and δ_B, the values X_A and S_i are computed. Then, these values are substituted in Eq. (19) to solve the equation system. Thus, the constant D, ℓ_a, f, η, and x_c are determined. The coordinate y_c is computed by $T_i = \eta(Y_1 - y_c) - \eta(Y_{i+1} - y_c)$ from the geometry of Fig.5. From this expression, the data η, t_i, Y_i are known and y_c should be determined. This procedure is performed by the next equation system

$$\begin{aligned}
T_1 &= \eta(Y_1 - y_c) - \eta(Y_2 - y_c). \\
T_2 &= \eta(Y_1 - y_c) - \eta(Y_3 - y_c).
\end{aligned} \qquad (20)$$

The values T_1, T_2, Y_1, Y_2 and Y_3 are collected from the camera orientation in y-axis. These values are substituted in Eq.(20) to solve the equation system and the value y_c is determined. In this manner, the vision parameters are determined based on the data provided by network and image processing.

Typically, occlusions appear in the scanning of the foot sole. These occlusions are shown in the reported results by lighting methods (Cortizo E., et. al 2003). To avoid the occlusions, the camera position should be moved away from the surface in z-axis or toward laser line in x-

axis. When an occlusion appears, the electromechanical device moves the camera away from the surface in z-axis. In this case, initial geometry has been modified and the displacement magnitude is different from the initial configuration. Therefore, the displacement should be converted to a displacement of the initial configuration. This procedure is carried out by the calibration of vision parameters via Eq.(19). Thus, the vision system provides online geometric modifications to overcome occlusions. The accuracy provided by the network and by the perspective projection is shown in section 5.

5. Results of foot sole scanning

The proposed foot sole scanning is performed based on Bezier networks without physical measurements. In the setup, the scanning is carried out a 15 mW laser line in x-axis in steps of 1.69 mm. The laser line is captured and digitized in 256 gray levels. From each image, the network computes the surface depth based on the line shifting s_i. The depth resolution provided by the network is around of 0.28 mm. Thus, the foot sole is reconstructed by the network.

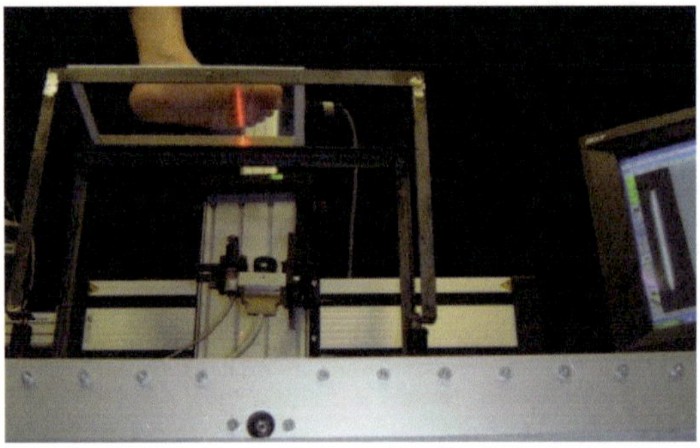

(a)

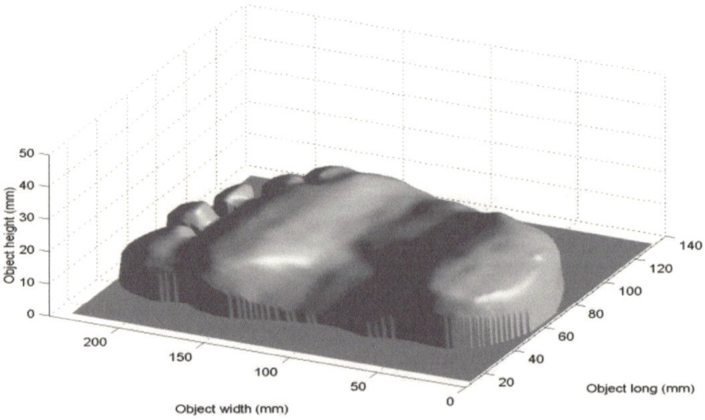

(b)

Fig. 6. (a)- Scanning of foot sole by the vision system. (b)-Three-dimensional shape of the foot sole.

The foot sole to be profiled by the vision system is the shown in Fig.6(a). To carry it out, the foot sole is scanned by a laser line. In this step, a set of images is captured by the CCD camera. From each image, the network computes the surface depth. This procedure is performed detecting line shifting via Eq.(1). Then, the shifting s_i is converted to a value u via Eq.(5). Also, the coordinate y of the laser line is converted to a value v via Eq.(6). Then, the values (u, v) are substituted in Eq.(9) to obtain the depth $H(u,v)$. Thus, the network produces a transverse section of the foot sole. Then, all transverse sections are stored in array memory to obtain the shape of the foot sole shown in Fig. 6(b). The scale of this figure is in *mm*. One hundred and forty six images are processed to obtain the whole foot sole. In this case, the number of data to compute the accuracy was n=490 and the error is a rms=0.137 *mm*.

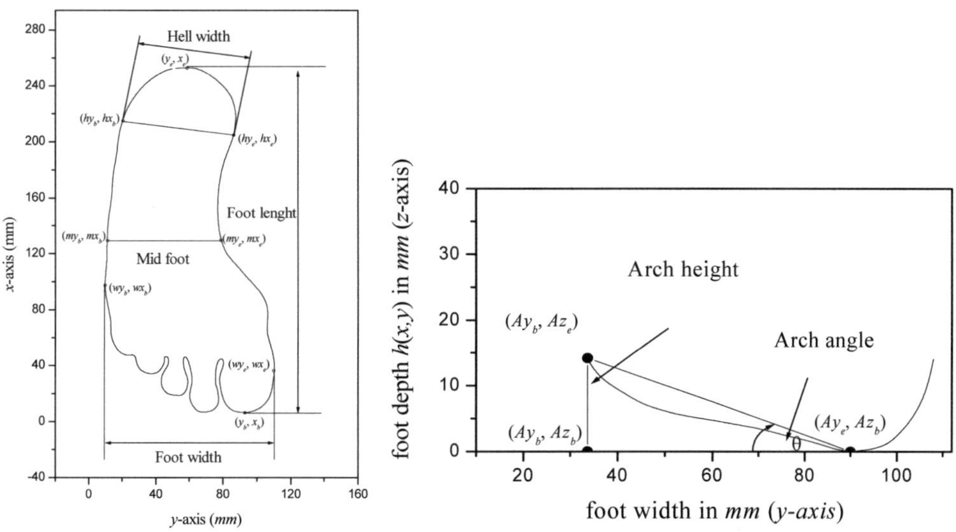

(a) (b)

Fig. 7. (a) Plantar contour of the foot sole, which include foot length, foot width, heel width, mid foot width, arch height and the arch angle. (b) Geometry of arch height and arch angle.

The morphological parameters that provide the model of a foot sole are the foot length, foot width, heel width, mid foot width, arch height and the arch angle (San-Tsung B. Y., Zhang M., Fan Y. B., Boone D. A., 2003). Therefore, these parameters should be computed from the foot sole to perform the fitting to the shoe mould. To compute these parameters, the contour of the foot sole shown in Fig. 7(a) is extracted via edge detection. The foot length is obtained by detecting the distance between the heel point to the maximum value of the contour in the direction of x-axis. Thus, the foot length is computed by the distance = x_e-x_b=247.8136 *mm*. The foot width is the breadth computed along the foot sole in y-axis. This distance is computed by detecting the maximum and minimum value in of the contour in y-axis. Thus, the foot width = wy_b - wy_e = 102.4685 *mm*. The mid foot width is the breadth of the foot sole at ½ of the length foot in x-axis. Therefore, the mid foot width is computed by detecting the coordinates of the two points of the contour in y-axis. Thus, mid foot width = my_b - my_e = 67.3522 *mm*. The heel width is the breadth of the foot sole at 1/6 of the foot length in x-axis. Therefore, in this position, the heel width is computed by detecting the coordinates of the two points of the contour in y-axis. Thus, heel width = hy_b - hy_e =66.0853 *mm*. The arch height is the distance from the reference plane to the apical arch in z-axis. Therefore, arch height is computed by detecting the maximum distance in z-axis. Thus, the arch height $=Az_e-Az_b=14.1472$ *mm*, which is the shown in Fig.7(b). The arch angle is the angle formed between the apical arch and the reference plane in z-axis. This angle is computed based on geometry shown in Fig.7(b). Thus, arch angle is computed as $\theta = tan^{-1}[(Az_e-Az_b)/(Ay_e-Ay_b)] = 14.7816$. The accuracy of the morphological parameters of the foot sole is around 0.1 *mm*. Typically, the accuracy required in foot sole measurement is 1 *mm* (Cortizo E., et. al, 2003). The accuracy provided by the Bezier network for the three-dimensional vision of the foot sole is around of a *rms*= 0.130 *mm*. This means an error minor than 1%. This result is good according to the reported results in the recent years (Lee Au E. Y., et. al, 2007).

The approach of this technique is to fitting a shoe last to the foot sole. This procedure is carried out by adjusting the shoe last to the plantar contour. The shoe last to be fitted is shown in Fig. 8(a). To carry it out, the shoe last is scanned in the vision system and a set of images is captured. The network computes the shoe last depth from each image. Thus, the network produces a transverse section of the shoe sole from each line image. Then, all transverse sections are stored in array memory to obtain the shoe sole shown in Fig. 8(b). The scale of this figure is in *mm*. The accuracy computed based on the error for the shoe sole mould is a *rms*=0.121 *mm*.

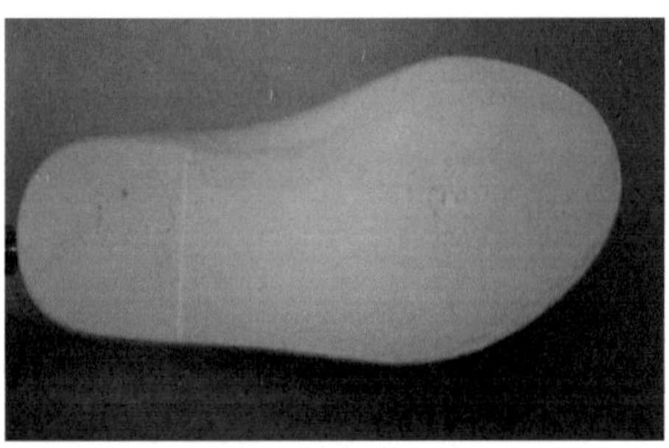

(a)

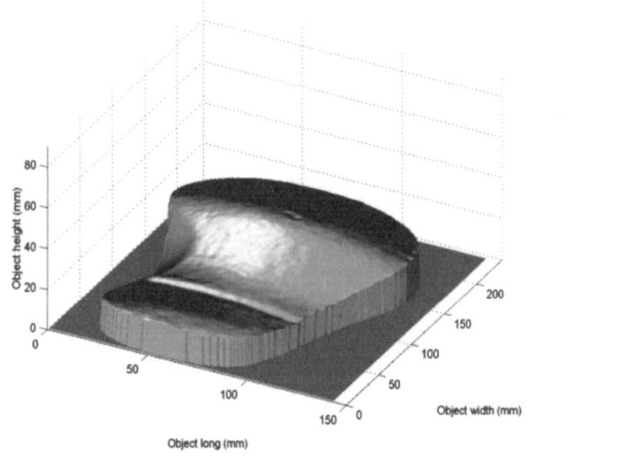

(b)

Fig. 8. (a) Shoe las to be profiled. (b) Three-dimensional shape of the shoe last.

Secondly, the morphologic parameters of the foot sole are compared with the parameters of the contour of shoe last. In this case, the parameters of the shoe mould are different to parameters of the foot sole. Therefore, the contour of the shoe last should be modified. To carry it out, the contours of the foot sole and the shoe last are overlapped. The procedure of this overlapping is shown in Fig.9(a). In this figure, the dot line corresponds to the foot sole contour and the continuous line corresponds to the shoe last. Thirdly, Bezier curves Eq.(3) and Eq.(4) are applied to perform the contour fitting. This procedure is carried out using the foot sole points, which are outside of the shoe mould contour. In this case, $x(u)$ corresponds to the coordinate in x-axis and $I(u) = y(u)$ corresponds to the coordinate in y-axis. Thus, the coordinates x are substituted into Eq.(3) and the coordinates y are substituted into Eq.(4).

These two equations are evaluated in the interval 0≤ u≤1 to fit the shoe sole mould. The result of the fitted contour of the shoe sole mould is shown in Fig. 9 (b). In this manner, the fitting of the shoe sole mould has been completed.

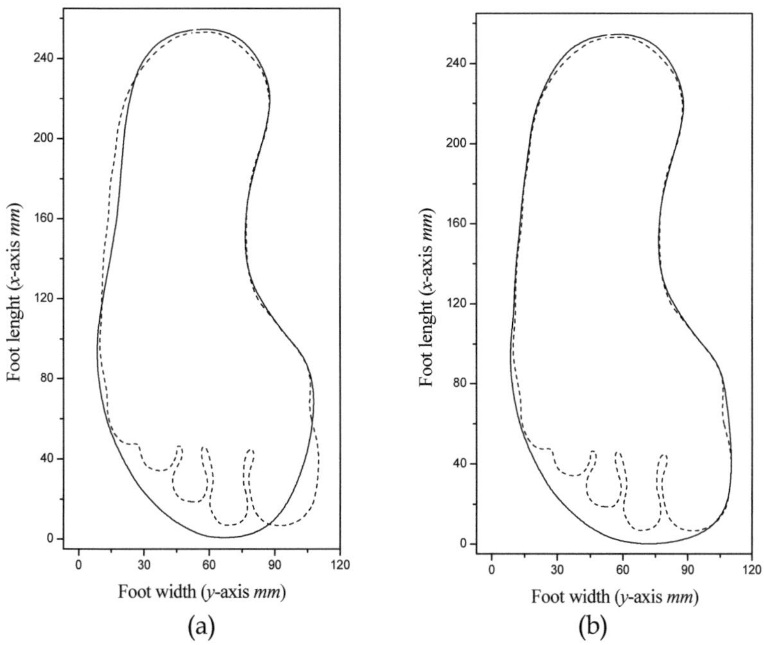

Fig. 9. (a) Overlapping of the foot sole contour on the shoe last contour. (b) Fitted contour of the shoe mould to the foot sole contour via Bezier curves.

The employed computer in this vision system is a PC to 1.8 GHz. The capture velocity of the camera is 34 fps. Also, the electromechanical device is moved at 34 steps per second. Each image of laser line is processed by the network in 0.010 sec. The shape of the foot sole is profiled in 5.75 sec. The time processing of the proposed system provides a fast method to perform the three-dimensional vision of the foot sole. In this procedure, distances of the geometry of the setup are not used to compute the surface depth. Therefore, the proposed reconstruction of the foot sole is easier than the reported techniques. The reported are based on the laser triangulation and perspective projection techniques and the vision parameters are collected by an external procedure. In this manner, the proposed technique is performed automatically by Bezier networks and the measurements on optical setup are avoided.

6. Conclusions

A technique for three-dimensional vision of the foot sole based on Bezier networks has been presented. The described technique here provides a valuable tool for the footwear manufacturing. In this procedure, a Bezier network provides the measurement of the foot sole and the vision parameters. This system performs the fitting of the shoe sole mould to the foot sole via Bezier curves. The automatic technique avoids the physical measurements on the setup. Also the proposed network provides the data to compute the geometric modifications due to the mobile setup. This procedure improves the performance and the accuracy of the results. It is because measurement errors are not passed to the vision system. The ability to detect the laser line with a sub-pixel resolution has been achieved by Bezier curves. This procedure is achieved with few operations. In this computational-optical setup, a good repeatability is achieved in each reconstructed surface. Therefore, this technique is performed in good manner.

7. Acknowledgment

J. Apolinar Muñoz Rodríguez would like to thank the financial support by CONACYT Mexico.

8. References

Chen Y., Medioni G.G., 1992. Objec Modeling by Registration of Multiple Range Images. Images and Vision Computing, 10(3): 145 - 155.

Chen M. J. L., Chen C.P.C., Lew H. L., Hsieh W. C., Yang W. P., Tang S. F. T., 2003. Measurement of forefoot varus angle by laser technology in people with flexible flatfoot. American Journal of Physical Medicine and Rehabilitation, 82: 842-846.

Cortizo E., Moreno Years A., Leopore J. R., Garavaglia M., 2003. Application of structured illumination method to study the topography of the sole of the foot during a walk. Opt. Lasers Eng. 40: 117 - 132.

Frederick H., Lieberman G. J., 1982. Introduction to operations research, McGraw- Hill, USA.

Lee Au E. Y., Ravindra Goonetilleke S., 2007. A qualitative study on the comfort and fit of ladies' dress shoes. Applied Ergonomics, 38: 687–696.

Masters T., 1993. Practical Neural Networks Recipes in C++, Academic Press, San Diego (CA) U.S.A.

Mclvor A. M., 2002 . Nonlinear calibration of a laser stripe profiler. *Opt. Eng.* 41: 205-212.

Mortenson M. E., 1997. Geometric Modeling, Willey, Second edition, U.S.A.

Muñoz Rodríguez J. A., Rodríguez-Vera R., 2003. Evaluation of the light line displacement location for object shape detection. Journal of Modern Optics, 50: 137-154.

Muñoz Rodríguez J. A., Rodríguez-Vera R., 2007. Shape detection using a light line and a Bezier approximation network. Imaging Science Journal, 55: 29-39.

Muñoz Rodriguez J. A., 2011. Online self-calibration for mobile vision based on laser imaging and computer algorithms. Opt. Lasers Eng. 49: 680-692.

San-Tsung B. Y., Zhang M., Fan Y. B., Boone D. A., 2003. Quantitative comparison of the plantar foot shapes under different weight bearing conditions. Journal of rehabilitation Research development, 40: 517-526.

Wang G., Hu Z., Wu F., Tsui H. T., 2004. Implementation and experimental study on fast object modeling based on multiple structured light stripes. Opt. Lasers Eng., 42: 627-638.

Xiong S., Zhao J., Jiang Z., Dong M., 2010. A computer-aided design system for foot-feature - based shoe last customization. Int. J. Adv. Manuf. Technol., 46: 11–19.

Zhang S., Huang P. S., 2006. Novel method for structured light system", Opt. Eng., 45: 083601.

Zhou F., Zhang G., Jiang J., 2005. Constructing feature points for calibration a structured light vision sensor by viewing a plane from unknown orientation. Opt. Lasers Eng., 43: 1056-1070.

2

Human Sensing in Crowd Using Laser Scanners

Katsuyuki Nakamura[1], Huijing Zhao[2],
Xiaowei Shao[3] and Ryosuke Shibasaki[3]
[1]*Central Research Lab., Hitachi, Ltd.**
[2]*Peking University*
[3]*The University of Tokyo*
[1,3]*Japan*
[2]*China*

1. Introduction

Human sensing is a critical technology to achieve surveillance systems, smart interfaces, and context-aware services. Although various vision-based methods have been proposed (Aggarwal & Cai, 1999)(Gavrila, 1999)(Yilmaz et al., 2006), tracking humans in a crowd is still extremely difficult. Figure 1 shows a snapshot of a railway station during rush hour, which is one of the hardest examples of human sensing.

Suppose a camera is diagonally set up on a low-ceiling, like the one in Fig. 1. Significant occlusion tends to occur in crowded places because pedestrians severely overlap each other. As a result, sufficiently high sensing performance cannot be achieved. On the other hand, if a camera is positioned to take measurements looking straight down in order to reduce occlusions, the viewing angle is limited. Furthermore, covering large areas using multiple

Fig. 1. Snapshot of a railway station during rush-hour

*This work was done while the first author was a Ph. D student in the University of Tokyo.

cameras is difficult due to the computational cost of data integration. These problems cannot be solved even using fisheye cameras or omni-directional cameras to expand the viewing angle. In addition, there are some cases in which cameras cannot be installed due to privacy concerns.

In this chapter, we propose a method to tackle these problems using laser scanners for human sensing in crowds. We especially focus on human tracking and gait analysis techniques. Our proposed method is well-suited to privacy protection because it does not use images but only range data. Moreover, because of the simple data structure of the laser scanner, we can easily integrate data even as the number of sensors increases, and real-time processing can be performed even when multiple sensors are used. Therefore, our method is especially suitable for crowd sensing in large public spaces such as railway stations, airports, museums, and other such facilities. That is to say, the above issues can be solved with our approach. We conducted an experiment of the proposed method in a crowded railway station in Tokyo in order to evaluate its effectiveness.

This chapter is structured as follows. Section 2 reviews existing research on human sensing. Section 3 proposes a method of tracking people in crowds using multiple laser scanners. Section 4 describes the gait analysis of tracked people. Section 5 is a performance evaluation in a crowded station.

2. Review

In this section, we briefly review the existing research on human sensing. The approaches are roughly classified into three types: vision-based, laser-based, and sensor fusion.

2.1 Vision-based approach

The first type is the vision-based approach using video cameras.

Much research has been done using this approach, although the number of people targeted for tracking has been relatively small. A well-known human detector was proposed by the article (Dalal et al., 2005). They used histograms of oriented (HOG) descriptors with a support vector machine (SVM). Felzenszwalb et al. extended this detector based on the deformable part-based model (Felzenszwalb et al., 2009). For tracking targets, mean shift trackers are widely used (Comaniciu et al., 2000).

In order to extend such approaches to track multiple targets in a crowd, we have to handle significant occlusion of each object. A typical solution is to utilize data association such as a Kalman filter, particle filter, or Markov chain Monte Carlo (MCMC) data association approach. Okuma et al. proposed a boosted particle filter that can track multiple targets by combining the AdaBoost detector and a particle filter (Okuma et al., 2004). Zhao et al. proposed a principled Bayesian framework that integrates a human appearance model, background appearance model, and a camera model (Zhao et al., 2008). The optimal solution is inferred by an MCMC-based approach. The result shows that up to 33 people are tracked in a complex scene. Kratz et al. models spatial-temporal motion patterns of crowds by using a hidden Markov model (HMM), and tracks individuals in extremely crowded scenes (Kratz & Nishino, 2010).

Other research on human sensing can be found in published surveys (Aggarwal & Cai, 1999)(Gavrila, 1999)(Yilmaz et al., 2006)(Szeliski, 2010).

2.2 Laser-based approach

The second approach is based on lasers. Fod et al. proposed laser-based tracking using multiple laser scanners (Fod et al., 2002). Their system measures a human's body, and tracks it by using a Kalman filter. More practical approaches for tracking people in crowds have been proposed by (Zhao & Shibasaki, 2005) (Nakamura et al., 2006). They measure pedestrians' feet to reduce the occlusion, and track individual pedestrians in crowds by recognizing their walking patterns. Experimental results showed that 150 people were simultaneously tracked with 80% precision in a railway station. Cui et al. combines laser-based tracking with a Rao-Blackwellized Monte Carlo data association filter (RBMC-DAF) to overcome tracking errors that occur when two closely situated data points are mixed (Cui et al., 2007). Song et al. have proposed a unified framework that couples semantic scene learning and tracking (Song, Shao, Zhao, Cui, Shibasaki & Zha, 2010). Their system dynamically learns semantic scene structures, and utilizes the learned model to increase the accuracy of tracking.

2.3 Sensor fusion

The third approach involves sensor fusion. Several techniques have been proposed to track multiple people by fusing the laser and vision approaches. Nakamura et al. used the mean shift visual tracker to support laser-based tracking (Nakamura et al., 2005). Cui et al. extended this approach by combining it with decision-level Bayesian fusion (Cui et al., 2008). Song et al. proposed a system of joint tracking and learning, which trains the classifiers that separate the targets who are in close proximately (Song, Zhao, Cui, Shao, Shibasaki & Zha, 2010). Trained visual classifiers are used to assist in laser-based tracking. Katabira et al. proposed an advanced air-conditioning system that combines laser scanners and wireless sensor networks (Katabira et al., 2006). The area that should be ventilated is determined by the humans' positions in the room and the temperature distribution.

2.4 Focus of this chapter

This chapter introduces the method of laser-based tracking and gait detection, which first emerged as a practical technique for sensing people in crowds that target more than a hundred people.

3. Laser-based people tracking

3.1 Sensing System

3.1.1 Human sensing using laser scanner

We use a laser scanner called SICK LMS-200. This sensor measures distance by using the time of flight (ToF) of laser light, and can also perform wide-area measurements (30-m distance). In addition, because the dispersion of the laser waves is minimal, resolution is high, and the angular resolution is 0.25° at the maximum. The wavelength of laser light is 905 nm

Fig. 2. Snapshot of human sensing using a laser scanner

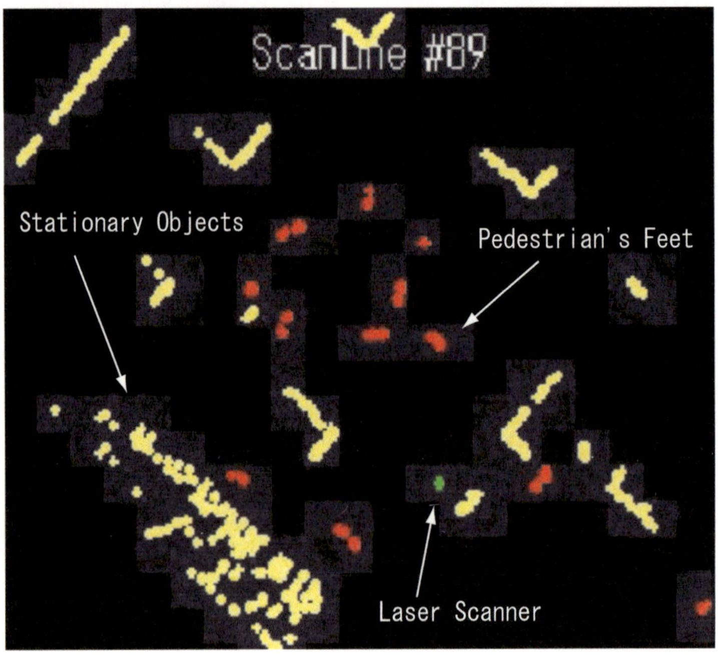

Fig. 3. Example of range data obtained with laser scanner

(near-infrared region), and it is a class 1A laser that is safe for peoples' eyes. The sampling frequency varies depending on the settings; it was 37.5 Hz in our case.

In the proposed method, a flat area about 16 cm off the floor is scanned with the sensors set on the floor. As a result, range data for ankles, including both static objects and moving objects, can be obtained. Figure 2 shows a sensing system, and Fig. 3 shows an example of the obtained range data.

3.1.2 Human sensing using multiple Laser scanners

We performed human sensing by using multiple laser scanners in order to minimize occlusions in the wide-area sensing. Suppose that each sensor obtains data at the same horizontal level; integration of multiple range data can be achieved using the following

Helmart transformation.

$$\begin{pmatrix} u \\ v \end{pmatrix} = m \begin{pmatrix} \cos\alpha & \sin\alpha \\ -\sin\alpha & \cos\alpha \end{pmatrix} \begin{pmatrix} x \\ y \end{pmatrix} + \begin{pmatrix} \Delta x \\ \Delta y \end{pmatrix} \tag{1}$$

where (x, y) represent a laser point in the local coordinate, (u, v) represent a transformed laser point in the global coordinate, m is a scaling factor, α is a rotation angle, and Δx and Δy are vectors shifted from the origin. These parameters are estimated by taking a visual correspondence using the rotation and shift of shared static objects (e.g. walls, pillars, etc.) measured by each sensor. The interface that performs this operation is built into the software. After the integration and synchronization, human tracking is conducted by the algorithms explained in Section 3.2.

3.2 Tracking algorithm

3.2.1 Tracking flow

Figure 4 illustrates the flow diagram of laser-based people tracking.

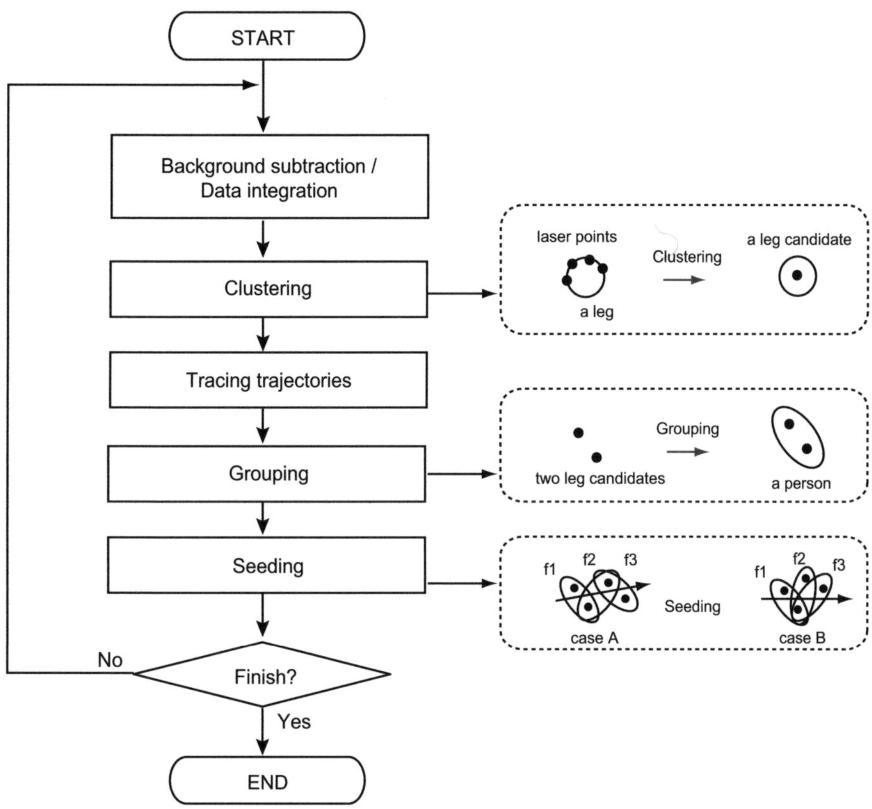

Fig. 4. Flow diagram of laser-based people tracking

First, background subtraction is conducted in each sensor in order to detect moving objects and integrate them to the global coordinates by using equation (1).

Second, several laser point strikes on the foot are clustered in order to extract one foot candidate. In this study, a group of points within a radius of 15 cm is clustered as a foot candidate. In practice, due to errors in the sensor calibration, there are several cases in which the foot of one human is not entirely within a cluster, or the feet of several humans are within the same cluster. However, such false positives or false negatives can be reduced during the subsequent stages, and there is no significant impact on tracking processing.

Third, the existing trajectories are extended to the current frame by using the Kalman filter. The details of this process are described in Section 3.2.2 to 3.2.4. By using a dynamic model of humans, the best foot candidate is integrated.

Last, if the foot candidate does not integrate into the existing trajectories, a new trajectory is created in the following steps of initial detection, and the initial state is set for the Kalman filter.

1. Grouping: When foot candidates not belonging to any of the existing trajectories are within 50 cm of each other, two foot candidates are grouped to create a human candidate. When there is a crowd, several human candidates can be created. Invalid human candidates are eliminated using the following seeding process.
2. Seeding: In consecutive frames, the candidates that satisfy the following two conditions are taken to represent the same human, and the connected centers of gravity for the two moving foot candidates represent a new trajectory. (a) At least one foot overlaps for a human candidate in consecutive frames (three or more frames). (b) The motion vector created by the other foot, which does not overlap, changes smoothly.

3.2.2 Walking model

When walking, pedestrians make progress by using one foot as an axis and moving the other foot. The two feet change roles alternately as they reach the ground and create a rhythmic walking motion. According to the ballistic walking model (Mochon & McMahon, 1980), muscle power acts when generating speed during the first half of the foot's movement, and the latter half of the foot's movement is passive. Figure 5 represents a simplified model of a pedestrian walking with attention given to the changes in the movement, speed, and acceleration of the feet.

In this research, the movement of the two feet is defined using four phases. Phase 1 is defined as going from a stationary state for both feet through the acceleration of the right foot alone, and to where the two feet are in alignment. Phase 2 is defined as when the right foot then decelerates and reaches the ground. In the same fashion, Phase 3 is defined as when the left foot accelerates, and Phase 4 is defined as when it decelerates.

The values v_L and v_R are the speed of the left and right feet respectively, a_L and a_R are their acceleration, and p_L and p_R are their positions. These variables are taken to have the values in the observed plane integrated using the process in Section 3.1.2.

Table 1 summarizes the transitions of state parameters in walking phases. When $|v_R| > |v_L|$, the right foot is in front, with the left foot serving as the axis. Here, the acceleration $|a_R|$ acting on the right foot can be taken to be a function of muscle power as based on the ballistic walking model. The authors define the acceleration of the right foot in walking phase 1 as $|a_R| = f_R\dot{v}$.

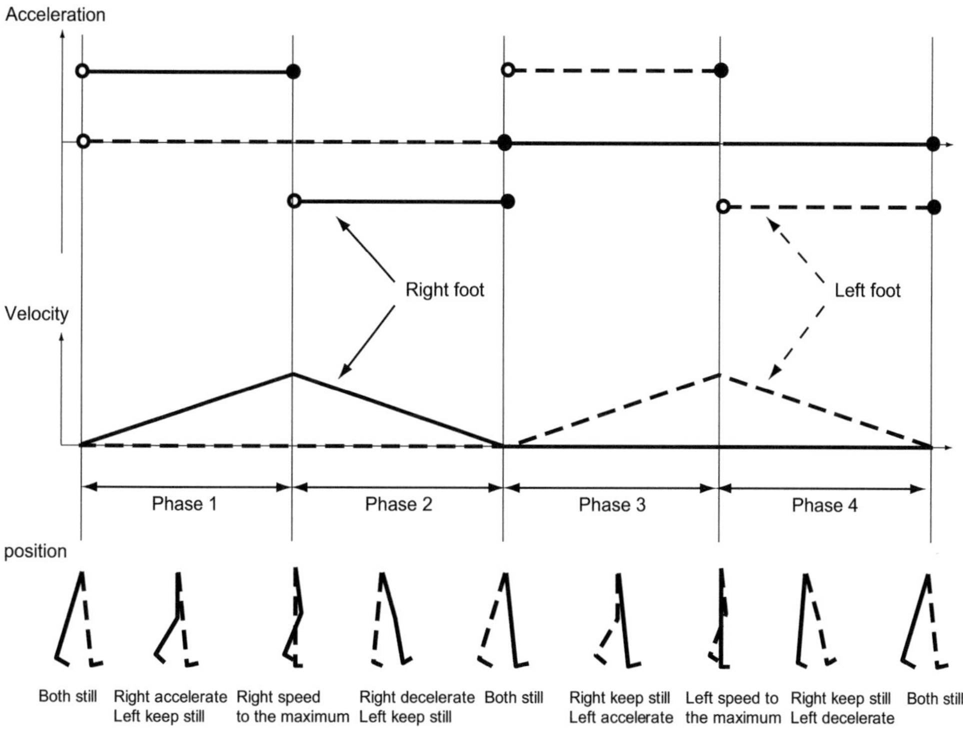

Fig. 5. Simplified walking model (top row: acceleration; middle row: velocity; bottom row: position)

Here, $f_{L/R}$ represents the acceleration function for the two feet defined in Equation (8), and $\dot{v}$ represents the unit direction vector. In walking phase 2, the right foot decelerates at a steady rate, and both feet are on the ground. The acceleration acting here has a negative value because of the effect of an external force other than muscle power. This is defined as $|a_R| = -f_R\dot{v}$. In walking phases 1 and 2, the left foot is virtually stationary, and thus, $|v_L| \approx 0$ and $|a_L| \approx 0$. When the right foot is the axis, the acceleration for the left foot in walking phases 3 and 4 for foot movement is $|a_L| = f_L\dot{v}$ and $|a_L| = -f_L\dot{v}$ and the velocity can be defined as $|v_R| \approx 0$ and $|a_R| \approx 0$. In the state in which both feet are on the ground, $|v_{L/R}| \approx 0$ and $|a_{L/R}| \approx 0$.

Phase	Phase1	Phase2	Phase3	Phase4
v_L	$\lvert v_L\rvert \approx 0$	$\lvert v_L\rvert \approx 0$	$\lvert v_R\rvert < \lvert v_L\rvert$	$\lvert v_R\rvert < \lvert v_L\rvert$
v_R	$\lvert v_R\rvert > \lvert v_L\rvert$	$\lvert v_R\rvert > \lvert v_L\rvert$	$\lvert v_R\rvert \approx 0$	$\lvert v_R\rvert \approx 0$
a_L	$\lvert a_L\rvert \approx 0$	$\lvert a_L\rvert \approx 0$	$\lvert a_L\rvert > 0$	$\lvert a_L\rvert < 0$
a_R	$\lvert a_R\rvert > 0$	$\lvert a_R\rvert < 0$	$\lvert a_R\rvert \approx 0$	$\lvert a_R\rvert \approx 0$

Table 1. Transitions of state parameters in the walking phases

3.2.3 Definition of Kalman filter

As was described in Section 3.2.2, the walking model proposed in this chapter has three state parameters: $v_{L/R}$, $a_{L/R}$, and $p_{L/R}$. As shown in Fig. 5, although the position and velocity of each pedestrian vary continuously, the acceleration varies discretely depending on the phase of the foot movement. Thus, the state parameters are divided into two vectors, and the Kalman filter is defined based on the dynamics for moving objects.

$$\mathbf{s}_{k,n} = \Phi \mathbf{s}_{k-1,n} + \Psi \mathbf{u}_{k,n} + \omega \tag{2}$$

Here, $\mathbf{s}_{k,n}$ represents the state vector for the position $p_{L/R}$, and the velocity $v_{L/R}$ for both feet for pedestrian n at the measurement time k. The vector $\mathbf{u}_{k,n}$ represents the state vector for the acceleration $a_{L/R}$, and ω represents the system noise. The subscripts x and y for each element represent the spatial coordinates.

$$\mathbf{s}_{k,n} = \begin{pmatrix} p_{L_{x,k,n}} \\ p_{L_{y,k,n}} \\ v_{L_{x,k,n}} \\ v_{L_{y,k,n}} \\ p_{R_{x,k,n}} \\ p_{R_{y,k,n}} \\ v_{R_{x,k,n}} \\ v_{R_{y,k,n}} \end{pmatrix} \tag{3}$$

$$\mathbf{u}_{k,n} = \begin{pmatrix} a_{L_{x,k,n}} \\ a_{L_{y,k,n}} \\ a_{R_{x,k,n}} \\ a_{R_{y,k,n}} \end{pmatrix} \tag{4}$$

The transition matrix Φ and Ψ are related to the state vector $\mathbf{s}_{k,n}$ and $\mathbf{u}_{k,n}$ from the past frame $k-1$ to the present frame k. Here, Δt is the interval for observations, and in this study $\Delta t \approx 26$ milliseconds.

$$\Phi = \begin{pmatrix} 1 & 0 & \Delta t & 0 & 0 & 0 & 0 & 0 \\ 0 & 1 & 0 & \Delta t & 0 & 0 & 0 & 0 \\ 0 & 0 & 1 & 0 & 0 & 0 & 0 & 0 \\ 0 & 0 & 0 & 1 & 0 & 0 & 0 & 0 \\ 0 & 0 & 0 & 0 & 1 & 0 & \Delta t & 0 \\ 0 & 0 & 0 & 0 & 0 & 1 & 0 & \Delta t \\ 0 & 0 & 0 & 0 & 0 & 0 & 1 & 0 \\ 0 & 0 & 0 & 0 & 0 & 0 & 0 & 1 \end{pmatrix} \tag{5}$$

$$\Psi = \begin{pmatrix} \frac{1}{2}\Delta t^2 & 0 & 0 & 0 \\ 0 & \frac{1}{2}\Delta t^2 & 0 & 0 \\ \Delta t & 0 & 0 & 0 \\ 0 & \Delta t & 0 & 0 \\ 0 & 0 & \frac{1}{2}\Delta t^2 & 0 \\ 0 & 0 & 0 & \frac{1}{2}\Delta t^2 \\ 0 & 0 & \Delta t & 0 \\ 0 & 0 & 0 & \Delta t \end{pmatrix} \tag{6}$$

Moreover, the state vector $\mathbf{u}_{k,n}$ is estimated by recognizing the walking phase using the algorithm 1. Here, $\cdot$ represents the inner product of the vector.

Algorithm 1 Predicting $\mathbf{u}_{k,n}$

if $|v_{L_{k-1,n}}| < |v_{R_{k-1,n}}|$ **then**
  if $(p_{R_{k-1,n}} - p_{L_{k-1,n}}) \cdot \dot{v}_{R_{k-1,n}} > 0$ **then**
    /* Right foot is the rear foot (Phase 1) */
    $a_{R_{k,n}} \leftarrow f_R \dot{v}_{L_{k-1,n}}$
    $a_{L_{k,n}} \leftarrow 0$
  else
    /* Right foot is the front foot (Phase 2) */
    $a_{R_{k,n}} \leftarrow -f_R \dot{v}_{L_{k-1,n}}$
    $a_{L_{k,n}} \leftarrow 0$
  end if
else if $|v_{L_{k-1,n}}| > |v_{R_{k-1,n}}|$ **then**
  if $(p_{L_{k-1,n}} - p_{R_{k-1,n}}) \cdot \dot{v}_{L_{k-1,n}} > 0$ **then**
    /* Left foot is the rear foot (Phase 3) */
    $a_{L_{k,n}} \leftarrow f_L \dot{v}_{L_{k-1,n}}$
    $a_{R_{k,n}} \leftarrow 0$
  else
    /* Left foot is the front foot (Phase 4) */
    $a_{L_{k,n}} \leftarrow -f_L \dot{v}_{L_{k-1,n}}$
    $a_{R_{k,n}} \leftarrow 0$
  end if
end if

The acceleration function $f_{L/R}$ is calculated with the equations below using the average step length $S_{L/R}$.

$$S_{L/R} = \frac{1}{N} \sum_{t=j+1}^{k} \left| p_{L/R,t,n} - p_{L/R,t-1,n} \right| \tag{7}$$

$$f_{L/R} = \frac{S_{L/R}}{(k-j+1)^2 \Delta t^2} \tag{8}$$

Here, N represents the number of walking phases recognized from frame j to k. Frame j is determined experimentally. Furthermore, the initial value for the acceleration is calculated based on the amount of movement in the new tracing.

The Kalman filter updates the state vector $\mathbf{s}_{k,n}$ using the following equation based on the observed value vector $\mathbf{m}_{k,n}$.

$$\mathbf{m}_{k,n} = \mathbf{H}\mathbf{s}_{k,n} + \epsilon \tag{9}$$

where, $\mathbf{H}$ represents measurement matrix, ϵ is measurement noise.

$$\mathbf{m}_{k,n} = \begin{pmatrix} p_{L_{x,k,n}} \\ p_{L_{y,k,n}} \\ p_{R_{x,k,n}} \\ p_{R_{y,k,n}} \end{pmatrix} \tag{10}$$

$$\mathbf{H} = \begin{pmatrix} 1 & 0 & 0 & 0 & 0 & 0 & 0 & 0 \\ 0 & 1 & 0 & 0 & 0 & 0 & 0 & 0 \\ 0 & 0 & 0 & 0 & 1 & 0 & 0 & 0 \\ 0 & 0 & 0 & 0 & 0 & 1 & 0 & 0 \end{pmatrix} \tag{11}$$

3.2.4 Tracing trajectories using Kalman Filter

Figure 6 shows the flow diagram of the tracing trajectories using the Kalman Filter.

First, the walking phase is recognized by using the algorithm described in Section 3.2.3, and $\mathbf{u}_{k,n}$ is estimated. Then, $\hat{\mathbf{s}}_{k,n}$ and $\hat{\mathbf{m}}_{k,n}$ are predicted. Next, the foot candidates in a search area S_{area} from among the predicted vector $\hat{\mathbf{m}}_{k,n}$ are searched. If a foot candidate is detected in the search area, then it is taken to be the foot $\mathbf{m}_{k,n}$ of the pedestrian candidate, and the state vector $\mathbf{s}_{k,n}$ is updated. If several foot candidates are found, then the one with the smallest Euclid distance for $\hat{\mathbf{m}}_{k,n}$ is taken to be the foot $\mathbf{m}_{k,n}$ of the pedestrian candidate. If no foot candidates are found, then because of the possibility of occlusion, this is allowed for only a set period of time T_{thd}. In this instance $\mathbf{m}_{k,n}$ cannot be obtained, and so only the state vector and the error covariance matrix are predicted. If the set period of time T_{thd} is exceeded, then the target is considered lost and the search is canceled. Tracking is performed by repeating this process until all tracings are completed.

4. Gait analysis for tracked people

4.1 Gait features

Gait refers to the walking style of humans, and is defined by several parameters such as walking speed, stride length, cadence, step width, ratio of stance phase, and swing phase. These parameters are useful not only in clinical applications, but also in research focusing on human identification, gender recognition, and age estimation (Sarkar et al., 2005). Gait detection is actively studied in the field of computer vision; for example, Bobick and Johnson used the action of walking to extract body parameters instead of directly analyzing dynamic gait patterns (Bobick & Johnson, 2001), BenAbdelkader et al. analyzed the periodicity of the width of an extracted bounding area, and then computed the period using autocorrelation (BenAbdelkader et al., 2002).

Stride length and cadence (number of steps per minute) are generally considered to be the most important gait features. This is because these features are easy to measure by visual observation. However, laser scanners can achieve more detailed analyses that have shorter time intervals than those observed visually. In this research, we extracted step length rather than stride length, and cycle time (walking cycle) rather than cadence, as depicted in Fig. 7.

4.2 Gait detection by using spatial-temporal clustering of range data

Generally, the movements of pedestrians' feet are periodic. If we put all the laser points into the spatial-temporal domain, we can see that some periodic spiral patterns are generated, as shown in Fig. 8. The cross points of this spiral pattern correspond to the axes of the feet. Therefore, we can detect gait features by using nearby cross points.

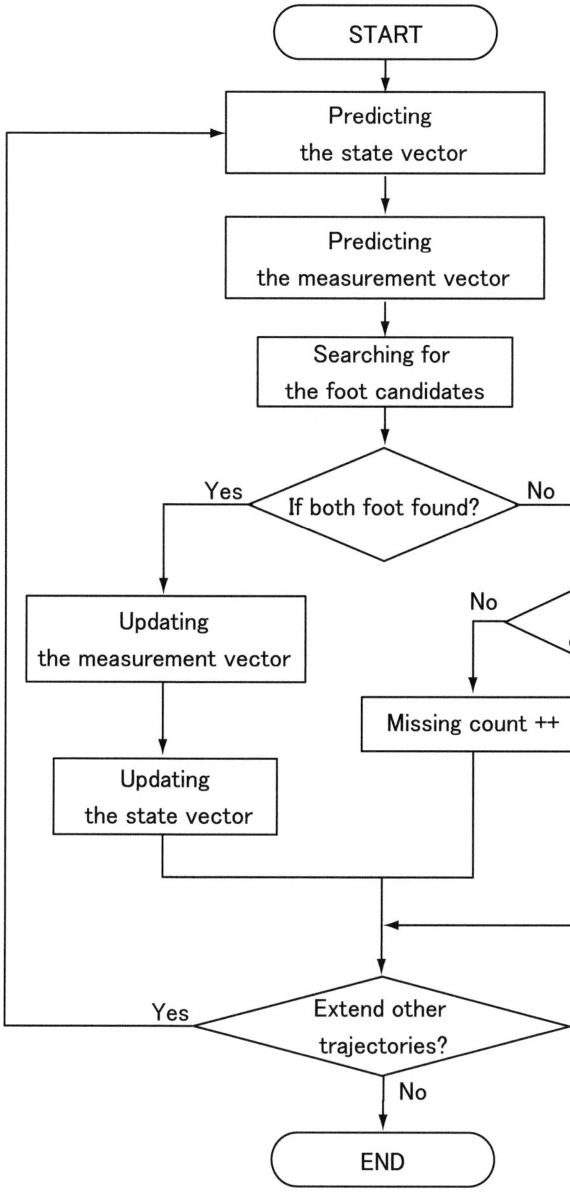

Fig. 6. Flow diagram of tracing trajectories using the Kalman filter

In this research, we used mean shift clustering (Comaniciu & Meer, 1999) to detect cross points. Mean shift is a well-known algorithm for finding the local maximum of an underlying density function. Here, a Gaussian kernel is used, where σ_s and σ_t stand for the kernel size in the space and time domains, respectively. We implement the mean-shift algorithm with $\sigma_s = 0.15$ m $\sigma_t = 0.5$ sec. The detected cross points are indicated by solid circles in Fig. 8. It can be seen that the cross points have been correctly detected. More details of this process can be found in our previous research (Shao et al., 2006).

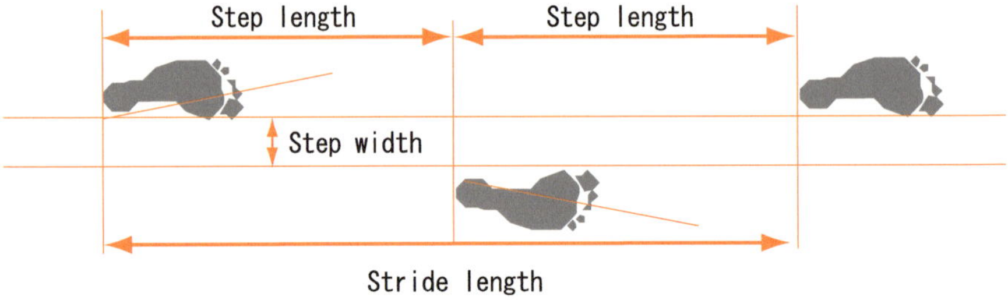

Fig. 7. Gait features

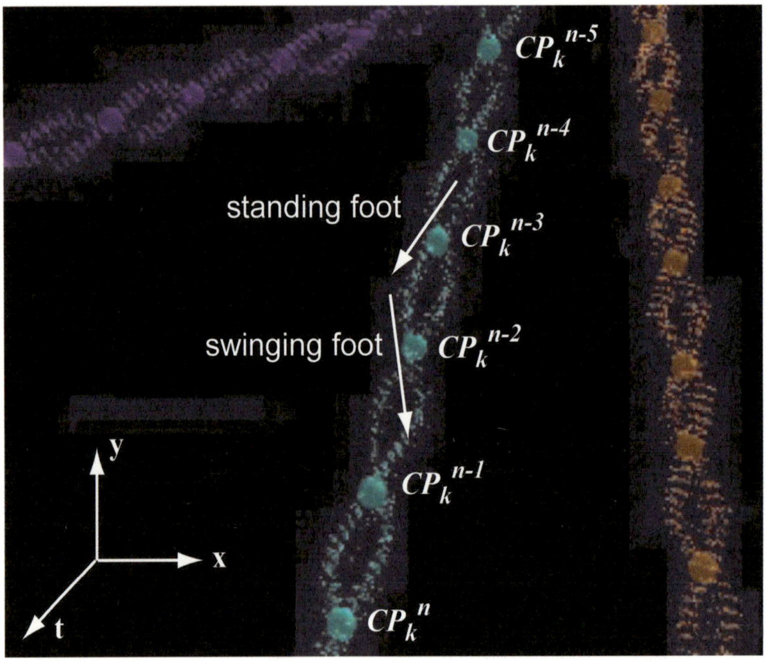

Fig. 8. Periodic spiral patterns of laser points in spatial-temporal domain. Detected cross points (CPs) representing axes of the feet are also marked by solid circles.

With the cross point $CP_k^n = (cx_k^n, cy_k^n, ct_k^n)$ for pedestrian k at measurement time n, step length s_k^n and cycle time ω_k^n can be computed by the following equations.

$$s_k^n = \left\{(cx_k^n - cx_k^{n-1})^2 + (cy_k^n - cy_k^{n-1})^2\right\}^{\frac{1}{2}} \tag{12}$$

$$\omega_k^n = ct_k^n - ct_k^{n-1} \tag{13}$$

Moreover, walking speed can be calculated by $v_k^n = s_k^n/\omega_k^n$. In this research, several cross points satisfying $v_k^n \leq 0.2$ m/s were eliminated, because a stationary human does not provide a spiral pattern in the spatial-temporal spaces, which leads to a detection error.

5. Experiment

5.1 Experimental conditions

We evaluated the effectiveness of the proposed method through an experiment conducted at a railway station in Tokyo, which is used by roughly 250,000 people per day. The station concourse is about 20 meters by 30 meters, and can hold over 150 passengers at a time.

Figure 9 shows a plane view of the concourse and the locations of the sensors. The shadow areas are indicated in the observation field. The darker the shadow, the greater the number of sensors observing the area. Eight laser sensors (#1 through #8) were set up around the most crowded area. Furthermore, in order to evaluate the proposed method with real-world conditions, the authors set up several cameras to obtain video. Six cameras (#C3 - #C8) were positioned on the ceiling to take video from directly above the concourse, and two video cameras (#C1 - #C2) were positioned to take video diagonally.

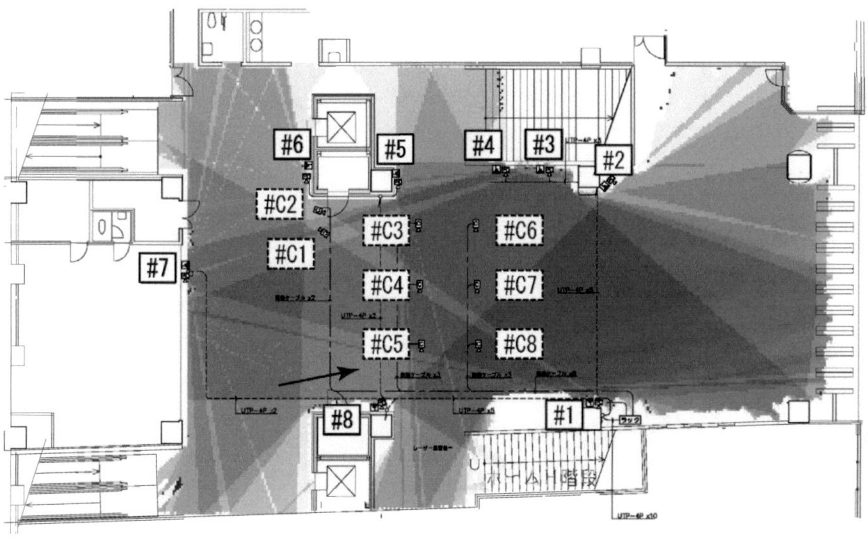

Fig. 9. Sensor alignments in a railway station, where #1 to #8 and #C1 to #C8 represent the position of laser scanners, and video cameras respectively. Shadow area shows the observation fields.

5.2 People tracking in crowd

Figure 10 shows the results of people tracking in crowds during rush hour. The red ellipses are recognized people, and yellow points are laser points[1]. Although significant occlusions occur, our proposed method can robustly track each pedestrian in the crowd. We found that a maximum of 150 people could be tracked at the same time and that tracking precision

[1] Calibration between cameras and lasers was done using Tsai's method (Tsai, 1987). Recognized people are approximated by a 170-cm height by 50-cm width, and back-projected to the image plane of the #C2 camera.

exceeded 80% during rush hour. The average pedestrian density at this time was roughly about 0.6 people/m^2.

The proposed method was more effective for tracking people in crowds in wide open areas than with vision-based methods. Because this method is also useful for protecting privacy due to using only range data, it can be used for sensing in areas where it is difficult to set up video cameras.

(a) frame 41

(b) frame 70

(c) frame 97

(d) frame 127

Fig. 10. Results of people-tracking in crowd during rush-hour

5.3 Gait analysis

Figure 11 plots the distribution of step length and cycle time of some different walking patterns. The mean and variance of step length, cycle time, and speed are listed in Table 2. The x-axis in Fig. 11 is the step length, and the y-axis is the cycle time. As we can see, pedestrians #2 and #3 have almost the same speed (average speed = 1.21 m/s and speed variance = 0.08 m/s) but they are well separated because they have different step lengths and cycle times. Also, pedestrian #4 walks stably, with a variance of step length of 3 cm and variance of cycle time of 40 ms. As another example, pedestrians #1, #2, and #3 have almost the same cycle times, but they are separable because of their different step lengths. Pedestrian #5 was running. He walked very fast and then began to run. From Table 2 we can see that his cycle time is short and stable, with a mean of 0.33 s and a variance of 20 ms, but his step length varies greatly, from about 0.6 m to 1.5 m, because of the change from fast walking to running. In this experiment, step length and cycle time of each step were extracted from our walking model and employed with speed to analyze different walking patterns. The results demonstrate that different walking patterns have their own distributions in the step length to cycle time space, and useful information about their behavior can be obtained. More information on activity recognition using gait features can be found in the reference (Nakamura et al., 2007).

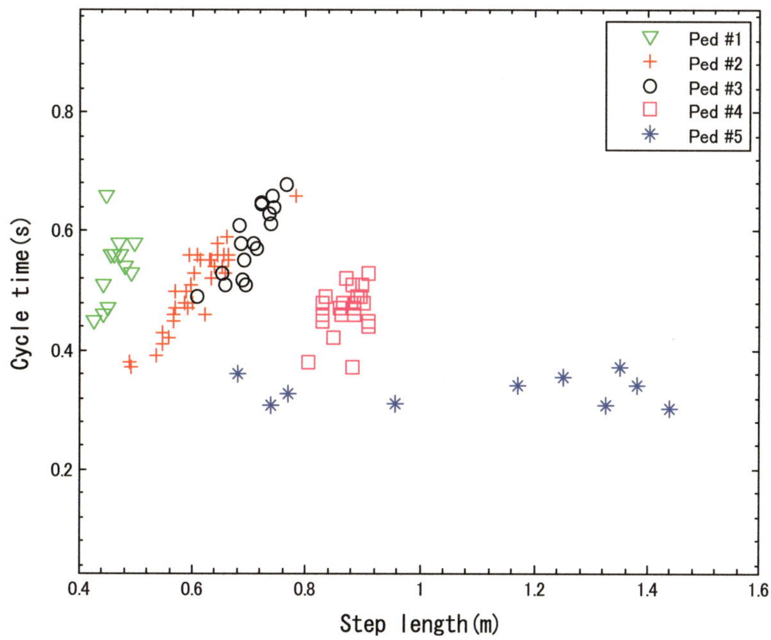

Fig. 11. Examples of detected gait features in different walking styles

Ped #	Step length (m)	Cycle times (s)	Speed (m/s)
1	0.46 ± 0.02	0.54 ± 0.06	0.87 ± 0.08
2	0.60 ± 0.06	0.50 ± 0.07	1.21 ± 0.08
3	0.70 ± 0.04	0.59 ± 0.06	1.21 ± 0.08
4	0.87 ± 0.03	0.47 ± 0.04	1.88 ± 0.17
5	1.11 ± 0.29	0.33 ± 0.02	3.34 ± 0.92

Table 2. Statistical features of different walking styles

5.4 Application to crowd-flow analysis

Our method can be applied to analyze both crowd flow and local activities. Figure 12 shows the results of visualizing crowd flow for one day. Blue lines indicate movement from right to left, and yellow lines are the opposite flow. The red points represent static people (e.g. moving at a speed < 0.3 m/sec), and white points show collision avoidance between two people (e.g.

two passengers get close to each other within 60 cm). Figure 13 shows a detected average number of train passengers during a day. We can see two peaks during the commuter rushes.

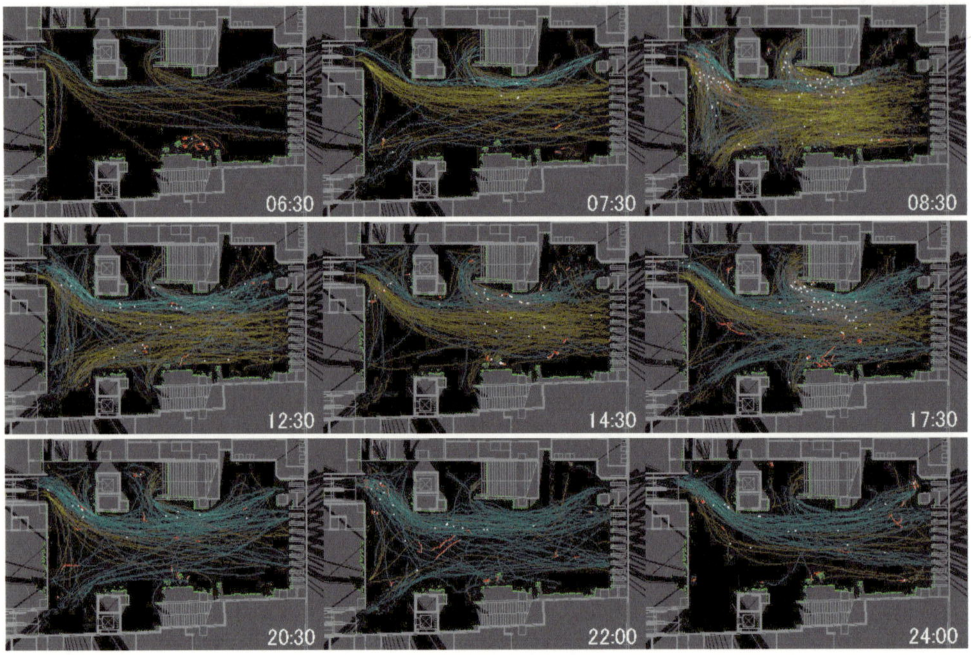

Fig. 12. Visualization results of crowd flow in one day. Blue lines are movement from right to left, and yellow lines are from left to right. Red points represent static people, and white points indicate collision avoidance between two people.

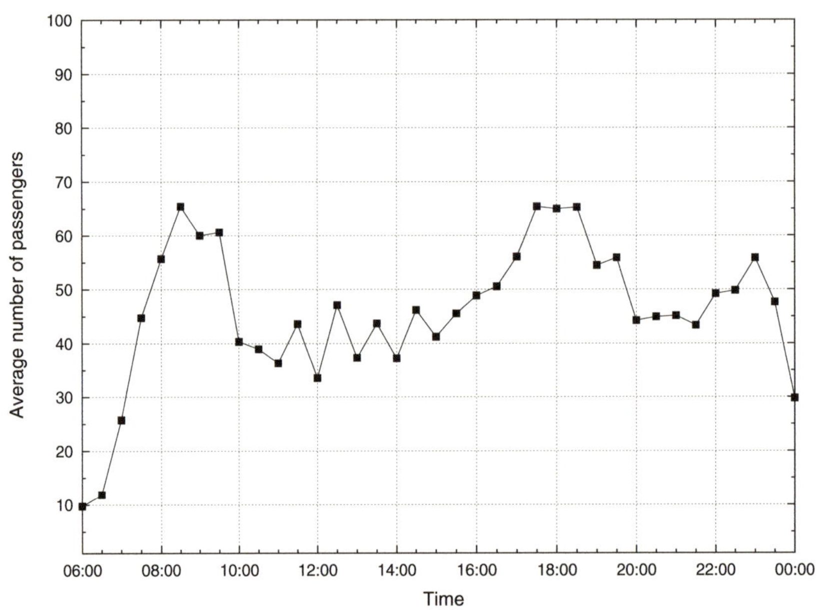

Fig. 13. Detected average number of passengers at a railway station in one day.

6. Conclusion

In this chapter, we described a method of human sensing in a crowd using multiple laser scanners. We evaluated the effectiveness of the proposed method for human tracking and gait analysis in a crowd through an experiment conducted at a large railway station. Our proposed method is well-suited to protect privacy because it does not use images but uses range data only. Therefore, our method is especially suitable for crowd sensing in public spaces such as railway stations, airports, and museums. We believe that this laser-based method is a necessary approach to complement vision-based methods that makes it possible to achieve a wider range of applications.

7. Acknowledgment

We would like to thank Dr. Sakamoto and Ms. Nakagawa for their invaluable assistance in the experiment at the railway station.

8. References

Aggarwal, J. K. & Cai, Q. (1999). Human Motion Analysis: A Review, *Computer Vision and Image Understanding* 73(3): 428–440.

BenAbdelkader, C., Cutler, R. & Davis, L. (2002). Stride and cadence as a biometric in automatic person identification and verification, *Proceedings of the IEEE International Conference on Automatic Face and Gesture Recognition (FGR)*, pp. 357–362.

Bobick, A. F. & Johnson, A. Y. (2001). Gait recognition using static, activity-specific parameters, *Proceedings of the IEEE International Conference on Computer Vision and Pattern Recognition (CVPR)*, pp. 423–430.

Comaniciu, D. & Meer, P. (1999). Distribution Free Decomposition of Multivariate Data, *Pattern Analysis & Applications* 2(1): 22–30.

Comaniciu, D., Ramesh, V. & Meeo, P. (2000). Real-Time Tracking of Non-Rigid Objects using Mean Shift, *Proceedings of the IEEE International Conference on Computer Vision and Pattern Recognition (CVPR)*, pp. 142–151.

Cui, J., Zha, H., Zhao, H. & Shibasaki, R. (2007). Laser-based detection and tracking of multiple people in crowds, *Computer Vision and Image Understanding* 106(2-3): 300–312.

Cui, J., Zha, H., Zhao, H. & Shibasaki, R. (2008). Multi-modal tracking of people using laser scanners and video camera, *Image and Vision Computing* 26(2): 240–252.

Dalal, N., Triggs, W. & Schmid, C. (2005). Human Detection using Oriented Histograms of Flow and Appearance, *Proceedings of the European Conference on Computer Vision (ECCV)*, Vol. 3952, Springer, pp. 428–441.

Felzenszwalb, P. F., Girshick, R. B., Mcallester, D. & Ramanan, D. (2009). Object Detection with Discriminatively Trained Part Based Models, *IEEE Transactions on Pattern Analysis and Machine Intelligence* 32(9): 1–20.

Fod, A., Howard, A. & Matarić, M. J. (2002). Laser-Based People Tracking, *Proceedings of the IEEE International Conference on Robotics and Automation (ICRA)*, pp. 3024–3029.

Gavrila, D. M. (1999). The Visual Analysis of Human Movement: A Survey, *Computer Vision and Image Understanding* 73(1): 82–98.

Katabira, K., Zhao, H., Shibasaki, R. & Ariyama, I. (2006). Real-Time Monitoring of People Behavior and Indoor Temperature Distribution using Laser Range Scanners and Sensor Networks for Advanced Air Conditioning Control, *Proceedings of the International Conference on Networked Sensing Systems (INSS)*, pp. BOF–11.

Kratz, L. & Nishino, K. (2010). Tracking with local spatio-temporal motion patterns in extremely crowded scenes, *Proceedings of the IEEE International Conference on Computer Vision and Pattern Recognition (CVPR)*, pp. 693–700.

Mochon, S. & McMahon, T. A. (1980). Ballistic Walking, *Journal of Biomechanics* 13: 49–57.

Nakamura, K., Shao, X., Zhao, H. & Shibasaki, R. (2007). Recognizing Non-Stationary Walking based on Gait Analysis using Laser Scanners (in Japanese), *IEEJ Transactions on Electronics, Information and Systems* 127(4): 537–545.

Nakamura, K., Zhao, H. & Shibasaki, R. (2005). Tracking Pedestrians using Laser Scanners and Image Sensors (in Japanese), *Proceedings of the Symposium on Sensing via Image Imformation (SSII)*, pp. 177–180.

Nakamura, K., Zhao, H., Shibasaki, R., Sakamoto, K., Ohga, T. & Suzukawa, N. (2006). Tracking pedestrians using multiple single-row laser range scanners and its reliability evaluation, *Systems and Computers in Japan* 37(7): 1–11.

Okuma, K., Taleghani, A., Freitas, N. D., Little, J. J. & Lowe, D. G. (2004). A Boosted Particle Filter : Multitarget Detection and Tracking, *Proceedings of the European Conference on Computer Vision (ECCV)*, Springer, pp. 28–39.

Sarkar, S., Phillips, J., Liu, Z., Vega, I. R., Grother, P. & Bowyer, K. W. (2005). The Human ID Gait Challenge Problem: Data Sets, Performance, and Analysis, *IEEE Transactions on Pattern Analysis and Machine Intelligence* 27(2): 162–177.

Shao, X., Zhao, H., Nakamura, K., Shibasaki, R., Zhang, R. & Liu, Z. (2006). Analyzing Pedestrian's Walking Pattern Using Single-Row Laser Range Scanners, *Proceedings of the IEEE International Conference on Systems, Man, and Cybernetics (SMC)*, pp. 1202 – 1207.

Song, X., Shao, X., Zhao, H., Cui, J., Shibasaki, R. & Zha, H. (2010). An Online Approach: Learning-Semantic-Scene-by-Tracking and Tracking-by-Learning-Semantic-Scene, *Proceedings of the IEEE International Conference on Computer Vision and Pattern Recognition (CVPR)*, pp. 739–746.

Song, X., Zhao, H., Cui, J., Shao, X., Shibasaki, R. & Zha, H. (2010). Fusion of Laser and Vision for Multiple Targets Tracking via On-line Learning, *Proceedings of the IEEE International Conference on Robotics and Automation (ICRA)*, pp. 406–411.

Szeliski, R. (2010). *Computer Vision : Algorithms and Applications*, Springer-Verlag New York Inc.

Tsai, R. Y. (1987). A Versatile Camera Calibration Technique for High-Accuracy 3D Machine Vision Metrology using Off-the-shelf TV Cameras and Lenses, *IEEE Journal of Robotics and Automation* RA-3(4): 323–344.

Yilmaz, A., Javed, O. & Shah, M. (2006). Object tracking: A Survey, *ACM Computing Surveys* 38(4): 1–44.

Zhao, H. & Shibasaki, R. (2005). A Novel System for Tracking Pedestrians Using Multiple Single-Row Laser-Range Scanners, *IEEE Transactions on Systems, Man, and Cybernetics - Part A: Systems and Humans* 35(2): 283–291.

Zhao, T., Nevatia, R. & Wu, B. (2008). Segmentation and tracking of multiple humans in crowded environments., *IEEE Transactions on Pattern Analysis and Machine Intelligence* 30(7): 1198–1211.

3

A Multi-Faceted Assessment of the Applications of Full Body Scanners at Airports

Angelos Lakrintis[1], Konstantinos Malandrakis[1] and Leo D. Kounis[1,2]
[1]*Halkis Polytechnic, School of Applied Sciences, Dept. of Aviation Technology*
[2]*State Aircraft Factory*
[1,2]*Greece*

1. Introduction

Performed research indicates that the number of passengers travelling by air is increasing. Likewise, the percentage of global air cargo is growing. For safeguarding established safety levels, the International Civil Aviation Organization, ICAO, in conjunction with national Civil Aviation Authorities has introduced over the years a number of rules and frameworks that aviation companies and airports need to adhere to. Indeed, it can be argued that although ICAO's main scope is to regulate international air travel, the means and methods implied to achieve safety levels in the aviation sector remains at the corresponding Civil Aviation Authorities and the associated airport management.

An escalating number of violations of security measures over the years have led some airport authorities to impose stricter aviation-specific security techniques, suggesting that current legal, and/or technology-related frameworks ought to be further optimised. In doing so, the quest for establishing higher quality and safety levels in the aviation sector has resulted in the development, marketing and operation of Whole Body Imaging, WBI technologies. Although regarded by airport authorities as a means of either supplementing already existing security checks, or even to replace metal, or/and explosive detectors, initial research indicates that public reaction is split, as it considers issues such as religious beliefs, health-related and ethical issues.

The preliminary study carried out so far indicates–primarily-the application of two different strategies per se. FAA requires the use of Full Body Scanners, FBS, at American airports. Within the European Union however, the use of FBS is not yet regulated, as it awaits the outcomes of an impact assessment review.

This work will acknowledge the current operating legal framework and will elaborate on contemporary implied practices, which will address the areas of quality/benchmarking, safety and security and will consider the time value of money taking into account the acquisition of such scanning machines. A potential sector-wide operation of Full Body Scanners, FBS may equally have an effect on associated flight operations and slot management.

Based on real-life case-studies, this research will address basic operating principles of FBS and will suggest best benchmarking practices within an applied Quality Management environment.

In view of the above, the aim of this study will be to suggest alternative means of checking methods, whilst maintaining set quality levels and enhancing benchmarking tools.

In doing so, this work will address the following objectives:

- to study current Legislation on safety matters in the aviation area
- to perform an in-depth analysis of the operating characteristics of the various types of FBS already in use
- to elaborate on a variety of potential drawbacks resulting thereof
- to develop questionnaires
- to review current quality levels regarding overall airport performance and handling-specific percentages
- to appraise benchmarking tools
- to conduct comparative analyses between airports already using WBI technology and airports that apply traditional means of checking points
- to validate alternative means of passenger checking
- to propose a viable and sustainable alternative

1.1 Initial hypotheses pertaining to safety levels in relation to quality

Initial research carried-out indicates that a wide-spread application of Full Body Scanners, FBS falls short. This may be attributable to the following reasons:

- their effectiveness has not yet been proven. This entails the capability of the scanning machine to effectively detect hidden ammunition on human body parts, whilst subjecting the scanned person to the least amount of radiation
- they are not regarded as being the panacea to airport and flight-specific safety.

It is the authors' view that both potential causes mentioned above should be viewed taking into consideration the "time value of money"-principle into account. This is defined as the amount of interest earned on an initial investment within a set time period. However, this term deduces a lean investment and a seamless payback period, not necessarily considering long-term "what-if" scenarios. The latter may include long-term radiation level exposure and its effects on travelers, in particular frequent flyers.

This study will emphasise the need for a multi-faceted assessment of the application of FBS at airports and major transportation hubs-such as international railway terminus-considering potential drawbacks on the long-term. In addition to this, it will suggest alternative scanning methods that will be able to effectively detect dangerous objects, whilst minimising the radiation dosage emitted. It is worth mentioning at this point that the latter forms part of current on-going research.

2. Literature review and methodology

Safety and security are amongst the most widely coined clauses in any airline's Mission Statement. Safety is defined as *"...freedom from danger, risk, or injury"*, whereas security is described as *"...freedom from risk, or danger; safety; anything that gives or assures safety; one who undertakes to fulfill the obligation of another; surety"* [1]. The latter of the term forms part of an airline ticket and is seen as a legal term of a contract agreed and signed between the passenger and the air carrier. Their mutual acceptance of this contract detailing potential claims and law suits in the event that either party may have failed to provide the service as stated in the Warsaw Convention of 1929 [2]. It is worth mentioning at this point that the latter was subject to two revisions, namely the Hague amendment of 1955 and that of Montreal in 1999.

Research performed by Topouris *et al*, (2011), has shown that global air travel exhibits a continuous growth in passenger numbers and cargo. This in turn, necessitated -amongst others-the introduction and development of safety procedures and built-in safety devices, such as metal detectors.

The Ontario Privacy Commissioner IPC, in collaboration with the Dutch Data Protection Authority established in 1995 the term Privacy Enhancing Technologies, PETs. This refers to "...coherent systems of information and communication technologies that strengthen the protection of privacy information systems by preventing the unnecessary or unlawful collection, use and disclosure of personal data, or by offering tools to enhance an individual's control over his/her data."[4].

The events of September 11th, 2001, led the Research, Engineering and Development Advisory Committee of the Federal Aviation Administration, FAA (2002), to introduce a number of approaches in order to minimise potential security threats. Among the suggestions launched, the Advisory Subcommittee proposed the adaptation of the following short-term means:

- hardening the cockpit door
- initiating integrated airport-wide security test beds
- further improving human performance
- demonstrating technology to screen people, and
- triage procedures for screening people and their belongings

Screening technology as proposed by the Advisory Subcommittee translates to Whole Body Imaging, WBI, technology. This refers to basically two different types of apparatuses, namely:

- Millimeter-Wave, and
- Backscatter x-ray

The FBS listed above, use an array of algorithms so as to provide the security controller with a detailed view of any illegal items carried by the traveler, while aiming at safeguarding personal data.

In the long-term, it is stated that the FAA would need to tackle the aviation security issue as a generic system and develop henceforth a number of technologies and techniques, capable of integrating potential threat information. It is the authors' view that this scenario would require continuous investigation. The latter ought to consider cutting-edge technology, processes and procedures alike, so as to cater for potential unauthorized "system-entries". Indeed, it can be argued that potential "loopholes" within the aviations' generic system are subject to misconduct.

The Transportation Security Administration, TSA, commenced the deployment of WBI technologies and procedures at airports. Opponents of the latter argue that it violates fundamental human rights.

However, the Securing Aircraft From Explosives Responsibly: Advanced Imaging Recognition Act, S.A.F.E.R A.I.R Act, actually imposes the positioning of WBI apparatuses at airports by 2013.

Performed research reveals that the number of FBS at airports is increasing. The United States are pioneering in this area, followed by the United Kingdom, Holland and Germany to name but a few. Bart (2011), states that "...system-wide there are more than 750 screening checkpoints and over 2.000 screening lanes across 450 airports in the USA, with TSA's goal being the installation of 1.800 units by the end of 2014."

It is worth mentioning at this point that the legal framework addressing the deployment of FBS has not been finalised yet.

The European Directive No 2320/2002 sets the platform for passengers and cabin baggage screening placing the emphasis on metal detectors. However, the European Resolution (2008) 0521 of October 23rd, 2008 states that:

"...essential information is still lacking and asks the Commission, before the expiry of the three-month deadline, to:

- carry out an impact assessment relating to fundamental rights;
- consult the EDPS, the Article 29 Working Party and the FRA;
- carry out a scientific and medical assessment of the possible health impact of such technologies;
- carry out an economic, commercial and cost-benefit impact assessment;

2. Believes that this draft measure could exceed the implementing powers provided for in the basic instrument, as the measures in question cannot be considered mere technical measures relating to aviation security, but have a serious impact on the fundamental rights of citizens;..."

Preliminary research reveals that WBI technology may have long-term effects primarily on the health of passengers going through these devices. This is a focal point that is acknowledged by the EC. The latter awaits the outcome of current on-going research. The data gathered will serve as a platform in order to establish an EU-wide legal framework addressing effective and efficient security measures at airports. It is the authors' view that the long-term effects of the widespread use of FBS ought to be further researched.

Another noteworthy point is the potential infringement of fundamental human rights, considering the fact that aforementioned Resolution calls for "strong and adequate safeguards" to "protect the right to personal dignity". Finally, the Resolution stresses the fact that the impact of a wide-area deployment of FBS devices ought to protect and cater for an unlawful processing of personal data. Indeed, the latter part of the sentence is addressed by the Obscene Publications Act 1959 of the United Kingdom. To accommodate for any unlawful processing of personal data, the Obscene Publications Act introduces the term "test of obscenity".

It is the authors' view-based on preliminary research performed- that advise was not sought for at the European Data Protection Supervisor (EDPS), as required by Article 28(2) of Regulation (EC) No 45/2001. In addition to this further research ought to be carried-out, so as to include feedback from a variety of agencies such as the Fundamental Rights Agency, FRA.

EU Commission Regulation 185/2010 acknowledges that: "... methods, including technologies, for detection of liquid explosives will develop over time. In line with technological developments and operation experiences both at Community and global level, the Commission will make proposals, whenever appropriate, to revise the technological and operation provisions on the screening of liquids, aerosols and gels". This is in par with the European Directive 95/46/EC concerning "...the protection of individuals with regard to the processing of personal data and on the free movement of such data". However, the Directive 95/46/EC coins the term of "necessary measure", allowing airport security to perform a detailed check, if the person is a suspect.

Owing to the multiple of applicable Laws, Regulations and the novelty of the area, this work is aimed at providing all potential users of WBI technologies with a fundamental platform, taking into consideration legal/ethical, technological, quality and health-related issues.

Initial research includes the study of a number of International, European and National Laws and Regulations, which revealed that the deployment of WBI technologies and apparatuses is distinctly split between opponents and proponents. The opponents primarily represent the EU, whereas the latter stand for the United States of America.

However, it is noteworthy to mention that within the EU, Member-States have decided to address aviation security by different means. As such, the United Kingdom has decided to implement FBS at all UK airports by the end of 2010, whereas in Germany, only Fühlsbüttel Airport in Hamburg is deploying such a scanner, albeit on a voluntary basis. Performed research has shown that backscatter x-ray scanners are illegal for use at German airports, with current research focusing on millimeter-wave scanners.

In order to provide a detailed view concerning public perception of a potential EU-wide deployment of FBS questionnaires were developed. These were targeted at random to travelling passengers from a number of Greek airports. The outcome thereof will be discussed as of chapter 3 on.

When searching for the corresponding legal framework, in particular the one dealing with "Health and Safety at Work", it was found that this subject area was not addressed. In view of the above, on-going research will focus on developing alternative WBI means, emphasising on effectiveness and efficiency, while minimising the emitted radiation.

3. Results

As already mentioned in chapter 2, the Obscene Publications Act 1959, OPA, of the United Kingdom sets the platform for safeguarding unlawful distribution of personal data. To this extend, it equally includes sections of the Broadcasting Act 1990 concerning the terms "programme" and "programme service". As such, the OPA introduces the so-called "test of obscenity" whereby *"...the effect of any one of its items is, if taken as a whole, such as to tend to deprave and corrupt persons who are likely, having regard to all relevant circumstances, to read, see or hear the matter contained or embodied in it.*

In this Act "article" means any description of article containing or embodying matter to be read or looked at or both, any sound record, and any film or other record of a picture or pictures.

For the purposes of this Act a person publishes an article who—(a) distributes, circulates, sells, lets on hire, gives, or lends it, or who offers it for sale or for letting on hire; or(b)in the case of an article containing or embodying matter to be looked at or a record, shows, plays or projects it [or, where the matter is data stored electronically, transmits that data.]

For the purposes of this Act a person also publishes an article to the extent that any matter recorded on it is included by him in a programme included in a programme service.

Where the inclusion of any matter in a programme so included would, if that matter were recorded matter, constitute the publication of an obscene article for the purposes of this Act by virtue of subsection (4) above, this Act shall have effect in relation to the inclusion of that matter in that programme as if it were recorded matter."

It is the authors' view that this Act should serve as the legal cornerstone regarding the lawful and legal handling of personal data, in association with the use of FBS. The latter is the outcome of further research by leading industries aimed at enhancing security checks at airports. As already mentioned in chapter 2, there are two different WBI apparatuses currently in use:

- Millimeter Wave, and
- Backscatter X-ray

Millimeter wave utilises non-ionising radio frequency energy in the corresponding mm-wave spectrum, so as to produce a 3D-picture of the object under study. Figure 1 shows the operating principle of such a device.

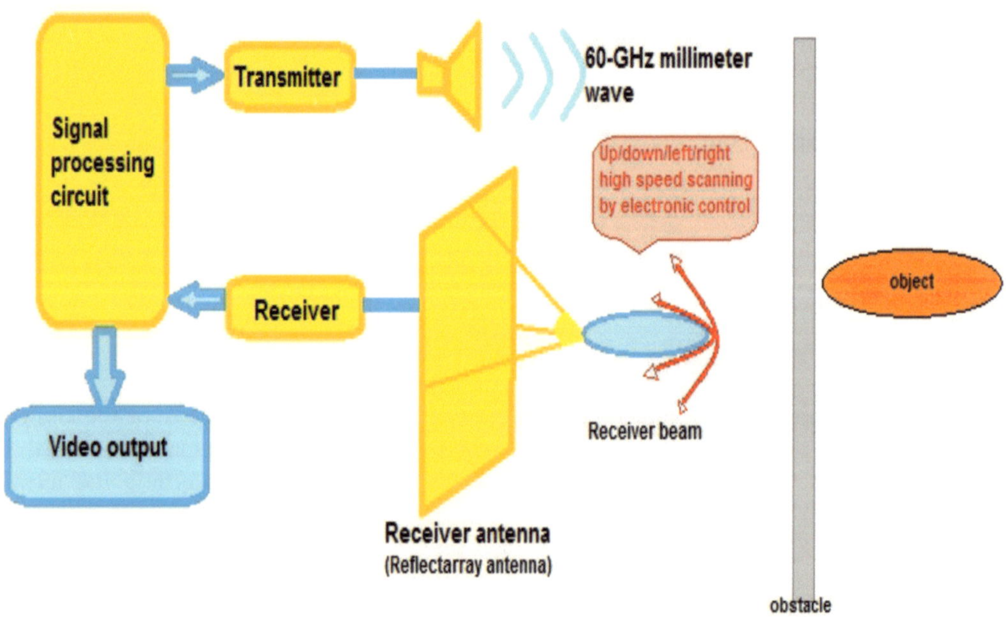

Fig. 1. Operating principle of a Millimeter wave device

The image is displayed on a remote monitor, subject to further analysis.

Backscatter X-ray devices use a narrow, low-energy x-ray beam that scans the surface under study at a certain speed. Their operating principle is depicted in figure 2.

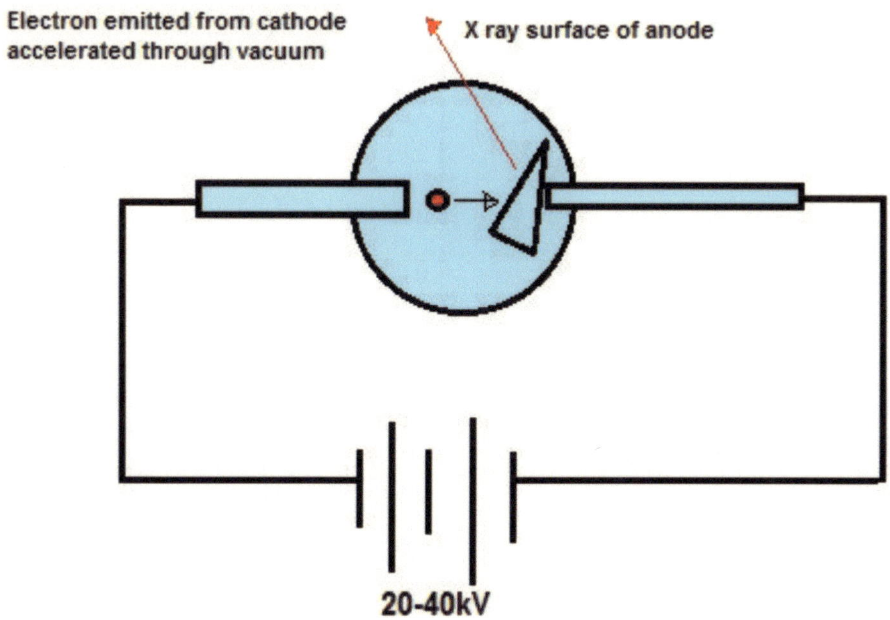

Fig. 2. Operating principle of a Backscatter X-Ray device

It thereafter generates a picture of the object that is projected onto a remote monitor for further analysis.

Preliminary analysis shows that the radiation dosage of the backscatter x-ray scanners is:

$H= E / wT = (0.0001 / 0.01)$ mSv per scan = 0.01 mSv per scan.

Supposing that the entire dose used in order to calculate the effective dose (E) is the skin, and the "manufacturer's worst case" estimate of E = 0.0001 mSv per scan, then the worst case equivalent dose (H) to the skin is as calculated above 0.01 mSv per scan.

Table 1 shows various radiation levels from a number of human activities.

It is worth mentioning at this point that performed research showed the legal ban of all backscatter x-ray devices in the Federal Republic of Germany. Further studies aim at optimising millimeter x-ray scanners, as the apparatus at Hamburg's Fühlsbüttel Airport was put out of operation. The latter was due to experienced bottlenecks, both in terms of overall passenger handling delay, as well as the number of complaints received by travelers.

Examples	mSv
Eating one banana	0.0001 mSv / banana
Sleeping next to a human for 8 hours	0.0005 mSv / night
Dental x-rays	0.005 mSv / visit
Backscatter X-ray	0.01 mSv/scan
Mammogram	3 mSv / visit
Chest CT scan	6-18 mSv / scan
Living near a nuclear power plant	0.0001-0.01 mSv / year
Living in San Diego (radiation from the sky)	0.24 mSv / year
Living in San Diego (radiation from the ground)	0.28 mSv / year
From your own body	0.40 mSv / year
Airplane trip	0.01 mSv / 1000 miles flown
Smoking a pack and a half a day	13-60 mSv / year
Start of level known to cause cancer	100 mSv / year
Standing in the middle of the Fukushima Daiichi nuclear power plant	100 to 400 mSv

Table 1. Radiation levels from a number of human activities

Indeed, a study by the German Radiation Protection Registry revealed some statistical data based on a study concerning the flight hours of pilots and crew members during the 5-year period 2004-2009 [14]. Authors Frasch G., Kammerer L., *et al* (2011) found that the average effective dosage of 2.35mSv per flight crew in the year 2009 was 20% higher as opposed to 2004. They justified this as being the result of "the cosmic radiation due to the altitude". Indeed, epidemiological studies point to increased health risks of flight personnel. However, an explicit correlation between altitude radiation and carcinosis is subject to further scientific research. It is therefore suggested that flight crew adhere to safety standards and optimise flight and route planning, so as to avoid high radiation areas.

In order to gauge public concern and evaluate the reactions concerning a potential use of FBS on European Airports and Transportation hubs, a questionnaire was developed and the opinion of a random sample of 50 travellers asked. The questions raised dealt with airport security matters. In particular, the query focused on how a potential introduction of FBS may affect their lives. The outcomes are shown in the following pie-charts below:

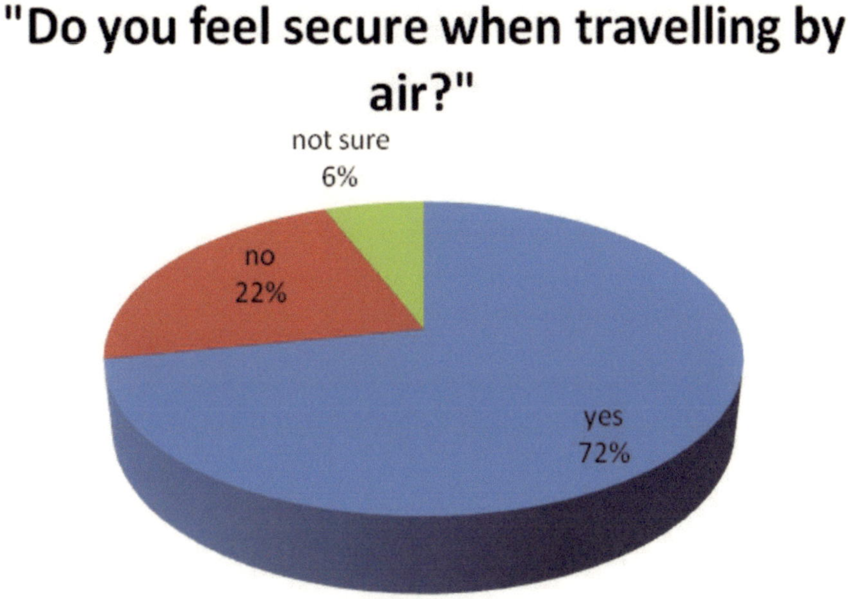

Fig. 3. Passenger replies concerning air travel

When passengers were confronted with the question, if they felt secure when travelling by air, 72% responded that they did. Only a 22% was feeling insecure. The latter may be subject to further research. Likewise, airports were regarded as inducing a feeling of security to 74% of passengers, as opposed to 4% who felt not secure. By comparing these two pie-charts one may observe that the positive responses do not show any significant variation.

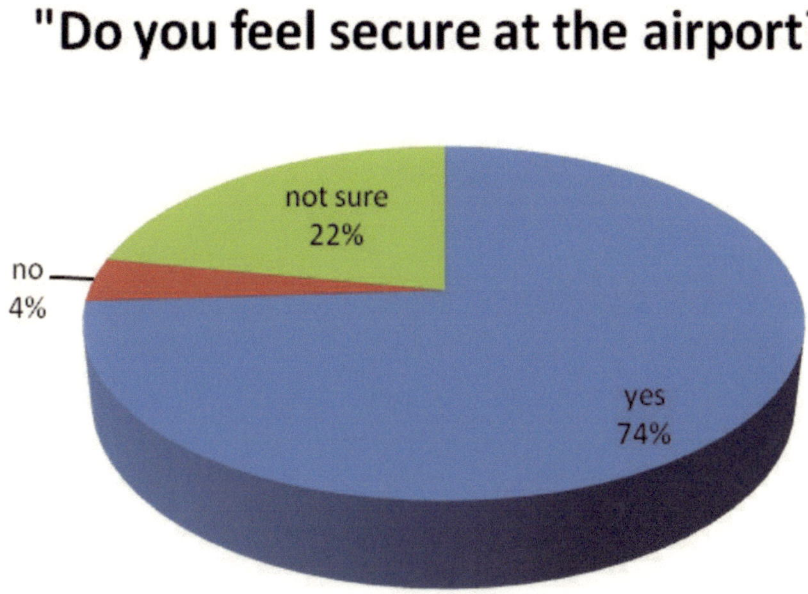

Fig. 4. Customer outcome regarding airport-related security

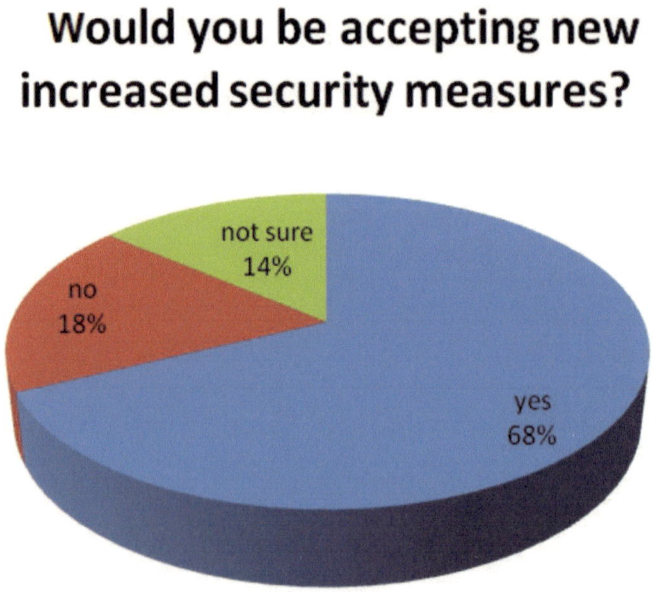

Fig. 5. Traveler results pertaining to security measures

68% of passengers travelling by air would accept new increased security measures. The outcome of the latter contradicts previous results, which underlie the fact that on average 73% of travelers feel both secure at the airport and on the aircraft.

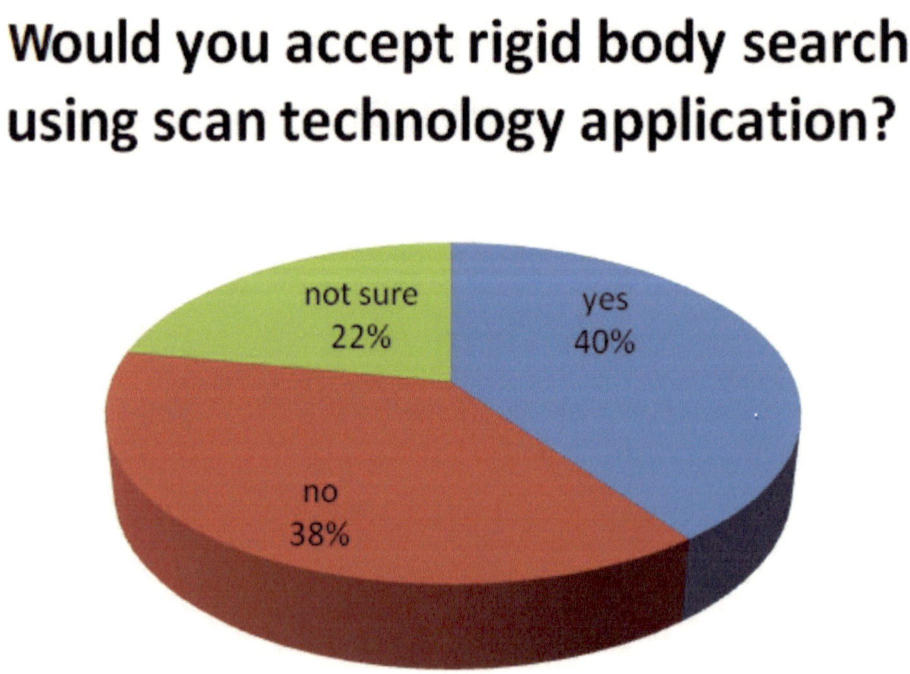

Fig. 6. Proponents vs. opponents of scan technology at airports

When further narrowing the question pertaining to the type of security device used, the responses show an approximate equal split. 40% of travellers would accept WBI technology, while 38% would not accept FBS. The remaining 22% of people quoting "not sure" needs further investigation as to the potential causes of their reply, past experience, social, moral and ethical beliefs to name but a few. The results thus far hand-shake with the initial hypotheses and indicate that:

- the majority of passengers would accept increased security measures, and
- that the deployment of FBS is not regarded as being the panacea to airport and flight-specific security
- a considerable percentage of people ranging between (14-22)% were not certain concerning increased security measures and the use of rigid body scan technology at airports, so as to enhance an overall security feeling

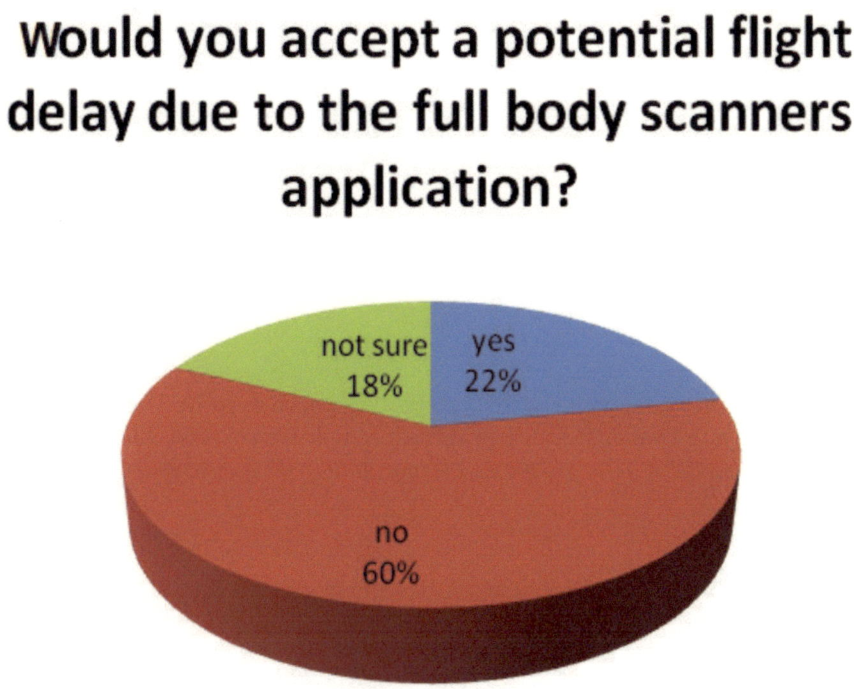

Fig. 7. Potential operational impact of FBS on slots

The pie-chart shown above suggests the following:

- 60% of passengers would not accept a delay in the slot-management of their flight, albeit having passed a full body scan. This empowers initial hypotheses concerning the development of optimised benchmarking tools and techniques. The latter would be the outcome of probable bottlenecks in passenger handling. It is the authors' view that this may also be the cause of past passenger-specific experience regarding the usefulness of FBS at countries already deploying them
- 22% of travelers would accept a likely delay of their flight in lieu of a full body scan
- 18% of responses were "not sure" about their acceptance of WBI apparatus

Based on the findings of the questionnaire, passengers do indeed opt for air travel as it is a fast, safe and effective transportation means. In their view, air travel has become cheaper. Modern airplanes manage to cover long distances by offering an improved level of comfort.

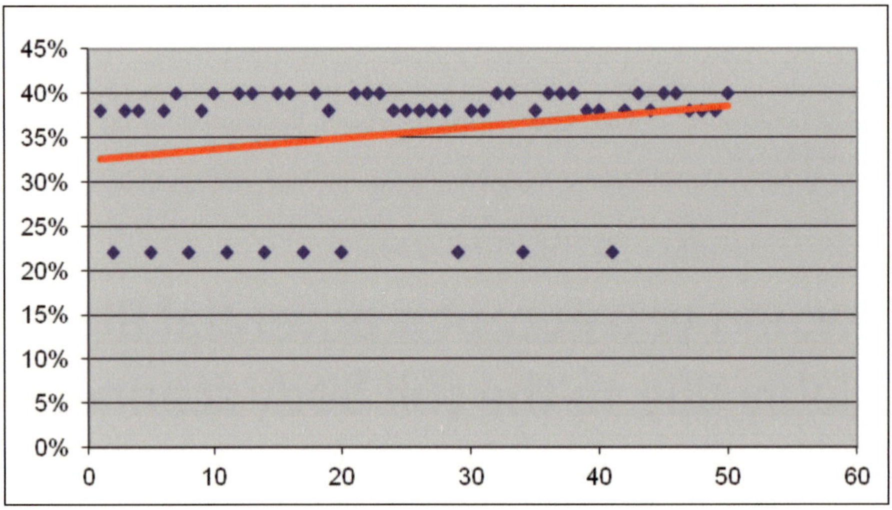

Diagram 1. Trend-analysis based on the sample of 50 passengers

4. Analysis

Initial research shows that airport authorities aim at satisfying passenger needs between (95-99)% of all times. In order for a statistical process to be regarded as being within set tolerances, the term Capability Index is introduced. Bergman and Klefsjö (1994) define capability as "*...the ability of a process to produce units with dimensions within the tolerance limits...*".

Capability Indices, Cp_ks, are a simple and widely used quality technique tool (Kounis, 2010). They measure the dispersion of a process in relation to the range of the set tolerance interval, by including the Upper Specification Limit, USL, and the Lower Specification Limit, LSL. Capability Indices may equally be applied to evaluate a process, or design. The higher a Cp_k value is, the lesser the amount of rejects. However, although a Cp_k shows a small variation, it is not always centered around the target value (Kounis *et al*, 2000).

As Cp_ks, were not readily available by airport authorities, or required detailed evaluation, a sample size of 50 was deemed appropriate. Indeed, research performed by Kounis *et al* (2000) have argued that a random sample of less than 50, may lead to making the resulting capability index Cp_k unacceptable. However, the authors suggest than in cases such as inadequate funding, or inhabitive volumes, Extreme Value Statistics, EVS, may be regarded as an alternative method to capability indices, since it focuses on the behaviour of the tails of the distribution. Indeed, the authors argue that EVS "...are capable of describing these situations asymptotically."

Diagram 1 shows a trend analysis of the received outcomes to the question concerning the acceptance of scan technology apparatuses.

It primarily highlights two different areas:

- the clear split between the people who are "not sure about accepting" such a technology, representing 22% of the sample, and
- a near 50:50 division of respondents standing for the acceptance of such a technology, as opposed to the ones not accepting it

The outcomes received thus far indicate that further research ought to be carried-out, as the application of WBI technologies affects a multitude of transport-related areas and entities. In addition to the technology implied, the results suggest a refinement of the corresponding legal framework, so as to:

- foster security measures at transportation hubs
- cater for cumulative radiation levels
- consider passenger demographics

With regards to increasing security measures at transportation hubs (airports and railway terminus), the "CRS Report for Congress" of May 26th, 2005, acknowledges an "*...inherently vulnerable*" state of passenger rail systems world-wide; the latter resulting from the Madrid 2004 rail events.

For an organisation to develop a thread-based risk management, Roper (1999), conceptualises it by means of the following equation:

$$\text{Vulnerability} + \text{Threat} + \text{Criticality} = \text{Risk}$$

As such, "vulnerability" is defined as a system's open, unprotected and exposed boundaries and/or areas to potential attacks.

Likewise, "threat" refers to the potential likelihood of an attack on a system, whereas "criticality" associates the potential consequences of an attack to the system's behaviour and performance.

5. Discussion

The outcomes of this preliminary study and basic Extreme Value Statistics, EVS concepts suggest that the introduction of more effective security means is indeed condoned by travellers. However, public perception is split when it comes to FBS, as they are deemed to either infringe, yet even violate basic human privacy rights, or to inadvertently cause health-related diseases.

The latter would have to form an area of on-going further research. It is the authors' view that no explicit scientific outcomes may be formulated, as the following factors ought to be taken into consideration:

- amount of cosmic radiation, acknowledging global warming
- travel frequency of passengers
- selected routes and associated level of cosmic radiation
- travellers' demographics and related health condition
- effect of other radiation-emitting sources on human health

It is worth mentioning at this point that long-term effects ought to be studied and various external and interrelating parameters be taken into consideration.

On the other hand, research is currently focused on the development of so-called Sound Amplification by Stimulated Emission of Radiation, SASER, or sound laser. A saser is in effect a photon laser that produces a highly consistent beam of ultrasound. One might argue that a saser is a slightly modified version of a laser; the former forcing sound waves to travel at THz frequencies, albeit on a nano scale. The basic operating principle is the employment of phonons to produce sound waves. Rather than emitting waves and forcing them to pass through an optical cavity-as is the case with a laser- a saser-produced wave travels via a superlattice structure. The latter consists of 50 ultra-thin and alternating layers of gallium arsenide and aluminium arsenide. Once the phonons are inside said superlattice, they rebound, multiply and escape in the form of an ultra-high frequency beam.

The University of Nottingham has pioneered in this area (http://www.rexresearch.com/kentsaser/kentsaser.htm). Kent *et al* argue that saser could find potential application in the manufacture of computers resulting in higher execution times.

Experiments also focus on understanding and assessing the behaviour of saser technology in spectral distribution. In particular, the detection of phonons in R_1 luminescence is studied so as to draw practical applications [IOP Science] (http://iopscience.iop.org).

Another application of saser technology is to suppress the quantum noise that drowns out very faint signals in ordinary conductors [New Scientist]. The aim of on-going research is to try and reduce potential noise levels in electronic devices, so as to track down weak signals.

This paper acknowledges current research work and aims at developing a device capable of producing sound waves. These would retract once a surface or an object is been hit. Using a suitable interface, the projection of said object could be facilitated. Likewise, areas of high density could be shown by means of peaks on an oscilloscope. The hypothetical apparatus making use of saser technology is shown in figure 3.

It is the authors' view that the proposed device be simple and easy-to-use. In addition to the above, it ought to be capable of identifying and showing dangerous materials and/or tools, whilst protecting human health and personal data. The latter should be stored in means that would serve a dual purpose:

- be fail-safe, and
- prevent a potential violation by other users, or third parties

In order to enhance flight security suggested imaging apparatuses might be linked on a European-wide area, providing the outcomes of further research sectors are satisfying. This however, necessitates either amendments to current legislation, and/or the development of new EU-wide Laws. The latter ought therefore to be based on "best practice"-principles. Additional ethical and religious matters ought to be accounted for in a potential EU-wide legal framework.

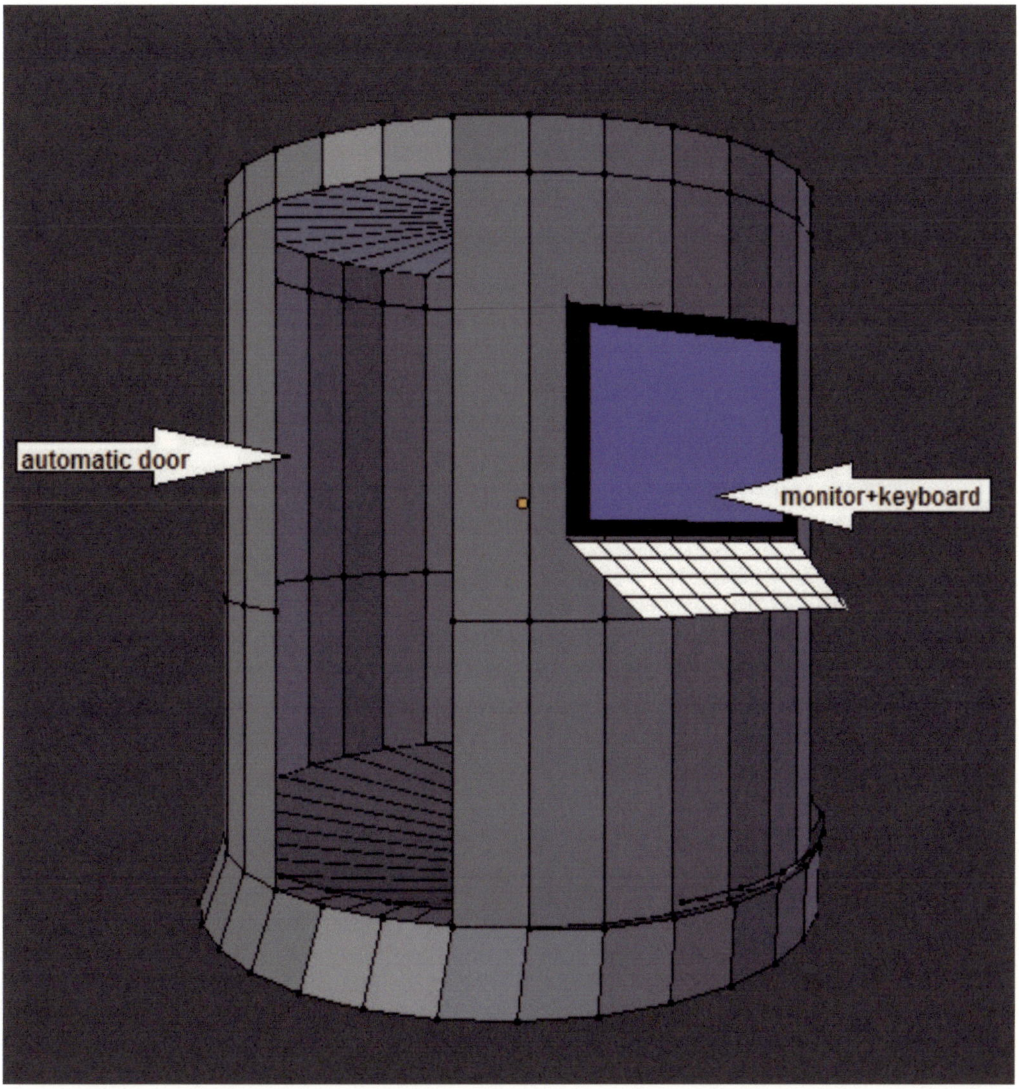

Fig. 8. Hypothetical scanning device based on saser technology

Finally, real-life data from airports already using FBS as opposed to the ones not applying such practices ought to be performed, so as to eliminate bottlenecks and optimise benchmarking techniques. The latter ought to consider the reasoning of airport authorities that have rejected the use of body scanners, based on "not satisfying outcomes", as postulated by their users.

Further research ought to account for the cumulative radiation levels for travellers as well as flight crew, in relation to the frequency at which these are subjected to WBI apparatuses. The outcome thereof would be compared to the percentage that would show potential health-related problems over a long-term period.

6. Conclusions and further work

This work has considered current legislative matters pertaining to safety and security levels in the aviation sector. It identified open-ended issues that require further research, both in the legal, as well as the technological area. To this end, it introduced basic operating principles of millimetre and backscatter x-ray scanners, detailing their characteristics and associated drawbacks.

Performed research indicates that the perception is not unanimous. A number of countries are aiming at a sector-wide deployment of WBI apparatuses, whilst other states either seek to develop alternative means, and/or to improve on already existing ones. This trend is equally mirrored by the preliminary outcomes of the questionnaire from a sample of travellers.

Up-to-date investigations concerning alternative means of passenger checking are likewise introduced and discussed. As such, owing to the novelty and the scientific wideness of this particular area it is the authors' view that further research ought to be conducted in areas such as benchmarking/quality and its effects on an airport's overall performance. In addition to the above, further work ought to be focused on grouping of potentially dangerous articles based on their physical and/or chemical properties, so as to develop effective scanning apparatuses.

In view of the above and acknowledging the fact that the FBS positioned at Hamburg's Fühlsbüttel Airport was taken out of operation, this study is serving as the platform for optimising security means at transportation hubs. Potential operating bottlenecks-as experienced by airport authorities already using WBI technologies- will be appraised and evaluated. The outcomes thereof will form the building platform of benchmarking models to optimise passenger handling.

It is also worth mentioning at this point that no specific and accurate measures can be derived thereof, as the time interval concerning the application of FBS in relation to the potential risk of carcinogenesis, is small and yet-to-be defined. The latter would include a diverse field of further potential influencing parameters as already outlined in the study of the German Agency for Radiation Protection. In particular, a clear split of environment-specific radiation as opposed to security-imposed radiation ought to be defined and boundaries established. For introducing a humanly-safe and secure apparatus, the effect of radiation on demographic groups ought to be further elaborated on. The latter forms part of further and continuous research aimed at identifying probable causes and interrelations.

7. References

Bart E., "Changes in Airport Passenger Screening Technologies and Procedures: Frequently Asked Questions", in *Congressional Research Service*, 7-5700, R41502, January 26th, 2011

Bergman B., Klefsjö B., „Quality: from Customer Needs to Customer Staisfaction", McGraw-Hill, 1994

Broadcasting Act 1990

Cavoukian A, "Whole Body Imaging in Airport Scanners: Activate Privacy Filters to Achieve Security *and* Privacy", in *Information and Privacy Commissioner of Ontario*, March 2009

Convention For The Unification Of Certain Rules Relating To International Carriage By Air, Signed At Warsaw On 12 October 1929 (revised at the Hague in 1955 and Montreal in 1999)

EU Commission Regulation 185/2010

European Directive 95/46/EC

European Directive No 2320/2002

European Resolution 0521 of October 23rd, 2008

Federal Aviation Administration, FAA, Research, Engineering and Development Advisory Committee, *Report of the Subcommittee*, February 2002

Frasch G., Kammerer L., Karofsky R., Schlosser A., Spiesl J., Stegemann R., „Die berufliche Strahlenexposition des fliegenden Personals in Deutschland 2004-2009", in *Bericht des Strahlenschutzregisters, Fachbereich Strahlenschutz und Gesundheit*, Salzgitter, August 2011, urn:nbn:de:0221-201108016029

IOP Science http://iopscience.iop.org

Kounis A.B., McAndrew I.R., O'Sullivan J.M., „Determining capability indices with small samples", in *Advances in Manufacturing Technology XIV*, 16th National Conference on Manufacturing Research, 5-7 September 2000

Kounis, L. D., McAndrew I.R., O' Sullivan J., M: *Modelling of parametric relationships in jet fans*, 3rdInternational Conference on Quality, Reliability and Maintenance, QRM 2000, pp.223-226, Ed: G. J. McNulty, University of Oxford, Oxford, United Kingdom, 30-31 March 2000

L.D. Kounis: Robustness and Capability Indices in the optimisation of an Airline's fleet. Bridging contradicting outcomes, in "Enterprise Networks and Logistics for Agile Manufacturing", pp. 359-397, Ed. Lihui Wang and S.C. Lenny Koh, Springer Publ., September 2010, ISBN 978-1-84996-243-8

New Scientist Journal reference: *Physical Review B*, in press

Obscene Publications Act 1959, United Kingdom

Peterman Randall D.: *Passenger Rail Security :Overview of Issues*, CRS Report for Congress, Updated May 26th, 2005, Congressional Research Service, The Library of Congress

Regulation (EC) No 45/2001, Article 28(2)

Roper Carl, A. *Risk Management for Security Professionals*, Butterworth-Heinemann, 1999

S.A.F.E.R A.I.R Act, Transportation and public works. 111th Congress, 2009-2010

The Heritage Illustrated Dictionary of the English Language, International Edition, Vol. II, Houghton Mifflin Company, 1979

Topouris S., Xenos-Kokoletsis A., Kounis L.D., "A feasibility study pertaining to the accommodation of the A380 at Athens International Airport", Paper ID2288, November 11-13, 2011 Zhengzhou, China

University of Nottingham Ref: http://www.rexresearch.com/kentsaser/kentsaser.htm

4

Laser Scanner: eSafety & ITS Applications

Nieves Gallego Ripoll
Universidad Politécnica de Valencia
Spain

1. Introduction

This chapter describes a proposal for a new type of Laser Scanner for use in traffic management. The chapter begins by outlining current traffic management systems, ongoing research areas and the perceived failures of present and some proposed implementations. Though the initial background presentation is long winded this is deemed necessary so the reader can gain a fuller insight into the new Laser Scanner technology being proposed.

The United Nations on its report "State of world population 2007" (UNFPA, 2007) declared that:*"In 2008, for the first time, more than half of the world's population will be living in urban areas. By 2030, towns and cities will be home to almost 5 billion people"*. This growth both in cities and inhabitants is directly proportional to mobility demands and requirements. In this situation, strategic traffic management is a key factor, even though it also involves different sectors such as economic (oil dependency), environmental (pollution, noise...), safety (both drivers and pedestrians) and health (breathing problems, circulation...).

Traffic authorities are encouraged to coordinate traffic management and monitoring focusing on Road Safety problems. The Intelligent Transport Systems (ITS) is a solution in traffic management and user mobility thanks to the development and use of different applications and technologies (Ertico, 2008; ITS, 2008). Intelligent Transport Systems and Services include a wide variety of sectors and areas that previously did not have a direct connection between them.

ITS systems increase road safety by incorporating breakthrough technologies in different levels of management and control. These subsystems (traffic flow information systems, travelers' information systems, highways monitoring and management systems, incident management, pedestrian detection...) rely on the traffic parameters provided by sensors measuring in real-time the situation on the road. The quality of the metrics received is a key for a correct network management and monitoring (García, 2000; Martin et al., 2003; Van Arem et al., 1993). The Utah Traffic Laboratory (UTL) on the report "Detector Technology Evaluation" carried out in November 2003 specifies that: *"The data collected must be plentiful, diverse, and accurate. These complex data requirements present a challenge to traffic detection systems"*(Martin et al., 2003). This statement demonstrates the importance of the traffic data and therefore of the sensor systems.

Traditionally, sensors have been used as mere measurement points, isolated from each other, that send the information to a central node which gathers the intelligence and information generated by the system (Sepulcres & Gozálvez, 2006). This node supports the applications, the processing and analysis of data received and also, in certain cases, executes the required remedial actions or control strategy.

Since there are a number of sensing technologies relevant to ITS, several studies and reports comparing them have been published (Fang et al., 2007; Klein et al., 2006; Mimbela & Klein, 2007; Skszek, 2001; Turner & Austin, 2000). Inductive loops and magnetic sensors require road closure for their installation and maintenance. The video image processor is vulnerable to weather and light conditions. Ultrasonic and infrared sensors lack stability in noisy environments. Microwave radars are not suitable for vehicle classification (Klein et al., 2006). Laser scanner sensors installation and maintenance do not require road closure, and are more robust and less vulnerable to weather conditions and noisy environments, and are thus the best option for traffic parameter detection.

To solve this situation, a solution is presented to optimize infrastructure sensors already installed in cities, focusing on the case of laser scanners, tackling the problem in two ways: on one hand by developing sensors able to provide more accurate and valuable traffic parameters and on the other by endowing the sensors with "intelligence". These innovations introduce an approach geared towards a new generation of sensors able to cooperate and communicate with the traffic control center (TCC).

The detection system used is a laser scanner, which provides information in real-time about: detection (presence), flow rate, counting and vehicle classification. The classification has been made following statistical methods in eight groups (8+1): powered two-wheels (motorbikes, ...), car, car with trailer, van, truck, truck with trailer, artic, bus + unclassified. The intention is to improve traffic management quality by providing supported information about the traffic parameters on the road network.

1.1 State of the art

Road safety became a global concern with the release of the, "World Report on Road Traffic Injury Prevention" (Peden et al., 2004). This report declares that, *"Of all the systems with which people have to deal every day, road traffic systems are the most complex and the most dangerous. Worldwide, an estimated 1.2 million people are killed in road crashes each year and as many as 50 million are injured. Projections indicate that these figures will increase by about 65% over the next 20 years unless there is new commitment to prevention"*. Some important mistakes were detected, "Insufficient attention to the design of traffic systems" (Peden et al., 2004). This statement confirms the reason to develop and design traffic monitoring systems to help reduce the actual traffic accident rate.

In April 2004, the General Assembly of the United Nations (UN) approved a resolution urging the improvement of road safety in the world (UN, 2004). This assembly invited the World Health Organization (WHO) to coordinate the safety activities within UN organizations. This was the beginning of the, "Road Safety Collaboration Group" with representatives of more than 42 organizations. Key provisions of the UN resolutions related with "Road Safety" are summarized in:

- May 2003 - UN General Assembly: Global Road Safety crisis. Called on governments and civil society to raise awareness and enforce existing road safety legislation (UN, 2003b).
- November 2003 - UN General Assembly: Global Road Safety crisis. Formalized, "Improving road safety" as an agenda item for the 60t̂h General Assembly Session (UN, 2003a).

- May 2004 - UN General Assembly: Improving Road Safety. Acknowledged the release of the World Report on Road Traffic Injury Prevention (UN, 2004).
- May 2004 - World Health Assembly: Road Safety and Health. Requested to encourage research to support evidence-based approaches for prevention of road traffic injuries and mitigation of their consequences (WHA, 2004).
- October 2005 - UN General Assembly: Improving Global Road Safety. Encouraged international community to lead financial, technical and political support to improve road safety (UN, 2005).
- March 2008 - UN General Assembly: Improving Global Road Safety. Encouraged private and public sector to implement policies to reduce crash risk for occupants and other road users (UN, 2008).

In this situation, strategic traffic management is a key factor in Intelligent Transport Systems applications. In urban areas, the traffic control center gathers parameters and information about real time traffic which are used to implement traffic policies and legislation.

1.2 Review of related work: Laser Scanner in ITS systems

In the literature, three main research lines are the most important: sensor development, applications and improvement of time-of-flight (ToF) techniques. Today, most of the effort is focused on development of the applications. The group of authors working on sensor development focuses on laser optical systems and laser telemetry. In the University of California-Davis, Cheng et al.(Cheng et al., 2005; 2001) developed a laser-based detector using a laser line projected on the ground and the reflected beam is collected and focused onto a photodiode array. Even though this sensor can provide information at speed, it uses two lasers. The weakness of the system is sensor costs are increased by using two transmitters and receivers. A slight error in the laser beam alignment during the installation of the sensors can result in significant errors and further, only one lane can be monitored at a time. The range sensor presented by the University of Central Florida cannot be applied in real-time (Hussain & Moussa, 2005).

Other research avenues consider it more important to use well designed and highly reliable commercial products than to work on the development of new sensors and work on improving acquisition and signal treatment without going into sensor technology. This group focuses on the application and represents the majority of the current research (Abdelbaki et al., 2001; Fuerstenberg & Dietmayer, 2004; Gallego et al., 2007; Harlow, 2001; Hussain & Moussa, 2005). Fuerstenberg and Dietmayer (Fuerstenberg & Dietmayer, 2004) investigate sensors installed in vehicles in order to detect moving pedestrians. Hussain and Moussa (Hussain & Moussa, 2005) develop a vehicle classification system called AVCSLII. The classification algorithm was produced by training neural networks and experimental results present five classes based on a database of 4995 vehicles. Harlow and Peng (Harlow, 2001) propose a solution to vehicle detection by processing range imagery. The solution produces different classification methods generated by two laser scan lines separated by 10°.

The authors do not take advantage of the two sweeping laser scan lines to exploit all the potential information, nor do they use the laser scans to measure vehicle speed. The proposal lacks in-depth analysis of speed, requires a larger database and further study under different weather conditions. Lastly, scientists have made a breakthrough in circuit design. It is worth mentioning the research of J. Kostamovaara of the University of Oulu in Helsinki (Palojarvi et al., 2005; Pehkonen et al., 2006), who investigated and optimized the receiver channel for a pulsed, ToF, laser range finder using BiCMOS technology.

2. System description

Present Laser Scanners are not designed specifically for traffic control applications, though the few which are do not provide the software needed for signal acquisition and treatment thus none of the actual systems can provide information in real time.

The proposed Laser Scanner System provides a set of sensor (laser scanner), communications and software modules able to detect and classify vehicles in real-time, acquiring and storing their silhouette and 3D image in addition to required traffic parameters.

2.1 Technical requirements

Laser Scanners measure the ToF of the coherent light coming from an emitter laser, and reflected by the vehicle for its detection. By means of the laser technology, parallel and coherent beams can be generated. A pulsed laser beam is emitted by the laser scanner and reflected by the vehicle. The reflection is registered by the scanner receiver. The time elapsed between the emission and reception of the pulse is directly proportional to the distance between the scanner and the object. For correct vehicle detections the minimum number of readings to identify a vehicle is 20 measurements. Laser scanners measure the time-of-flight of the laser pulse reflected by the vehicle. The laser works in the infrared range, approximate wavelength 0,9 um, which is outside the visible range thus avoiding possible distraction to road users. Furthermore the chosen laser frequency also lies outside the bandwidth of sunlight thereby minimizing this source of interference.

The laser scanner is located at a suitable height above the road to cover all lanes. Therefore the system must be located vertically above of the road with a minimum height of 5 meters, Fig. 1(a). In order to provide the required cross-sectional readings on each vehicle, it is advised to work with angular steps of 1°, Fig. 1(b). This angle can be increased as long as direct vision to the lanes is guaranteed.

(a) Laser Scanner Positioning

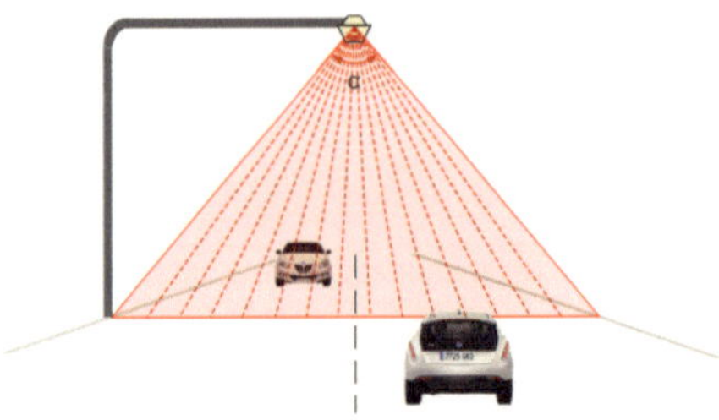

(b) Operation Principle

Fig. 1. Operation Principle

The communications between sensor and control unit included in the laser scanner system will be done via RS232 using a proprietary protocol. This communication can be implemented using the Ethernet protocol. By means of these specifications, the laser system should be able to detect and classify vehicles; as well as, acquire and store their silhouette.

After the preliminary studies carried out laser scanner system requirements for its use are shown in Table 1.

System Requirements	
General	
located vertically above of the road (m)	5
Maximum Consumption (W)	40
Laser Class	I (safe)
Emitter type	IR Laser diode
Communications	
Interface	RS232 optional RS485/422
Baudrate (KBd)	9,6 - 500
High speed option (MBd)	1,5
Scanner	
Scanning frequency (HZ)	15 - 100
Response time (ms)	53 - 10
Scanning angle (°)	100 - 180
Angular resolution (°)	0,25 - 1
Resolution (mm)	10
Functional	
The laser scanner system should be able to detect and classify vehicles	
The laser scanner system should be able to allow the acquisition and storage of silhouette of vehicles	
The laser scanner system should be located in the vertical of the road to a minimum altitude of 5 meters	
The laser scanner system must have direct vision to the lanes	
The laser scanner system should be able to communicate to the local traffic control center using Ethernet Communication	

Table 1. Technical Requirements

2.2 Laser Scanner measurement and data principle

Laser scanner is used in measuring mode (of the distance values). This means that the laser scanner outputs each distance value in two data bytes. In the standard measuring configuration, data bits 0 to 12 are used to represent the distance, while data bits 13 to 15 are not used. These 13 data bits enable 213-1= 8191 coding options to be represented. As we have selected a measured value resolution of 1 mm, this will result in a maximum measuring distance of 8191 mm.

The communications with the laser scanner are done through commands. Commands are a set of predefined instructions which carry out different operations like the configuration of the system. The following remarks are important to know how the laser scanner communicates, thus it is also important to understand the hardware developed:

- Data format is set as follows: 1 start bit, 8 data bits and 1 stop bit (8N1).

- Laser scanner outputs data in ascending angular steps. The angular values themselves are not transmitted; instead, the data field only comprises distance values.
- Laser scanner is configured by means of commands called telegrams. A send telegram always contains only one control command, and a response telegram contains only one response from the laser scanner.

The next instructions set must be followed in order to configure the laser scanner: baudrate configuration (9600, 38400 or 500K), range configuration (100°, 180°), resolution (1°, 0.5°, 0.25°) and measurement units (mm, cm).

2.3 Architecture hardware

The detector is divided into three parts: the sensor, the control unit, and the software classification program (Gallego et al., 2009). The sensor is installed in the pole of the traffic light. The control unit comprises the hardware and communications module. And lastly, the classification program running on a computer.

The communications between the sensor and the control unit is via RS422. At high speed baudrate communication it is not possible to use serial communication between the control unit and the computer, therefore we have chosen standard USB. This way connection to the computer would be effected using the USB port.

The control unit must: establish communication with the laser scanner, modify the configuration settings and carry out the signal pre-treatment using DSP microcontrollers (Digital Signal Processor). The signal processing is done in two stages which are in different physical locations. The vehicle detection is done on the control unit and the vehicle classification is done on the computer. The final control unit designed is shown in Fig. 2.

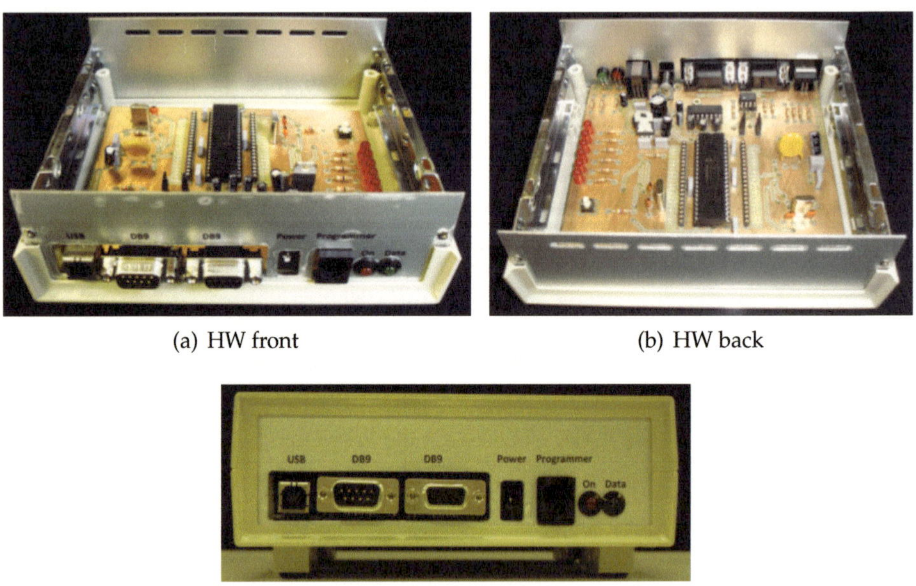

(a) HW front

(b) HW back

(c) HW connectors

Fig. 2. Hardware Design

The first task was to establish communication with the laser scanner sensor. The sensor sends data using the communication standard RS422; it provides a service for data transmission, using balanced or differential signaling. Therefore the first requirement was to develop hardware able to read this information using the computer serial port. For this purpose the hardware unit designed included differential drivers and converters. The differential driver and receiver pair was designed for balanced transmission-line and it is used to convert from RS422-TTL. Later on, another converter from TTL to RS232 level is needed to communicate using the serial port of the computer. The communication was established successfully with the laser scanner sensor. The bidirectional communication required for setting the configuration was facilitated by means of the sensor commands.

DSP microcontroller is the only solution, because the sensor sends 361 measurements per scan each 13 ms. The DSP firmware was developed for:

- signal pre-treatment,
- laser scanner configuration,
- and data configuration.

3. Data normalization

The signal processing is done in two stages which are in different physical locations. The vehicle detection is done on the DSP and the vehicle classification is done on the computer, using decision trees. Prior to vehicle classification data detected by the laser scanner needs to be pre-treated. This pre-treatment consist of geometric corrections and detection and elimination of static objects on the road. The goal of data normalization is obtain a 3D matrix representing detected vehicle.

Data received from the laser scanner must be normalized in three steps:

- Geometric corrections, due to data measure by laser scanner.
- Elimination of static objects on the road.
- Lost reflections.

3.1 Geometric corrections

Data sent by laser scanner is the distance between the sensor and the vehicle or the detected object. This distance is not perpendicular to the road and is related with the angle, α, at which the measurement is taken. Several authors refer to this consideration as geometric correction (Harlow, 2001). Graphically, in Fig. 3, value measured is A, whilst the information needed is the height h and the position x. By simple mathematics the height of the object h and the position x can be calculated.

3.2 Static objects

Detection and elimination of static objects on the road is easily implemented by storing an offset vector in absence of vehicles, Fig. 4. This offset vector is subtracted from the laser scanner data thus ensuring permanent fixtures and road anomalies are eliminated.

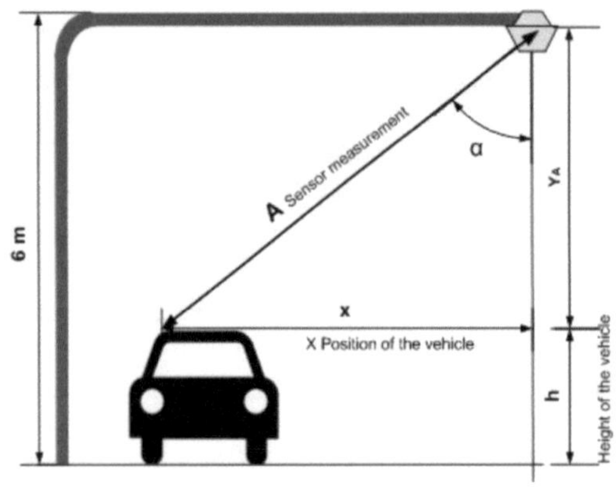

Fig. 3. Geometric Corrections

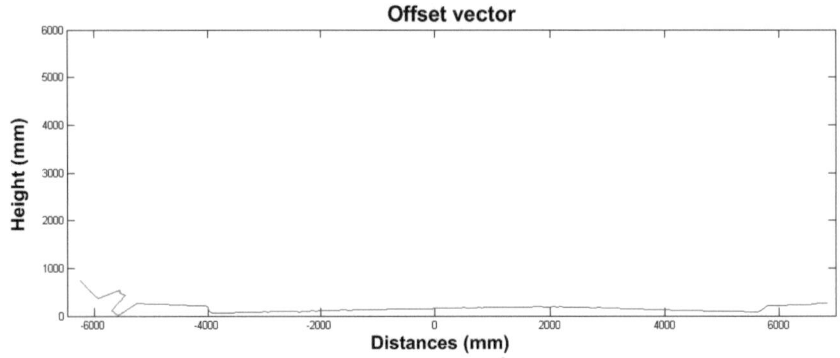

Fig. 4. Offset Vector

3.3 Lost reflections

The intensity of reflected sunlight from the road is a critical parameter that affects sensor measurements. Clouds blocking the sun are the most important cause for fast and/or large changes in intensity.

As explained previously, the laser scanner principle of operation is based on the measurement of the ToF. The changes in light intensity produce reflections too weak thus they are not received within the maximum detection time set by the sensor. These changes generate false measurements that can cause the sensor to report misleading data.

This effect is a characteristic of the vehicles most reflective parts i.e. the front and back windows. The light of the laser pulse emitted undergoes a reflection at an angle that does not agree with the detector position, thus the diffuse reflection does not take place in a direction that is advantageous to the detector. Fig. 5 shows the image of a vehicle with lost reflections that can be seen in the front and lateral windows because they show height values among 0 and 200 mm, corresponding to the blue color. In the case of the front window, it is possible to verify the data as the expected received pulse values should be between 1.000 and 1.400 mm, while what it is actually received are lower values due to the lost reflections.

In order to be able to correct these errors due to the intrinsic nature of the operation of the sensor it is necessary to do a depth revision of the detected values. After several studies, we conclude that in this situation all the erroneous values had a typical characteristic of height lower than one meter. Therefore, by filtering the information the erroneous data, so called lost reflections, are eliminated.

The following algorithm was used to eliminate the lost reflections:

- For each of the rows of the vehicle matrix:
  - Step one: If there are unknown values between well known values, these known values are copied instead of each of the unknown values.
  - Step two: If there are unknown values between two data sets with well known values, the greater values are copied instead of each of the unknown values.
  - If there is a row in the matrix containing only unknown data, the row is left as it is.
- The matrix is transposed.
- For each row repeat steps one and two.

This algorithm is programmed in the DSP firmware. To reduce the processing time only the valid vehicle detection data is sent to the computer.

The corrections after executing the algorithm are shown in Fig. 5.

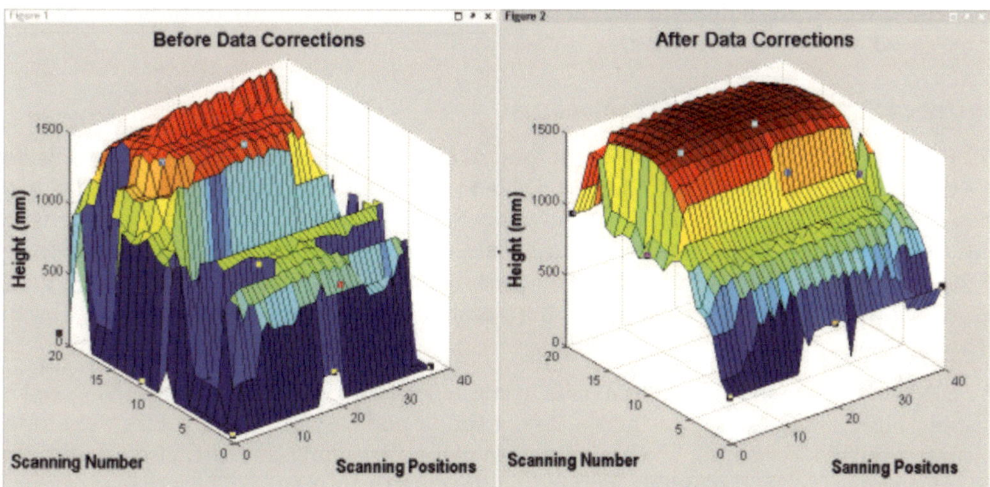

Fig. 5. Lost Reflections

3.4 Data anomalies

Laser scanner sensors that are mounted to the side or over a road may experience two types of data anomalies: lateral views and the effect of the speed (Klein et al., 2006).

Lateral view is an effect that can appear when a large vehicle hides another smaller one. The effect of tall cars blocking other lanes is shown graphically in Fig. 6. In this case a passenger car is almost hidden by a taller bus driving in an adjacent lane. The bus covers the passenger car from the sensor. The sensor detects the car, measurement A, but the next value detected, measurement B, is for the bus, so both detections overlap and cause an undercounting. In order to determine overlaps and solve errors, it is necessary to analyze the matrix with the isolated vehicle information.

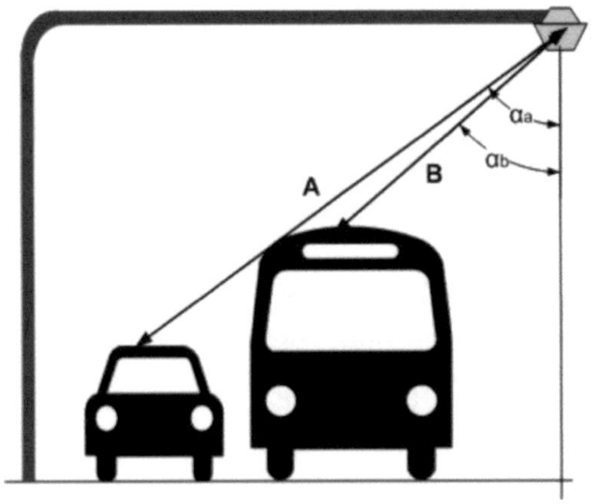

Fig. 6. Data Anomalies: Taller cars blocking angle's view

Also, the effect of speed has to be eliminated. Not all vehicles travel at the same speed on the road, so the information we have from each of them varies. This affects the silhouette obtained. By means of decimation and interpolation techniques, the matrices corresponding to all the vehicles which determinate their size are standardized. Thus, distortion effects due to the speed of travel are corrected.

3.5 Detection techniques: string contour

The detector covers not only one lane but four, so that more than one vehicle can be detected at a time, see Fig. 7(a). It is important to be able to isolate the different vehicles detected even though they have been detected at the same time. It is easy to reconstruct the vehicle when only one car crosses over the detection area. In this case all the information received belongs to the same car, and stream-by-stream the vehicle can be reconstructed. But in general, under normal circumstances this would not be the situation when several vehicles cross under the detector at the same time.

We used string contour techniques to isolate and reconstruct vehicles (Wilson, 1997), see Fig. 7(b) and Fig. 7(c). This technique makes it very simple and easy to implement the monitoring of contours in binary images. The basic function of the algorithm is to search, isolate and store the information of vehicles in independent files in a matrix format.

3.6 Conclusions

After a depth data normalization process information gathered by laser scanner can be isolated in a matrix. This matrix represents the vehicles detected once isolating one from others. Thus, the software requirements defined are shown in Table 2.

4. Methods

The classification algorithm aims to define a set of standards or group of patterns that allow the classification of the study group into two or more categories (Bishop, 2006; Jain et al., 2000). It was decided to use statistical recognition of standard systems because, from a set of numerical measures with well known probability distributions, recognition and classification is possible.

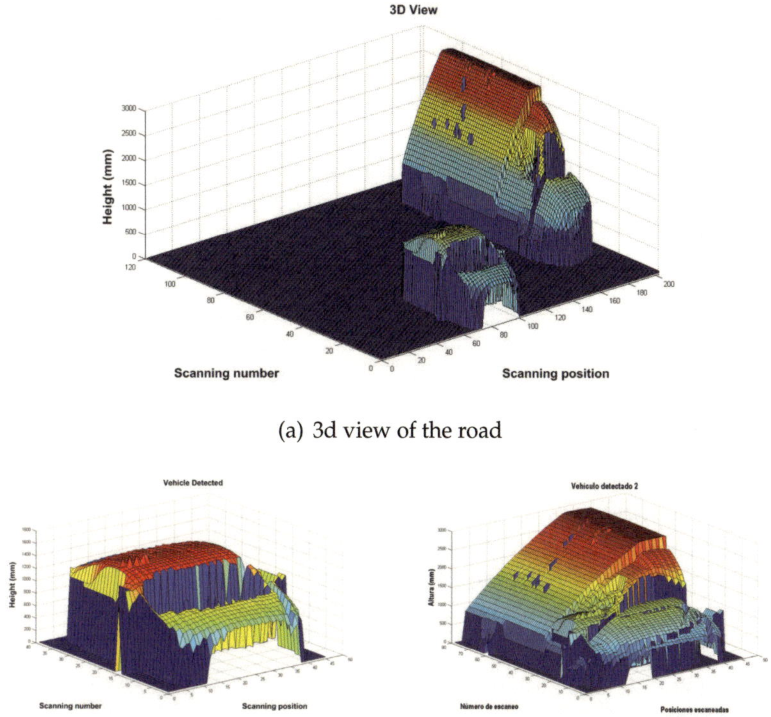

(a) 3d view of the road

(b) Vehicle detected 1

(c) Vehicle detected 2

Fig. 7. Detection techniques

Software Requirements
Data Acquisition
The system must be able to treat correctly the information sent by the laser scanner. DSP firmware must be able to carry data normalization by setting an offset line. The system, after the correct data normalization, must sent these information to the computer for its processing via USB port at 1.5MBd.
Data Treatment
Information received must be correctly treat in order to get valuable information to be sent to the traffic control centers. After this process the following information is due to obtain: vehicles detection and all the statistical and values that can be inferred of them: intensity of the traffic (veh/h) and density mainly. The processing of the signal must follow the algorithms defined of data normalization (geometric corrections, static objects and lost reflections) and data anomalies (lateral views and the effect of the speed). Information provided must be reliable and valuable.

Table 2. Software Requirements

The classic techniques of patterns recognition and classification can be parametric (Bayesian, linear discriminant, support vector machine. . .) or nonparametric (histograms, K neighbours, decision trees . . .). Since the 90's it is being use in addition techniques based on artificial neuronal networks. In the traffic area the neuronal networks have been used habitually in the case of artificial vision, with important congresses of IEEE dedicated to the subject. For example, the congresses on *"Computer vision and Pattern Recognition"* has already being celebrated since 1985. Multinomial pattern matching has been used in the detection of vehicles (Koch & Malone, 2006) by means of IR or acoustic sensors (Cevher et al., 2007a;b).

After the data recollection, the first stage in the pattern recognition is the learning or training stage. In this phase it is used a group of data which, a priori, the class is known which they belong and this group serve to train to the system. This strategy denominates supervised learning, when knowing itself the patterns; if they were not known it would deal with learning non supervised (Friedman & Kandel, 1999; Ripley, 1997).

In the cases of supervised learning two situations can be given: that the objective is to assign each value of entrance to one of a finite number of discreet categories, which is called classification problem; or that the objective is to assign one or more variable continuous, being then regression. The cases of supervised learning can have different objectives: to detect groups of similar examples within the data, clustering; to determine the distribution of the data in the entrance space, density estimation; or to project the data in spaces of smaller dimension to visualize them (Bishop, 2006).

The group of data known like training group, defines the patterns corresponding to each one of the categories that are desired to classify, and allows to determine the function that will be used for the classification (Bishop, 2006). The best the training group would be, the best the patterns will define and better discriminations will allow to carry out (Jain et al., 2000).

Once the model has been trained begins the second and last stage that is the classification or recognition of new data. This procedure is realized with the test group. In some practical applications where data are complex it is needed to pre-process itself to transform them into a new space of variables, so that the problem of recognition of pattern is easier to solve. The purpose of this stage of pre-processed (also called extraction of characteristics) is to find useful characteristics that are easy to calculate and that they maintain information for the discrimination in classes. That is to say, one is in charge to extract a set or vector of characteristics that will locate the data in points of the n-dimensional space of classification (Bishop, 2006; Sobreira & Rodríguez, 2008). In this case the vector of characteristics is used like entrance of the algorithm of recognition of landlords. Also the pre-processed is realized if it is desired to increase the speed of calculation in applications of real time (Bishop, 2006).

4.1 Classification: Decision trees & boostraping

Given a collection of records they are divided into two subgroups: a training group and a test group (Bishop, 2006). The classification algorithm is designed used decision trees. These trees are constructed beginning with the root of the tree and proceeding down to its leaves. Using decision trees we developed a model for each class attributes as a function of the values of other attributes (Matlab, 2009; Teknomo, 2009). Therefore future and unseen records could be assigned a category or class as accurately as possible. The test set is used to determine the accuracy of the model.

The construction of decision trees is the method of supervised inductive learning more used. Its dominion of application is not restricted to a concrete scope but they can be used in diverse areas, from applications of medical diagnosis to games like the chess or systems of weather forecast (Díaz, 2007).

The decision trees adjust perfectly, as anticipated, to the data of the learning group, but they can realize predictions not very right in the case of new values. The branches inferiors, especially, can be seen strongly affected by atypical values. Simpler trees can offer often better results avoiding overfitting it (Matlab, 2009).

The technique of bootstrap is a resampling method that was proposed by Bradley Efron in 1979. It is used to approximate the distribution in the sampling of a statistical one. We used Boostraping techniques since boosting is a general method for improving the accuracy of any given learning algorithm (Michien et al., 1994). The concept of boosting, i.e. adaptive resampling, applies to many learning methods (Correa, 2004; Schapire & Singer, 1999). However, resampling is particularly advantageous when used in conjunction with decision trees, because of two key reasons; decision tree algorithms are relatively efficient in high dimensions, and decision trees tend to have a bigger component of variance than other methods like nearest neighbors or neural nets (Apte et al., 1998).

4.2 Classification standards

Generally, the standards regarding vehicles classification do not have entity by themselves, but they are part and included in more extensive norms about ITS systems (Middlenton et al., 2002). The Department of American transport, by means of the family of standards NTCIP (acronym of National Transportation for Communication ITS Protocol) has defined the communication protocols and the vocabulary necessary to allow the interoperability between commercial equipment of traffic control of different manufacturers (NTCIP, 2009). Standard NTCIP 1209 deals with traffic sensors. It is in this standard where it will be included all regarding detection, count and classification of vehicles. In Europe, different standards for equipment have been developed from capture of data, but at national level, in countries like: France, Holland and Germany (Middlenton et al., 2002).

The number of axes of the heavy vehicles is also the base of the American standard defined by the FHWA (acronym of Federal Highway Administration) member of the American Department of Transport. The FHWA in its report "Traffic Monitoring Guide' ' it states: *"for many of the carried out analyses by the traffic agencies the simple schemes of three groups (vehicles of passengers, articulated trucks of a unit and trucks) are valid, but other times a more sophisticated classification is required"* (FHWA, 2001). At the moment, from the North American states are few that completely use the classification in 13 groups defined by the FHWA, although they use variations of the same following the vehicle park that circulates around each one of the states. In addition, the majority of the American manufacturers of commercial equipment already provides the classification of vehicles in the categories defined by the FHWA (Metrocount, 2009). The American standard is focused towards a classification centered in trucks, since of the 15 classes that define 6 of them they base on trucks with and without trailer and with diverse axes.

Finally, on Europe a great part of the counting equipment and commercial systems of classification are based on German standard TLS (defined by the federal institute of investigations of highways, Bundesanstalt fur Strassenwesen) (Bundesanstalt, 2008).

Like in the American case, the most important companies of the sector also have adopted this standard. An example of it is: Weiss Electronics with models like the MC2024 (Electronic, 2009); Efkon with models like the AE TITAN 3000 (Efkon, 2009); and Xtralis with models like the DT 350 or TT290 (Xtralis, 2009). Also, European projects of VII Program Frame like project TRACKSS, use the same standard for applications of vehicle classification.

Although these classifications of vehicles are used by the administration, exist other that are not based on ITS applications. It is the case of the automobile companies, that use different classifications from their models in classes like: microcar, sedan, station wagon, sport cars, grand tourers ... They exist, also, the famous classifications of security Euro NCAP (acronym of European New Car Assessment Programme) that group among others to the vehicles in: supermini, family car, executive, MPV, 4x4 ... (EuroNCAP, 2009).

The classification we used is done based on the German standard TLS. The classification is done in eight groups (8+1): powered two-wheels, passenger car, car with trailer, van, truck, truck with trailer, artic and bus, + unclassified. See Fig. 8.

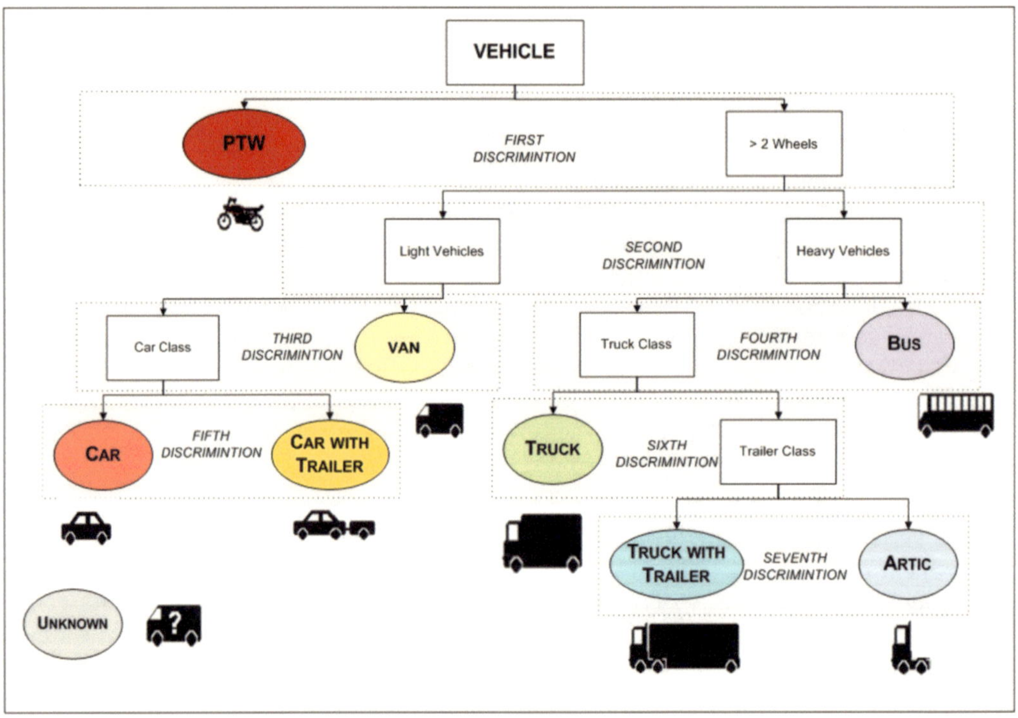

Fig. 8. Classification Tree

5. Experimental results

5.1 Laboratory tests

The detection and classification abilities of the prototype system were tested using a laboratory model in an environment as similar as possible to the road environment. The detector was placed 1.5 meters above the road instead of the 6 meters used in real conditions. Different vehicle prototypes were made on a 1:60 scale to the real ones.

Initial detection tests were carried out with prototypes crossing the detector area every 2 seconds. In this situation, the vehicle detection worked successfully. The confusion matrix shows the counts of the actual versus predicted class values. It shows not only how well the model predicts, but also presents the details needed to see exactly where things may have gone wrong, and it is represented by:

- TP (True Positive) - number of vehicles detected = 495.
- FN (False Negative) - number of vehicles not detected = 4.
- FP (False Positive) - number of objects different from vehicles that have been detected = 1.
- TN (True Negative) - number of objects different from vehicles that have not been detected = 0.

Evaluation from laboratory tests			
Confusion Matrix			
Detection System		D	$\overline{D}$
	D	495	1
	$\overline{D}$	4	0
Detection Rate: 99.2%			
Precision: 99.8%			
Accuracy: 99%			

Table 3. Laboratory Tests - Vehicle Detection

The system achieved high detection accuracy and precision in the case of vehicle detection with rates of 99.2% and 99.8%, respectively. When the prototypes speed was increased the detection was also successfully.

The classification tests were done using the following protocol:

- The prototype A crosses the detection area every 10 seconds at a constant speed, for two minutes.
- Idem for prototype B.
- Prototypes A and B cross the detection area at the same time with similar speed every 10 seconds for two minutes.

The protocol was followed with different prototypes at different positions to verify the correct classification of vehicles. Five detections were not successful due to the fact that two cars were detected as one. Although this case was solved with modifications in the signal processing software, in terms of detection, it has to be considered as an error in the preliminary tests.

When prototypes are used, besides the problems caused by them all having the same scale, there is also the problem of the discrimination equations having to be modified proportionally to the scale used in them. Height and weight should be reduced for the correct use of the data extraction method therefore the classification was tested using a suitable database from previous projects.

5.2 Experimental tests

The system was installed in the city of Valencia, Spain, on an urban road with two lanes of traffic in both directions. Thus, the system was tested with real urban traffic conditions. Also installed was a fixed camera next to the sensor to verify the information received and also to provide video images. The sensor is located on a post exclusively for this purpose. After the installation, minor adjustments were made to adapt the system to the characteristics of the road and the detection zone was delimited. The sensor was calibrated to correct possible misalignments during the installation. The system was installed outdoors for a period of six months. This period allowed testing the system in sunny conditions and at different times of the day.

First experiments were conducted at two different times of the day, early morning time between 8am and 11:30am, for "normal light value", and early afternoon period between 12noon and 2pm, for "high light intensity". The early morning is used for a reference when the sensor is known to be accurate. The early afternoon is used for testing the signal pre-treatment algorithm.

Several tests were carried out both in the "in the early morning" and "early afternoon" . All tests were successful at the communication level together with the signal acquisition and treatment software. Therefore, we can conclude that the detector work as expected in both light conditions.

The results of the experimental detection tests are shown in Table 4. The indicators employed to evaluate each classification group are the standard evaluation metrics in case of detection given by: detection rate 97.9% (probability of vehicles from a particular group classified as such), precision 99.7% (probability of vehicles classified from a particular group that belong indeed to this group) and accuracy 97.6% (probability of vehicles classified correctly). The indicator precision penalizes false positive, it is to say 3 detections of irrelevant items in this case. Meanwhile detection rate penalizes the false negative of the system; that is, 21 vehicles not detected.

Confusion Matrix			
Detection System		D	$\overline{D}$
	D	976	3
	$\overline{D}$	21	0
Detection Rate: 97.9%			
Precision: 99.7%			
Accuracy: 97.6%			

Table 4. Experimental Tests - Vehicle Detection

Even though several experiments were carried out for classification, the results present are based on 972 vehicles detected: 51 powered two-wheels (2W), 675 passenger cars (PC), 178 vans (V), 37 trucks (T), 7 trucks with trailer (TT) and 24buses (B). The results in Table 5 show high accuracy in the case of vehicle classification with a rate of 94.24%. Analysing the results one can conclude:

- The powered two-wheels are always classify correctly: 100% recall and precision.
- The best detection ratio, after powered two-wheels, is the passenger car vehicles. Thus, 97.6% of the vehicles from this group have been detected correctly.

- Errors in classification in the case of passenger car and van categories are due to the similarity between minivans and some passenger cars, mainly crossovers or category equivalent to the EuroNCAP class "Small Off-Roaders".
- The precision rate in the case of bus category is low, 81.5% due to the lack of sufficient samples.
- Because of the misclassifications of buses and trucks the recall rate in truck category is relatively low, 73.0%.
- There were no car and trailer combinations in the tests.

Evaluation from experimental tests						
Confusion Matrix						
	2W	PC	V	T	TT	B
2W	51	0	0	0	0	0
PC	0	659	15	0	0	0
V	0	15	151	4	0	1
T	0	1	12	27	0	1
TT	0	0	0	2	6	0
B	0	0	0	4	1	22
Accuracy: 94.2% based on 972 vehicles detected						
Classes	2W	PC	V	T	TT	B
Recall(%)	100	97.6	84.8	73.0	85.7	91.7
Precision(%)	100	97.8	88.3	65.8	75.0	81.5

Table 5. Experimental Tests - Vehicle Classification

6. Conclusion

Current Traffic Monitoring relies on information provided by the sensors installed in the road. This information is used both in real-time, to know what is happening on the roads, and in logs stored over long periods to have a global vision of the traffic over periods: of months or even years. The important aspects of the sensors is not only the parameters measured but also their reliability.

A laser scanner system for vehicle detection and classification, with high reliability, has been developed with a field of view able to cover up to 4 lanes. An original methodology for the treatment of the problem has been developed, with statistical rigor, from the normalization of the signals to silhouette generation. Preliminary tests presented good results with precision rate in detection of 97.9% and accuracy rate in classification of 94.2%.

The system and method developed presents a valid option which is both simple and with a reduced computational cost. The system offers the possibility of using sensors that use the vehicles area detection principle with high accuracy in both detection and classification. Although the laboratory and field results validate the design, further tests will be performed to assess the impact of different weather conditions i.e. snow or rain. Meanwhile, a prototype is working at the Traffic Control Center of Valencia (Spain) to use in different traffic control applications such as cooperation between Laser Scanner and traffic lights for a better traffic distribution or restriction of vehicles to the city center.

7. References

Abdelbaki, H., Hussain, K. & Gelenbe, E. (2001). A laser intensity image based automatic vehicle classification system, *Intelligent Transportation Systems, 2001. Proceedings. 2001 IEEE* pp. 460–465.

Apte, C., Damerau, F. & Weiss, S. (1998). Text mining with decision trees and decision rules, *in* C.-M. University (ed.), *Conference on Automated Learning and Discovery*.

Bishop, C. M. (2006). *Pattern Recognition and Machine Learning*, Information Science and Statics, Springer, New York.

Bundesanstalt für StraSSenwesen (2008). Technical delivery terms for route stations, TLS, Bundesanstalt für StraSSenwesen. [Online; Last Access October 2008]. URL: *http://www.bast.de/nn43710/EN/eHome/ehomepagenode.html?nnn=true*

Cevher, V., Chellappa, R. & McClellan, J. (2007). Joint acoustic-video fingerprinting of vehicles, part i, *Acoustics, Speech and Signal Processing, 2007. ICASSP 2007. IEEE International Conference on* 2: 745–748.

Cevher, V., Guo, F., Sankaranarayanan, A. & Chellappa, R. (2007). Joint acoustic-video fingerprinting of vehicles, part ii, *Acoustics, Speech and Signal Processing, 2007. ICASSP 2007. IEEE International Conference on* 2: 749–752.

Cheng, H. H., Shaw, B. D., Palen, J., Lin, B., Chen, B. & Wang, Z. (2005). Development and field test of a laser-based nonintrusive detection system for identification of vehicles on the highway, *Intelligent transportation systems, IEEE transactions on* 6(2): 147–155.

Cheng, H., Shaw, B., Palen, J., Larson, J., Hu, X. & Van katwyk, K. (2001). A real-time laser-based detection system for measurement of delineations of moving vehicles, *Mechatronics, IEEE/ASME Transactions on* 6(2): 170–187.

Correa, J. C. (2004). Método bayesiano bootstrap y una aplicación en la estimación del percentil 85 en ingeniería de tránsito, *Revista Colombiana de Estadística* 27(2): 99–107.

Díaz, Z. (2007). *Predicción de crisis empresariales en seguros no vida, mediante árboles de decisión y reglas de clasificación*, Colección: Línea 3000, Editorial Complutense.

Efkon (2009). Ae titan 3000, EFKON AG. [Online; Last Access August 2009]. URL: *http://www.efkon.com/docs/TDsiAETITAN01E03.pdf*

Electronic, W. (2009). Classification detector mc2024, Weiss Electronic GmBH. [Online; Last Access August 2009]. URL: *http://www.weiss-electronic.de/englisch/03-komponenten/03-03-03-*

Ertico (2008). ITS Europe. [Online; Last Access November 2008]. URL: *http://www.ertico.com/en/welcome_to_ertico_its_europe.htm*

EuroNCAP (2009). Euro ncap vehicle types, European New Car Assessment Programme. [Online; Last Access January 2009]. URL: *http://www.euroncap.com/*

Fang, J., Meng, H., Zhang, H. & Wang, X. (2007). A low-cost vehicle detection and classification system based on unmodulated continuous-wave radar, *Intelligent Transportation Systems Conference, 2007. ITSC 2007. IEEE* pp. 715–720.

FHWA (2001). Traffic monitoring guide. executive summary, *Technical report*, US Department of Transportation. Federal Highway Transportation. Office of Highway Policy Information. Report Number: FHWA-PL-01-021. URL: *http://www.fhwa.dot.gov/ohim/tmguide/*

Friedman, M. & Kandel, A. (1999). *Introduction to Pattern Recognition: statistical, structural, neural and fuzzy logic approaches*, Vol. 32 of *Machine Perception and Artificial intelligence*, World Scientific, Singapore.

Fuerstenberg, K. & Dietmayer, K. (2004). Object tracking and classification for multiple active safety and comfort applications using a multilayer laser scanner, *Intelligent Vehicles Symposium, 2004 IEEE* pp. 802–807.

Gallego, N., Mocholí, A., Menéndez, M. & Barrales, R. (2007). Explotación de las infraestructuras actuales en entornos urbanos para aplicaciones its, *VII congreso español en Sistemas de Inteligentes de Transporte, Valencia.* .

Gallego, N., Mocholí, A. & Menéndez, M. (2009). A real-time laser scanner intelligent sensor for its applications, *Intelligent Transportation, 6th International Workshop on* pp. 21–26.

García, L. A. (2000). *Diseño e implementación de una arquitectura multiagente para la ayuda a la toma de decisiones en un sistema de control de tráfico urbano*, PhD thesis, Departamento de Informática. Universidad Jaime I, Castellón de la Plana. España. Director: Dr. Francisco Toledo Lobo.

Harlow, C. (2001). Automatic vehicle classification system with range sensors, *Transportation research. Part C: Emerging Technologies* 9(4): 231–247.

Hussain, K. & Moussa, G. (2005). Automatic vehicle classification system using range sensor, *Information Technology: Coding and Computing, 2005. ITCC 2005. International Conference on* 2: 107–112.

ITS (2008). ITS Spain. Ministry of Science and Innovation. [Online; Last Access November 2008].
URL: *http://www.itsspain.com/itsspain/*

Jain, A. K., Duin, R. P. W. & Jianchang, M. (2000). Statistical pattern recognition: A review, *Pattern Analysis and Machine Intelligence, IEEE Transactions on* 22(1): 4–37.

Klein, L. A., Mills, M. K. & Gibson, D. R. (2006). Traffic detector handbook, *Technical report*, US Department of Transportation. Federal Highway Transportation. Publication No. FHWA-HRT-06-108.

Koch, M. W. & Malone, K. T. (2006). A sequential vehicle classifier for infrared video using multinomial pattern matching, *Computer Vision and Pattern Recognition Workshop, 2006. CVPRW '06. Conference on* pp. 127–127.

Martin, P. T., Feng, Y. & Wang, X. (2003). Detector technology evaluation, *Technical report*, Department of Civil and Environmental Engineering University of Utah Traffic Lab. MPC Report number 03-154. [Online; Last Access November 2008].
URL: *http://www.mountain-plains.org/pubs/html/mpc-03-154/*

Matlab (2009). *Statistics Toolbox 7. User's Guide*, version 7.1 (released 2009b) edn.

Metrocount (2009). Traffic executive software, MetroCount. [Online; Last Access August 2009].
URL: *http://www.metrocount.com/products/mte/index.html*

Michien, D., Spiegelhalter, D. & Taylor, C. (1994). Machine learning, neural and statistical classification.

Middlenton, D., Gopalakrishna, D. & Raman, M. (December 2002). Advances in traffic data collection and management, white paper, *Technical report*, Office of Policy. Federal Highway Administration, Washington, DC. Work Order Number BAT-02-006.

Mimbela, L. E. & Klein, L. A. (2007). A summary of vehicle detection and surveillance technologies used in intelligent transportation systems, *Technical report*, The Vehicle Detector Clearinghouse and US Department of Transportation. Federal Highway Transportation.

NTCIP (2009). National transportation communications for its protocol. [Online; Last Access August 2009].
URL: *http://www.ntcip.org/info/*

Palojarvi, P., Ruotsalainen, T. & Kostamovaara, J. (2005). A 250-mhz bicmos receiver channel with leading edge timing discriminator for a pulsed time-of-flight laser rangefinder, *Solid-State Circuits, IEEE Journal of* 40(6): 1341–1349.

Peden, M., Scurfield, R., Sleet, D., Mohan, D., Hyder, A. A., Jarawan, E. & Mathers, C. (2004). *World Report on Road Traffic Injury Prevention*, 1st edn, World Health Organization, Geneva, Switzerland.

Pehkonen, J., Palojarvi, P. & Kostamovaara, J. (2006). Receiver channel with resonance-based timing detection for a laser range finder, *Circuits and Systems I: Regular Papers, IEEE Transactions on* 53(3): 569–577.

Ripley, B. D. (1997). *Pattern Recognition and Neural Networks*, Cambridge University Press, Cambridge, UK.

Sobreira, M. & Rodríguez, A. (2008). Clasificación automática de fuentes de ruido de tráfico, *Acústica 2008, V Congreso Ibérico de Acústica, 39ž Congreso Español de Acústica, Coimbra, Portugal* .

Schapire, R. E. & Singer, Y. (1999). Improving boosting algorithms using confidence-rated predictions, *Machine Learning* pp. 297–336.

Sepulcres, M. & Gozálvez, J. (2006). Dimensionado de sistemas de comunicaciones móviles entre vehículos para aplicaciones de seguridad, *Libro de Actas -URSI 2006. XI Simposium Nacional de la Unión Científica Internacional de Radio, 2006* pp. 1.158–1.161.

Skszek, S. L. (2001). State-of-the-art report on non-traditional traffic counting methods, *Technical Report FHWAAZ-01-503*, Arizona Department of Transportation.

Teknomo, K. (2009). Tutorial on decision tree. [Online; Last Access August 2009]. URL: *http://people.revoledu.com/kardi/tutorial/decisiontree/*

Turner, J. D. & Austin, L. (2000). A review of current sensor technologies and applications within automotive and traffic control systems, *Institute of the Mechanical Engineers, 2000. Proceedings IMechE 2000* 214(D): 589–614.

UN (2003a). Resolution adopted by the general assembly: Global road safety crisis. 56th plenary meeting, United Nations. A/RES/58/9.

UN (2003b). Resolution adopted by the general assembly: Global road safety crisis. 86th plenary meeting, United Nations. A/RES/57/309.

UN (2004). Resolution adopted by the general assembly: Improving road safety. 84th plenary meeting, United Nations. A/RES/58/289.

UN (2005). Resolution adopted by the general assembly: Improving global road safety. 38th plenary meeting, World Health Assembly. A/RES/60/5.

UN (2008). Resolution adopted by the general assembly: Improving global road safety. 87th plenary meeting, World Health Assembly. A/RES/62/244.

UNFPA (2007). State of world population 2007.unleashing the potential of urban growth, United Nations Population Fund.

Van Arem, B., Van der Vlist, M., Blonk, J. & Van den Berg, L. (1993). Demonstration of a general european road data information exchange network - gerdien, *Vehicle Navigation and Information Systems Conference, 1993., Proceedings of the IEEE-IEE* pp. 163–168.

WHA (2004). Resolution adopted by the 57th assembly: Road safety and health, World Health Assembly. WHA57.10.

Wilson, G. (1997). Properties of contour codes, *Vision, Image and Signal Processing, IEE Proceedings* - 144(3): 145–149.

Xtralis (2009). Traffic solutions, Xtralis. [Online; Last Access August 2009]. URL: *http://xtralis.com/p.cfm?s=22p=381*

5

Laser Scanning Technology for Bridge Monitoring

Shen-En Chen
Department of Civil and Environmental Engineering
University of North Carolina at Charlotte
USA

1. Introduction

After the collapse of the Minnesota I-35 bridge (August 1, 2007), there has been a renewed interest in the US to enhance bridge infrastructure monitoring (Liu et al., 2009). Other than developing traditional inspection and material testing techniques, there has been also increased discussions about possible applications of Commercial Remote Sensing (CRS) technologies for civil infrastructure monitoring (Al-Turk & Uddin, 1999, Shinozuka & Rejajaie, 2000, Chen et al., 2011). Laser scanning techniques are one of the remote sensing technologies that play significant role in environmental and infrastructure evaluation and monitoring. However, there are different sensing requirements for monitoring physical structures such as bridges, than conventional geospatial applications such as air quality, environmental impact and transportation operations, etc. The most important of which is the sensor resolution requirement.

This chapter discusses critical bridge monitoring issues and provides examples of applications of two laser scanner technologies that are currently being developed for bridge monitoring: 1) range finding laser (static) and 2) scanning laser vibrometer (dynamic). Both laser systems are terrestrial and single point systems that utilize mechanical or optical scanning mechanisms to create a field of view (FOV) of the optical receiver.

The range finding laser, also called LiDAR (for Light Detection and Ranging), is based on the transmission and receiving of pulsed lights. By determining the heterodyne laser beam phase shifts, scanning LiDAR can detect the distance information from a plane of data points, called point cloud. The point cloud information, which basically consists of the physical positions of any surface that the laser "sees", can then be used to detect useful critical information about a structure including the elevation (underclearance), surface (damage quantification) and deformation under loading (deflection), etc. Contrast to conventional analysis of photographic images, relatively simple algorithms can be used to manipulate the geometric point cloud data to retrieve the afore-mentioned information. Other bridge-related issues including validation of new constructions and comparisons before- and after critical event, can also be extracted from LiDAR scans.

Based on the measurement of Doppler effects of a returning continuous laser beam from a moving target, the scanning laser vibrometer (SLV) is a laser system that can detect the vibration of a subject. By covering the entire surface of a subject, SLV can not only detect the vibration frequencies but is able to separate the different vibration mode shapes of the subject (Oliver, 1995). This makes the SLV a very useful tool in isolating vibration-induced problems and in some cases, detect system or component level damages.

Because of the non-contact nature and the ability of sensing from a distance away, scanning lasers have the advantages of limited disruption to traffic, low labor requirements and providing permanent electronic documentations of the temporal changes of a structure. Scanning Laser is ideal as a bridge field inspection tool and can help reduce the costs of inspection and at the same time, enhance the accuracy in field inspections.

However, to implement laser scanning systems into bridge evaluation, one needs to understand the basics of bridge inspection practices and issues, in particular, recognizes the fact that some bridge issues are not necessarily associated with condition assessment, but with serviceability requirements such as adequate bridge underclearance, excessive bridge movements or traffic-induced vibrations.

Scanning lasers alone will not provide the critical information associated with bridge inspection, additional evaluation methodologies usually are needed to extract the necessary information associated with the bridge problems. The examples provided will demonstrate some additional physical theories that can be used to identify critical bridge information that affiliate with actual structural conditions.

2. Scanning laser technologies

The scanning laser technologies described herein can be best described as mid-range, terrestrial (ground-based) laser scanning systems that have found significant bridge health monitoring applications. The 3D scanning laser or LiDAR is a static laser that is a close cousin of the airborne LiDAR systems that generates large landscape footprints (Rueger, 1990). The SLV is the 2D dynamic laser systems that measures motions of specific position points based on Doppler shift measurements (Drain, 1980). Figure 1 shows the basic system components for both systems which include the ranging unit, the scanning mechanism, the laser controller and the data recorder.

Most laser scanners use servo-controlled rotating mirrors to reflect the laser beams on the target surface and usually allow the coverage of two-dimensional or three-dimensional areas. The servo-controller or galvanometers can be either moving-iron or moving-coil types and can be multiple-axial systems. The movable mirror system can be either through the use of a hexagonal mirror or the use of multiple-axis rotating flat mirrors. Laser beams bounce off the mirror and travel to different positions on the target and returns to the same mirror system. The returned laser beam help create the position data from the target surface. Most servo-controlled, rotating mirror has a fixed scanning speed. Hence, depending on the demand of data points, the duration of scan can vary from few minutes to several minutes. Figure 2 shows an example of a schematic moving-iron type galvanometer.

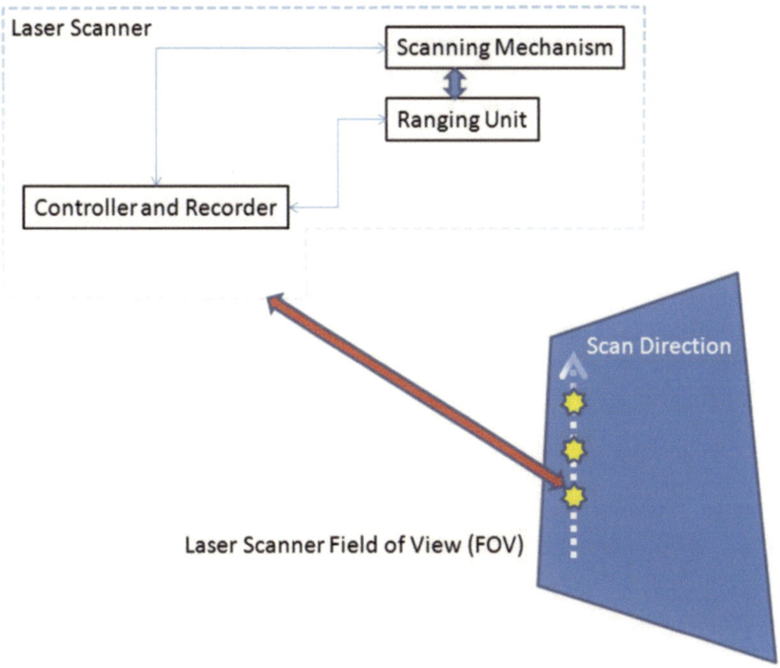

Fig. 1. Terrestrial Laser Scanning System Components

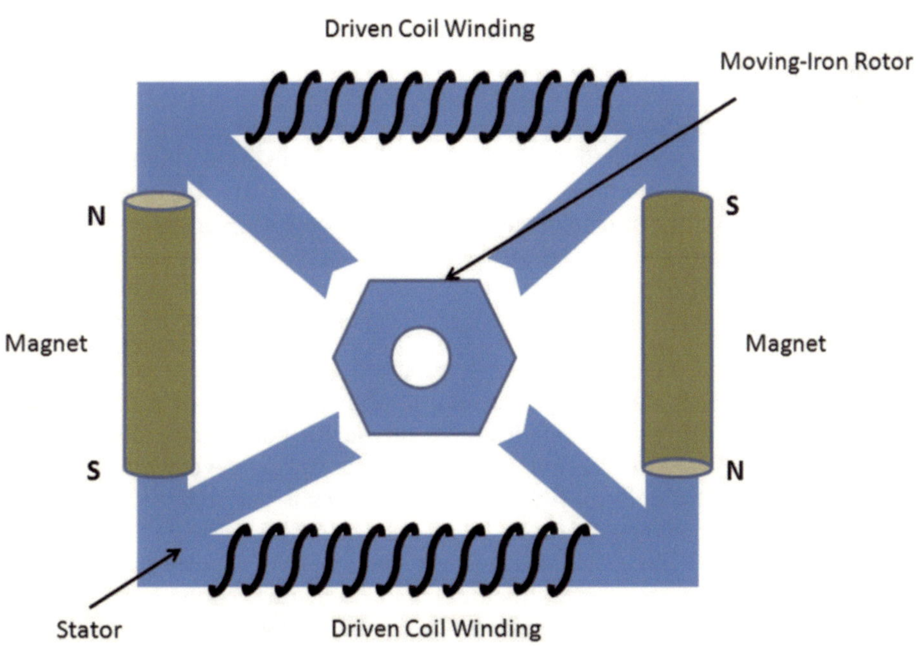

Fig. 2. Moving-Iron Galvanometer

2.1 3D scanning LiDAR

Figure 3 shows a scanning LiDAR and basics of the detection of returning signals. There are two approaches to the detection of the position data: 1) time of flight differences between emitted pulse and returned signals and 2) phase differences between the two signals (Jelalian, 1992). In a typical five to ten minute scan, the scanning LiDAR unit can collect millions of data points that include the XYZ position of each scan point. Applications of LiDAR scans are multi-facet: Airborne, long range LiDARs have been used in terrain mapping, ground canopy detection and environmental impact studies. Smith et al. (1997) reported using a Lawrence Livermore National Laboratory (LLNL) LiDAR for lunar surface (topography) measurements from the Clementine spacecraft. The solid state Nd:TAG (wavelength of 1.064 mm) laser has a maximum target range of 640 km.

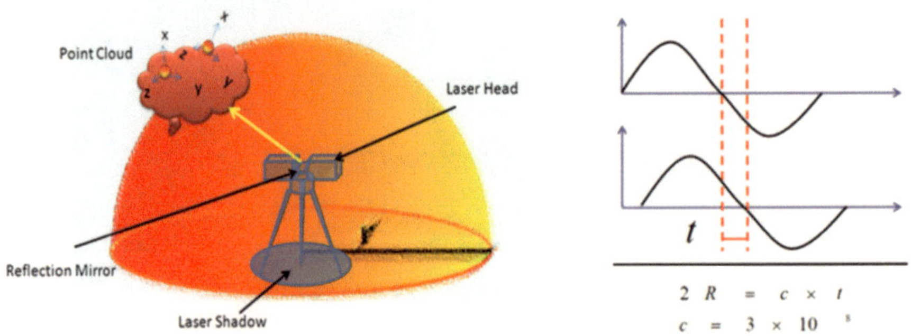

a) 3D LiDAR and FOV b) Range Determination

Fig. 3. Optical Principles of LiDAR: a) 3D LiDAR and Idealized Field of View (FOV); b) Basic Range Detection from Laser Wave Signals.

Scanning laser technology for structural monitoring really took off since the 1990s (Fritsch & Kilian, 1994): For infrastructure monitoring applications, Al-Turk and Uddin (1999) reported using airborne LiDAR for terrain and roadway mapping with the intent of assessing infrastructure inventory; for structural geometric measurements, Curless and Levoy (1996) reported using laser range finder to construct 3D structural geometry of historical structures; subsequently, several reports described using scanning LiDAR for detection of structural changes (Lichti & Gordon, 2004, Girardeau-Montaut & Roux, 2005, Pieraccini & Parrini, 2007).

For bridge applications, Lefevre (2000) first reported using radar for measuring bridge clearances. By comparing the position change of the scan points at each measurement location, deflection of bridge component can be measured. Fuchs et al. (2004a and 2004b) reported using LiDAR for displacement measurements during several bridge static load tests. However, to monitor multiple lines of a bridge during a load test, their laser needed to be placed at multiple locations manually. The accuracy of this measurement method was indicated to be at ±0.76 mm. There have also been reports of using vehicle-mounted scanning systems for bridge clearance measurements: when traveling at traffic speed, this technique can significantly reduce the time for bridge inspection (Kim et al., 2008). Liu et al. (2010a, 2010b and 2011) described several applications of scanning laser system for bridge monitoring applications.

On a tripod, the 3D scanning laser can be imagined to scatter the laser beams covering a sphere around the scanner (Figure 3). Depending on the design, there may be a "blind spot" where the laser will not be able to "see" (the laser shadow). Application of 3D LiDAR to image monitoring relies on the placement of the bridge to within the FOV and the construction of a dense point cloud image of the bridge or bridge components. Since terrestrial LiDAR scans from a single position, depending on the application, there may be the need to move the LiDAR to different physical positions in order to establish a complete image. Table 1 summarizes the different LiDAR applications for bridge monitoring and also the resolution requirements associated with each application.

Essential to LiDAR point cloud analysis is the appreciation of the geometric complexities of the scanned scene and how it ties to the position differences for the subject-of-interest. The position differences can be the calculation of the physical distances for subjects within the same scan or the differences between different scans (deformations) of the same subject. Before a bridge scan, the scanner should be calibrated such that each scan point represents the relative point position (X, Y, Z) to the scanner. Two approaches to the valuation of the scanned XYZ data are presented: 1) the Distance and Gradient Criterion (DGC) based method (Liu et al., 2010a) and 2) the Mean Sum Error and Triangulation (MSE&T) based method (Bian et al. 2011).

Applications	Geometric Dimensions	Resolution Requirements (not verified)
Bridge damage detection	L^2, L^3	±0.0001 m^2, ±0.000001 m^3
Clearance measurements	L	±0.001 m
Bridge displacement	L	±0.001 m
Accident study	L, L^2, L^3	±0.1 m, ±0.01 m^2, ±0.001 m^3
Pre- and Post-construction/event	L, L^2, L^3	±0.1 m, ±0.01 m^2, ±0.001 m^3
Traffic loading quantification	L, L/T	±0.1 m, ±0.01 m^2, ±0.001 m^3
Temperature effect	L	±0.1 m, ±0.01 m^2, ±0.001 m^3
Furniture detection	L^2	±0.01 m^2
Abuse (graffiti, homeless) detection	L^2	±0.01 m^2

Table 1. Potential LiDAR Applications

The DGC method depends on a two-criterion qualifier that defines different portions of the recorded point cloud. A reference plane for the selected point cloud is first defined, which is used to compare with the recorded data to identify the actual area of interest. The validity of each point within the area is then checked by comparing their coordinate value to the surrounding scan points using a search algorithm. For damage detection on a surface, irregular scan points of the selected area are identified by comparing the coordinate differentials between any neighbouring points and comparing the changes in gradient value of the scan points. These two criteria help to determine whether a scan point belongs to the defective part/parts.

Since the selected study area has been rotated and is parallel to the XY plane, D, the distance between the scan points to the reference plane can be easily obtained as

$$D = |Z_P - Z_{REF}| \tag{1}$$

where Z_P is the Z coordinate values of the selected points, and Z_{REF} is the Z coordinate value of the reference plane. The gradient of a certain irregular scan point, which has a column number C and row number R, can be represented as:

$$G = \left| \frac{Z(C+\Delta,R) - Z(C-\Delta,R)}{\sqrt{(X(C+\Delta,R) - X(C-\Delta,R))^2 + (Y(C+\Delta,R) - Y(C-\Delta,R))^2}} \right| + \left| \frac{Z(C,R+\Delta) - Z(C,R-\Delta)}{\sqrt{(X(C,R+\Delta) - X(C,R-\Delta))^2 + (Y(C,R+\Delta) - Y(C,R-\Delta))^2}} \right|, \quad (2)$$

where $X(C,R), Y(C,R), Z(C,R)$ is the X, Y, Z coordinate values of the selected point with a column number C and row number R. Δ is the number of points in each pre-established interval.

In the second method, Mean Sum Error (MSE) and the Delaunay triangulation calculations are used: two-variable regression is used to find an optimal reference plane with the least MSE. Linear reference plane for the selected area can be any surface that contains no anomalies. The MSE regression of each point is calculated against the reference surface and is defined as:

$$MSE = \frac{\sum_{i=1}^{n} (d_i)^2}{n} \quad (3)$$

where n is the number of the selected cloud points, and d_i is the distance of point i to the reference plane (i = 1, ..., n, respectively):

$$d_i = |\varepsilon_i| \quad (4)$$

where ε_i is the error, which is the identified distance to the reference plane.

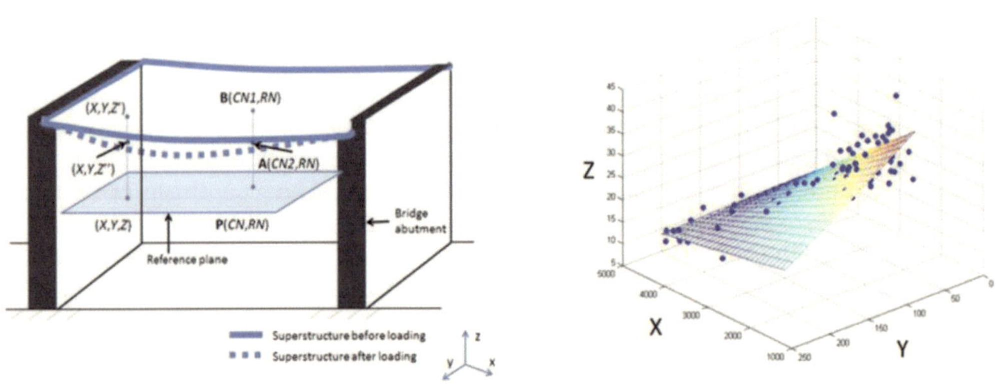

a) DGC Method b) MSE&T Method

Fig. 4. Conceptual Differences between DGC and MSE&T Methods (Bian et al., 2011)

MSE&T method uses the distance from scan points to the reference plane as the criterion to identify the defective parts. Therefore, a pre-determined tolerance value needs to be manually assigned before detecting the defective areas. After that, the Delaunay's triangulation algorithm is used to aggregate the projected points on X-Y plane and the z value is then assigned back to the projected point set, thus forming a 3D surface. Figure 4 shows the conceptual differences between the two methods: a) DGC and b) MSE&T.

Boehler and Marbs (2002) investigated the accuracies of 3D scanning technologies and identified several factors that can influence the scanner accuracies including temperature, atmospheric (lighting) and interfering radiations. The scanning angles between the laser and the target can result in significant scattering of the laser energy and reduce angular accuracy of the laser.

2.2 Scanning laser vibrometry

Laser Doppler vibrometer (LDV) functions by emitting a continuous laser beam to the target surface and measure the vibration of the surface using the Doppler shift between the incident beam and the returned beam (Drain 1980). Figure 5 shows one of the simpler arrangements of a LDV sensor, showing the laser beam being split into a target beam and a reference beam. The two beams are then coupled at the detector. If the laser has a wave length, λ, and the target is moving at a velocity, v, then the moving velocity can be determined by computing the frequency shift between the reference and the target beam:

$$v = \frac{f\lambda}{2} \tag{5}$$

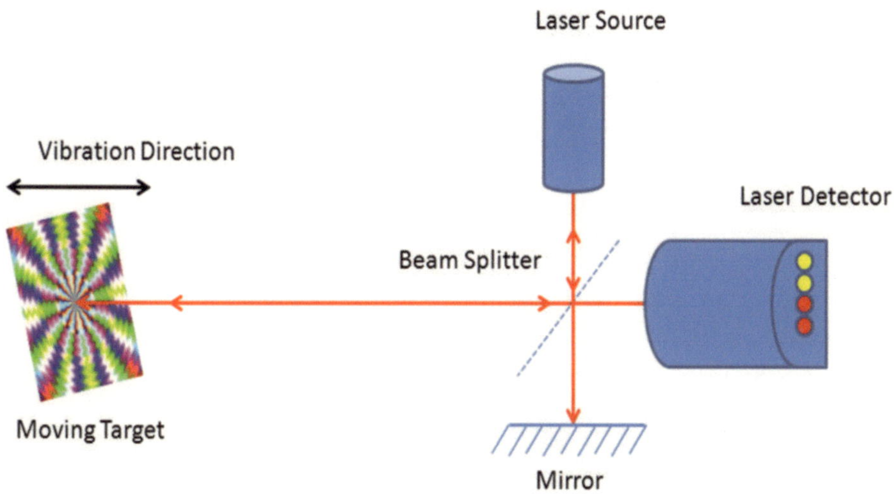

Fig. 5. LDV Sensor Principle

Two optical systems are commercially available: 1) Bragg Cell modulators and 2) Michelson interferometry. One of the key features of the SLV is that it potentially has no data density limitation and the sensitivity of the sensor improves as spatial data density is increased. Contrast to contact sensors, such as accelerometers, there is a limit to how many spatial points can be practically measured. SLV has been used extensively in the automobile industry for studying dynamic behaviors of cars (Junge, 1994). Other advantages of using a SLV for inspection include:

- no added mass on the structure,
- full-field measurement capability,
- speed test setup,
- accurate measurements,
- central remote operation (Oliver, 1995)

3. Examples of LiDAR applications for bridge monitoring

3.1 3D LiDAR for bridge damage evaluation

Surface damages in concrete members are common as a result of either excessive loading or environmentally-induced internal stressing (such as erosion or corrosion of rebars). Early detection of these surficial damages can enhance the durability and the preservation of the structures. Figure 6 shows typical damages to bridge concrete girders in the form of mass losses. The scanned image consists of three girders with girders 2 and 3 showing significant mass losses. If the mass loss can be repaired in time, corrosion resulting from the exposed rebars can be prevented. Using LiDAR scans, four defective areas were identified on two of the four scanned girders and the mass loss areas and volumes for each defective area are quantified (Liu et al. 2010a).

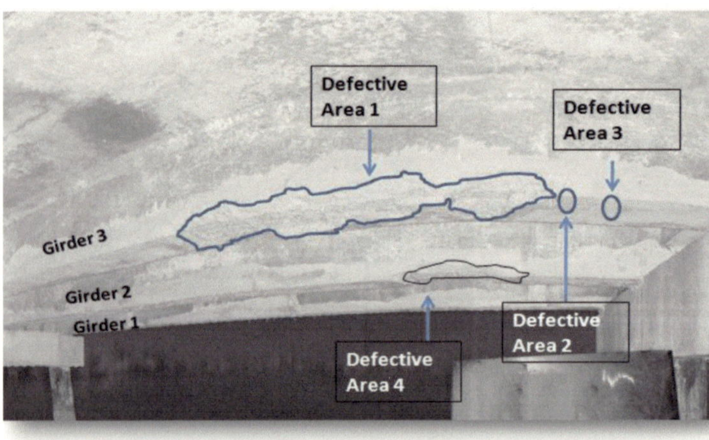

	Area (m2)	Volume (m3)
1	0.507	0.0285
2	6.62E-4	2.63E-5
3	2.13E-4	7.11E-6
4	0.225	0.0156

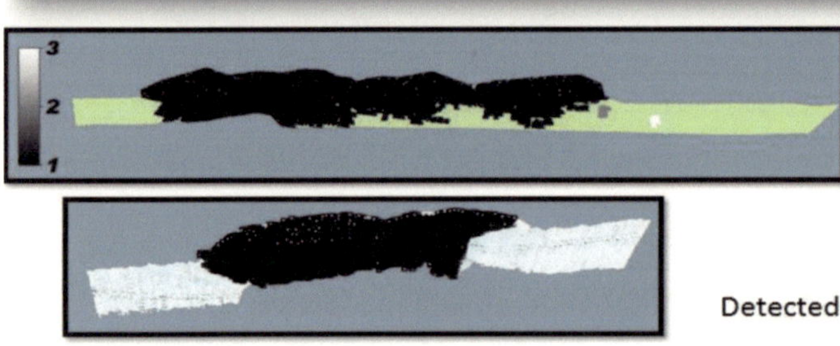

Fig. 6. Damage (Mass Loss) Quantification on Concrete Bridge Girders Using 3D LiDAR (Liu et al., 2010a)

Such damages can also occur when a bridge does not have the required clearances and resulted in vehicle collisions to the bridge superstructures. Hence, the underclearance measurements for a bridge are very important.

3.2 3D LiDAR for bridge underclearance measurements

Conventional clearance measurements are performed using surveying equipment and usually several measurement points are needed to determine the lowest point underneath a bridge, which is a time consuming process. LiDAR systems can provide bridge vertical clearance information for the entire bridge with accuracies in the order of millimeters. The display of clearance change over the entire bridge coverage area can be useful to assess damages and help engineers to devise bridge improvement planning. Using truck-mount LiDAR system, it is even possible to determine bridge clearance on the fly without stopping ongoing traffics (Kim et al., 2008).

a) Harris Boulevard Bridge

b) Laser Scan of Both Spans

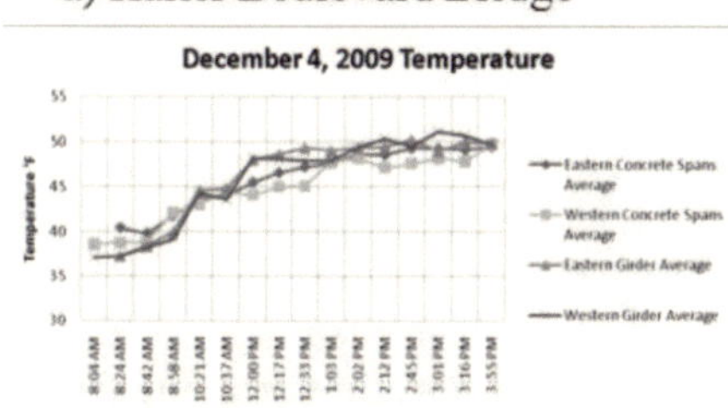

c) Temperature Deviation (Winter Scan)

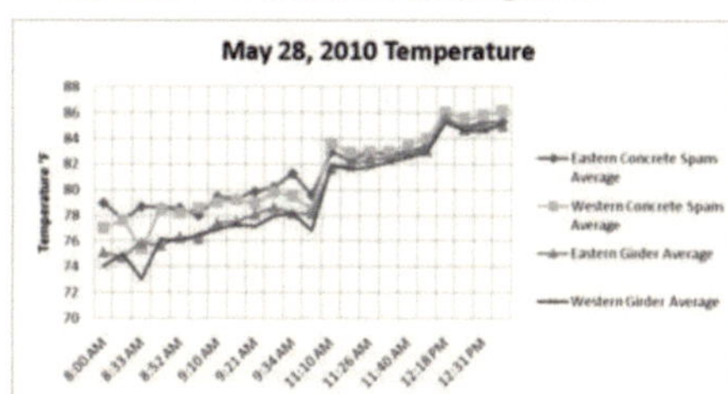

d) Temperature Deviation (Spring Scan)

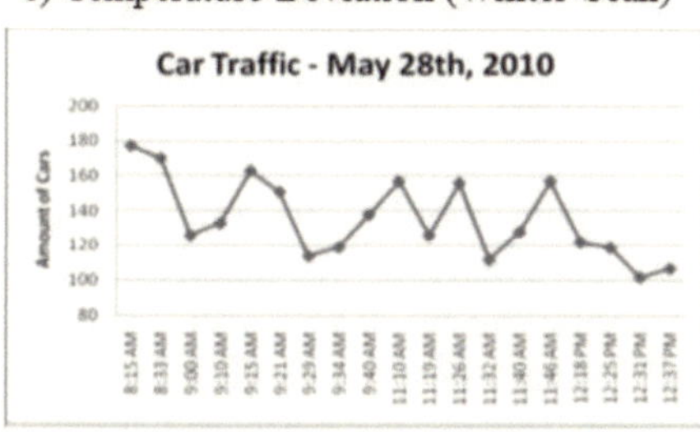

e) Traffic Data (Winter Scan)

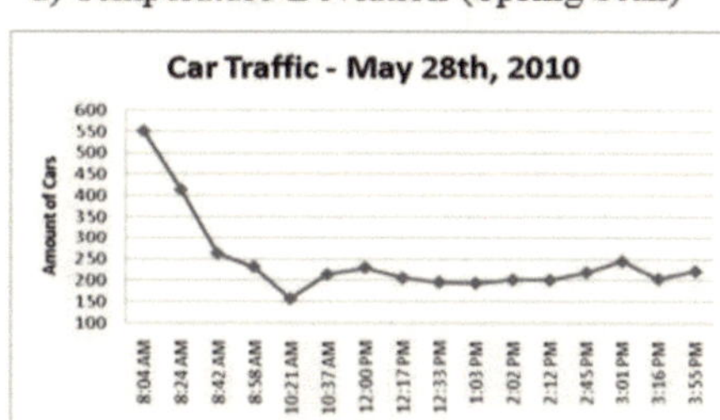

f) Traffic Data (Spring Scan)

Fig. 7. Laser Scan of Harris Boulevard Bridge: a) Harris Boulevard Bridge; b) Laser Scan Showing Both Spans; c) Temperature Deviation during Winter Scan; d) Temperature Deviation During Spring Scan; e) Winter Scan Traffic; and f) Spring Scan Traffic. (Watson et al., 2011)

However, it is necessary to establish the effects of traffic loading over the bridge to ensure no large displacements occur due to vehicle crossing. A temperature and traffic effects study is conducted on the Harris Boulevard Bridge, Charlotte, NC (Watson et al. 2011). Multiple scans were conducted during a day in the winter of 2009 and a day in the late spring of 2010. Figure 7 shows a) the bridge, b) example laser scan, c) winter temperature deviation, d) spring temperature deviation, e) winter traffic and f) spring traffic. Statistical analysis and a hypothesis testing were conducted on the test results to determine if the measurement deviations can be tied to either temperature effects or the passing traffics on the bridge. The statistical analysis and hypothesis testing indicated that LiDAR scans were not influenced by the weather or traffic effects.

3.3 3D LiDAR for post blast assessment

The Colony Road Bridge is a concrete culvert with a two-lane road above. The abutments of the culvert are backfilled with earth, which is retained by large trees, shrubs, and large granite gravel to protect the embankments and foundation culvert from erosion. At both openings of the culvert are wing walls that angled out from the culvert. The culvert crosses the Briars Creek which is about ten meters wide. Figure 8 outlines the basic geometry of the culvert, which has a width of 10.4 m and a height of 5.7 m.

For the Colony Road Bridge, there were several concerns with respect to a construction blasting project: A layer of rock had to be removed in order to lay a new sanitary pipe, but the construction area is nearby to the reinforced concrete culvert, a school, and family homes. The structure of most concern was the concrete culvert as the blasting would occur only 11 meters away from the structure, which is less than the allowable distance of a blasting from the City Ordinance (Charlotte City Council, 2011).

The blast plan called for 3.67 kg of high explosives including: 2x16 dynamite and 2 1/2x16 unimax blasting agent. The drill pattern was 1.54 x 1.83 with 20 to 30 holes drilled. The diameters of bore holes were approximately 0.089 m with a depth of 0.762. The blasting was done to remove a 3.96 layer of rock below 5.18 of earth so that a new 1.52 diameter sanitary sewer could be constructed. Figure 8(a) shows the location of the blast site and the location of the concrete culvert. Also shown are the blast records indicating blast less than 50 mm/sec limit.

The geophones placed on the Colony Road bridge detected vibrations up to 37.08 mm/sec. As a follow up to assessing the structure, a laser scan and post blasting analysis of the culvert were performed to determine if any noticeable permanent damage could be detected using terrestrial LiDAR. A set of LiDAR scans from before blasting was taken and compared to a set of scans after the blasting event. Figure 8(b) contains a black and white rendering top image of the reflectivity image of the Colony Road Bridge after the blasting. As shown the plants around the bridge have since grown to much higher and obscured much of the bridge. Figure 8(c) shows a comparison of the two scans indicating a difference value. Although small areas of the comparison result may show positive or negative displacements, this is expected as individual points may fall outside of the expected accuracy deviation. However, there is no general trend in the data to suggest that the structure has been moved or has been damaged by the blasting.

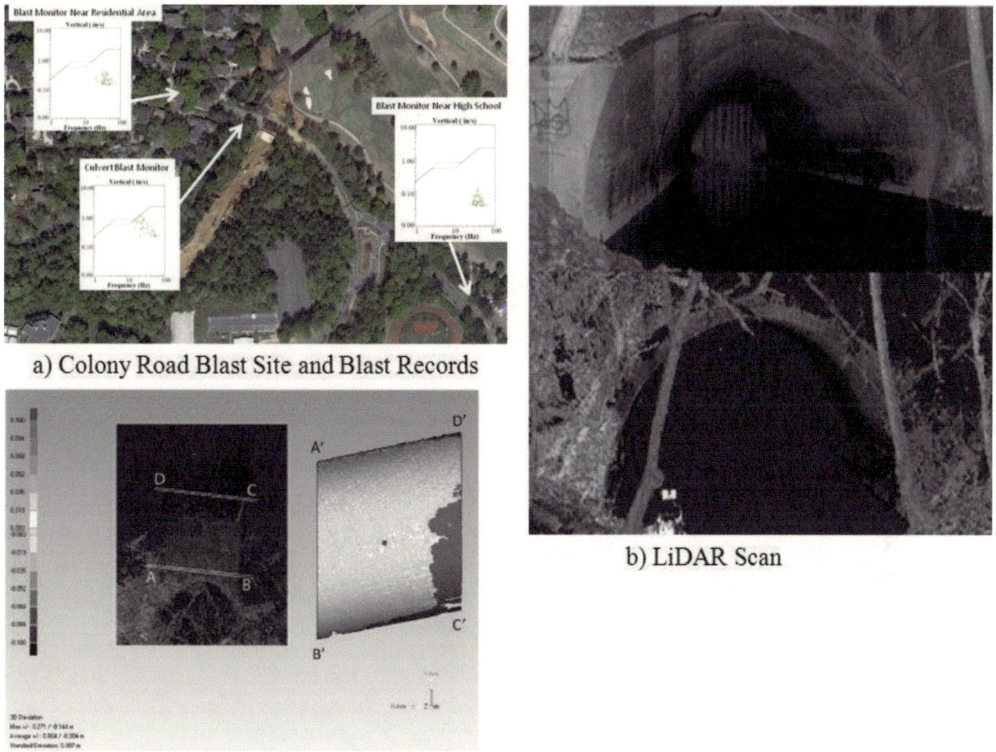

a) Colony Road Blast Site and Blast Records

b) LiDAR Scan

c) Pre and Post Blast Geometric Comparisons

Fig. 8. Colony Road Blasting: a) Colony Road Bridge Location; b) LiDAR Scan; c) Position Before and After Blasting Compared

3.4 Scanning laser for bridge deflection measurements

Load testing of bridges has been well recognized as a practical method to study the condition of bridges. Abnormal behavior of a bridge under a load test is a sign for the need for more frequent inspections and maintenance works. Bridge under given static load is often measured for displacement, stress, or strain of selected points using contact sensors such as strain gauges and displacement transducers. A skewed hybrid high performance steel (HPS) bridge located on SR1102 (Langtree Road) over I-77 in Iredell County, NC, have been studied using LiDAR scans and truck loading (Figure 9(a)). The bridge consists of two 46m spans with stay-in-place concrete decking. The length between the two abutments is around 90m and the width of the bridge is around 26m (Liu et al., 2010b).

For the load test, two heavy trucks were used and placed at designed locations on the bridge. The weight of Truck A was 25,237 kg and the weight of Truck B was 24,865 kg. The distance between the center of the front axle and the center of the rear axle is 6.2 m and the distance between the outer edges of the each pair of rear axles is 2.5 m. Three loading cases have been carried out as displayed in Figure 9(b). Since I-77 is a heavy traffic route, traffic control for strain gage placement is not allowed. The physical constraints inspired the use of 3D LiDAR scanner for static load deflection measurement (Figure 9(c)). Figure 9(d) shows the deflection measurements for position 1 where deflection below truckload is clearly shown.

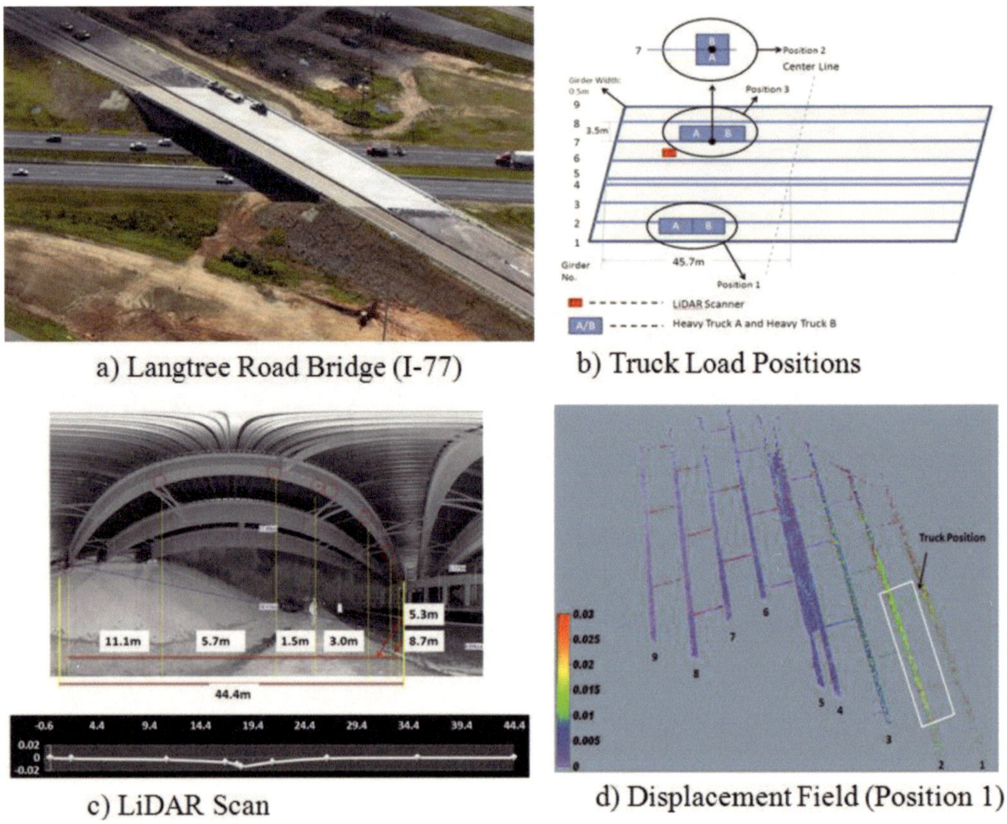

Fig. 9. Langtree Road Bridge over I-77 Highway: a) Bridge under Load Test; b) Load Truck Positions; c) LiDAR Scan with Clearance Measurements; d) Deflection Values under Each Girder (Liu et al. 2011b)

4. Examples of SLV applications for bridge monitoring

To determine vibration-related problems or inverse engineer the vibration measurements to determine damage within a bridge, the natural modal behaviors of the bridge must first be studied. Typically, contact transducers such as accelerometers or geophones are used in the vibration study of bridges. However, because of the sizes of a bridge, large grids of sensor placements typically resulted in long cables in place and result in extensive time and efforts for test preparation. SLV would have solved such problems. The following describes the modal testing on a military bridge.

4.1 SLV for full-scale bridge vibration mode identification

The Armored Vehicle Launched Bridges (AVLBs) are mobile military bridges used for tank crossing during military operations. Designed for Military Load Capacity (MLC) of 60 to 70 tons, these bridges are unique due to their structural complexity and to being light-weight. Constructed of high-strength aluminum alloy and steel, AVLBs are scissor-type structures usually built with two treadways supported on four hinged girders with tapered approaches. The bridge is typically mounted on a launcher and has a hydraulic mechanism to open the bridge. The average AVLB weight is approximately 13 tons with a span of 18.3 m. Figure 10 shows a typical AVLB bridge (10a) and the close up of the center hinge for mechanical opening (Chen et al., 2002).

a) Armored Vehicle Launched Bridge

b) Details of Center Hinge

Fig. 10. AVLB: a) Entire Bridge View and b) Close Up of Center Hinge [34] (Chen et al. 2002)

AVLBs are designed to resist cycles of launching impact and heavy tank loads, hence dynamic impact dominates the design limit states. As a result, the most common damages found on the AVLBs are cracking from high-stress, low-fatigue cycle loads (Cho, 1994). An understanding of the dynamic behaviors of the AVLB is therefore critical to the design and analysis of the bridge.

To simulate free-free conditions, the AVLB was suspended by four airbags underneath the four girders. Figure 11 shows the experimental setup which includes: a) an airbag underneath each of the girders and b) mechanical shakers. The airbags were used to separate the rigid body modes from the flexible modes and to minimize the effect of the supports on the bridge dynamic characteristics. Since the structure is hinged, to ensure uniform excitation, two shakers were placed symmetrically underneath each inner girder. The shakers were synchronized to provide the same amount of stroke at the same time.

The vibration data were measured using an Ometron VP4000 SLV which uses a "lock-in" approach for single frequency scans. To scan the AVLB, a crane was used to lift the laser to 18 m above the bridge (Figure 12). However, even at such height, it is difficult to collect the complete modal information. Figure 12(c) shows the stitched mode shape of the AVLB at 9 Hz excitation. To scan the entire AVLB with an allowable scan angle of 20 degrees would mean that the laser has to be at 64 m height.

Since the four girders act as loosely connected members, several torsion and out-of-phase bending modes have been observed. Figure 13 shows the first three bending modes of the AVLB (Nessler & Lovelace, 1997). The results indicated that the AVLB behaves like a loosely coupled structure where the two treadways act independently.

a) AVLB Scan with SLV on a Crane

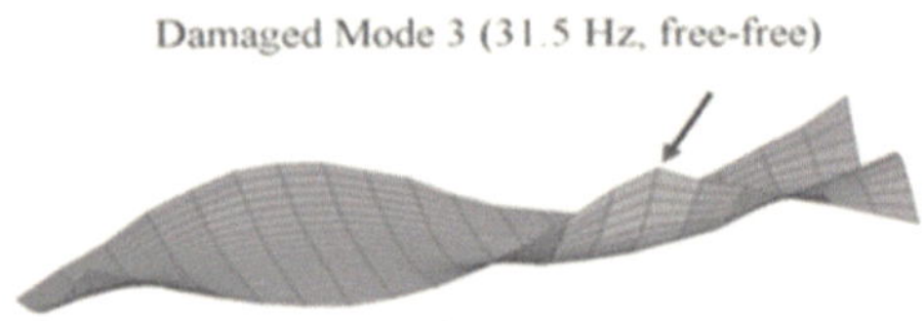

b) Damage Mode Showing Pin-Removal

c) Stitched Mode Shapes at 9 Hz Excitation

Fig. 11. SLV Scan of AVLB Bridge

a) Airbag Support

b) Shaker

Fig. 12. Testing Setup: a) Airbag and b) Shaker

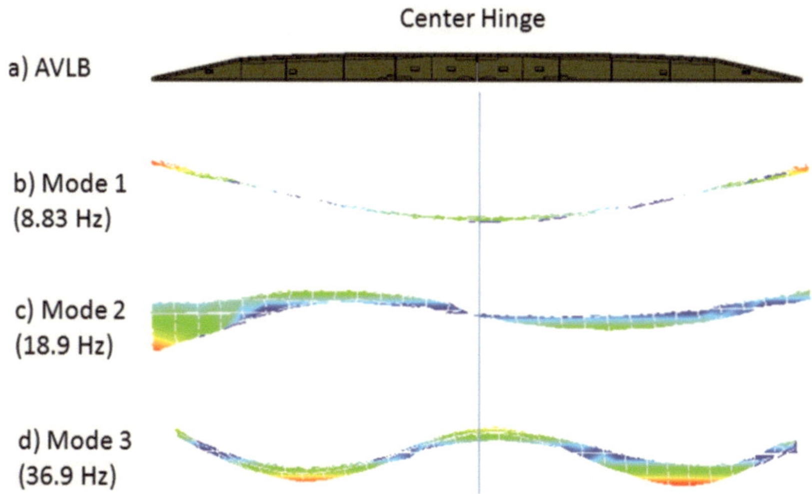

Fig. 13. Modal Behaviors of AVLB: a) AVLB; b) Mode 1; c) Mode 2; d) Mode 3

4.2 SLV for damage detection on bridges

Crack detection using vibration method has been investigated as early as early 1900s (DiPaquale et al. 1990, Hearn & Testa, 1991, Hausner et al. 1997,). Using multiple axial mode frequencies, Adams et al. (1978) were able to identify damage location on a straight bar. Cawley and Adams (1979) presented a method using the ratio of frequency changes in two modes obtained both experimentally and numerically to locate damage. To accommodate SLV vibration scans, a damage detection algorithm based on the strain energy distribution (SED) approach is presented (Pandey et al., 1991, Park & Stubbs, 1995). One of the key features of the strain energy distribution approach is that it is readily implementable to SLV for non-contact damage detection.

The SED method is based on changes in curvature of the mode shapes of a vibrating structure. Theoretical computation of SE is determined by taking the double integration of the derivatives of the measured mode shapes as follows

$$\left[U_{i,abcd}(x,y)\right]=\frac{D}{2}\int_a^b\int_c^d\left\{\begin{matrix}\left[\psi''_{i,x}(x,y)\right]^2+\left[\psi''_{i,y}(x,y)\right]^2+2\nu\left[\psi''_{i,x}(x,y)\right]\left[\psi''_{i,y}(x,y)\right]\\+2(1-\nu)\left[\psi''_{i,xy}(x,y)\right]^2\end{matrix}\right\}dxdy \qquad (6)$$

where D is flexural rigidity and $\psi(x,y)$, is the mode shape function or modal operational shape function and can be determined using modal testing with the scanning laser which generates the mode shape in x and y directions.

Using SLV, the measurements can be either frequency response functions at single frequency or operating deflection shapes at individual frequencies. The double primes ($\psi''(x)$) in Equation (8) denote the second derivatives of the shape function. The SED takes its integration *over small intervals*, i.e. *(a - b)* along the horizontal axis (i.e., *x*-axis) and *(c - d)* along the vertical axis (i.e. *y*-axis), which is the distance between two sampling points.

For damaged cases, it is assumed that the flexural rigidity does not change and the damaged strain energy can be expressed as

$$\left[U_{i,abcd_d}(x,y)\right] = \frac{D}{2}\int_a^b\int_c^d \begin{Bmatrix} \left[\psi''_{i,x_d}(x,y)\right]^2 + \left[\psi''_{i,y_d}(x,y)\right]^2 + 2\nu\left[\psi''_{i,x_d}(x,y)\right]\left[\psi''_{i,y_d}(x,y)\right] \\ +2(1-\nu)\left[\psi''_{i,xy_d}(x,y)\right]^2 \end{Bmatrix} dxdy \quad (7)$$

where the lower level subscript, *d*, denotes the strain energy of the damaged structure.

Local stiffness reduction as a result of cracking or damage would then be reflected in a local increase of SE. This local SE increase is present in different deflected shapes. The SED is computed on a mode-by-mode and element-by-element basis. It has been shown that any local stiffness reduction would lead to a concentration of the local curvature near the damaged cross section, hence producing higher SE. As a result, a comparison of the SEDs for damaged and intact structures can reveal locations of defects. Realistically, SE computations should take into consideration the reduction of cross sections. However, since crack locations are generally unknown beforehand, it is assumed that the effect is much smaller than the element sizes considered and is hence ignored. To detect damage, the numerical difference between damaged and undamaged SED values is used (Chen et al., 2000)

$$\Delta U_{i,abcd}(x,y) = U_{i,abcd}(x,y) - U_{i,abcd_d}(x,y) \quad (8)$$

Since SED is a function of curvature, its sensitivity improves with increased spatial data densities. Again, using SLV, the problem is easily addressed. Figure 12c shows damage detection on the AVLB by removing one of the hinges at the quarter point on the bridge.

4.3 SLV for cable tension measurements

To ensure the safety and integrity of tied arch bridges, it is crucial that tension levels in cables be monitored and do not exceed their design levels. One possible approach would be to correlate the vibration frequencies with the tensions in these cables. However, due to their long length, access to these cables for mounting contact sensors is not easy. Again using SLV, this problem can be solved.

Traditionally, cable vibration measurements of natural frequencies are used to predict cable tension using the taut string model, where the natural frequencies, ω_n, for the out-of-plane motion of a suspended cable are given as (Leonard, 1988):

$$\omega_n = n\pi\sqrt{\frac{T}{mL^2}} \quad (9)$$

where n = n^{th} mode of vibration, m = distributed mass per unit length, T = cable tension, and L = total or effective length of the cable. If the vibration frequencies can be clearly identified, then tension level can be back-calculated with exact measurements of the three unknowns, ω_n, m and L:

$$T = \left(\frac{\omega_n}{n\pi}\right)^2 mL^2 \quad (10)$$

The distributed mass, *m*, usually does not alter significantly, unless the metal has been heavily corroded. *L* is a function of the boundary condition, which typically was assumed to be pinned-pinned.

The approach is tested on the I-470 Bridge, which is also known as the Vietnam Veteran Memorial Bridge and is located in Wheeling, West Virginia (Chen & Petro, 2005). The bridge has a total length over 457.2 m and a central span of 237.4 m long. The four-lane highway bridge spans across the Ohio River with eastbound traffic towards West Virginia, and westbound traffic towards Ohio. The arch or rib and the tie of the bridge are both box sections. The load on the bridge deck is transferred to the rib cage through 16 hangers with intervals at 13.7 m between the two piers on each side of the bridge. The cables are fabricated from class A zinc-coated steel structural strands. 203 mm PVC pipe has been used as protectors on the lower part of the cables. The nominal diameter of the cable is about 57.2 mm. Each strand is pre-stretched under tension to about 155 tons. The modulus of elasticity of the braided cable is about 165.47 x 10^9 N/m^2. The strand has a minimum breaking strength of 310 tons. To prevent the cables from impacting each other, elastic spacers have been placed between the cables. The hanger acts as stress transmitter, for the box arch to carry the dead load of the bridge deck. The design tension levels in each cable were assumed not to exceed 10% of the cable's breaking strength (31 tons). Hence, this limit was used as a check on cable safety.

The testing was conducted with the SLV placed on the shoulder of the bridge deck (Figure 14). The vibration measurements were made under regular traffic conditions. The direct velocity data were collected at a sampling rate of 1 kHz with 50,000 data points. Spectrum analysis is then conducted on the time domain signals. The environmental condition on the test date is sunny with mild breezes.

Fig. 14. Cable Vibration Measurements from the Bridge Deck [41] (Chen et al. 2005)

Figure 15 shows the basic structure of the I-470 Bridge with the measured vibration frequencies and estimated tension loads. The tension levels were determined using equation (11). If each strand carries the same load, the total tension in each hanger would be four times the tension level in the strands.

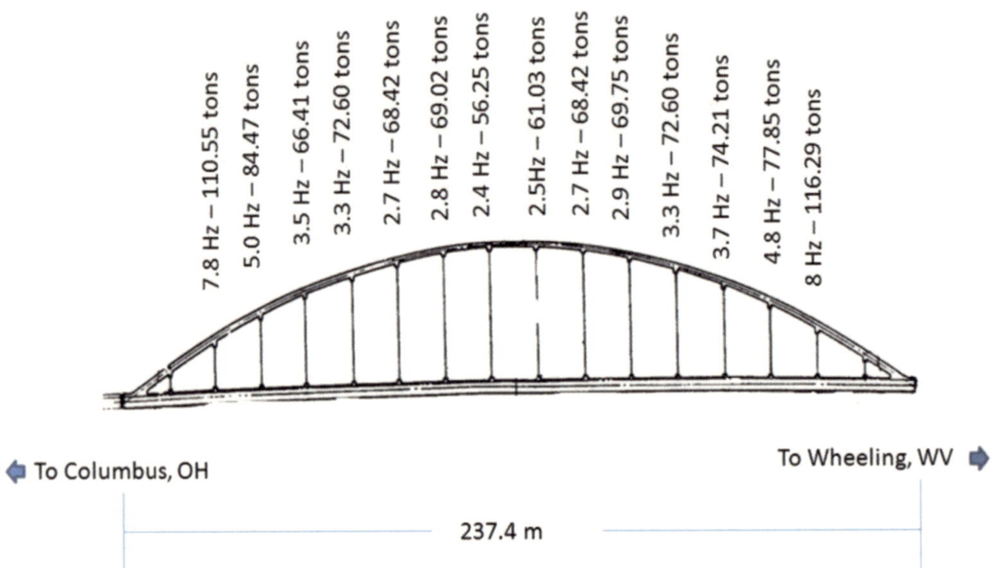

Fig. 15. Cable Load Capacity and Measured Vibration Frequencies (Chen et al. 2005)

5. Discussion

This chapter discusses several potential applications of scanning lasers for bridge monitoring. Other potential applications that are currently gaining popularity include high density, multi-dimensional geospatial model construction using LiDAR and photo imaging techniques for applications such as accident reporting and management, furniture (sign and luminaires) management, disaster mitigation and management, etc. Applying scanning laser technologies to bridge monitoring signifies a critical shift in bridge management paradigm - it naturally integrates conventional visual inspection and paper reporting with digital computing via geoscience technologies such as Geological Information Systems (GIS) and advanced image analysis.

However, before the industry catch on with the technology, there is a need to carefully evaluate the potential impacts to the industry, in particular, in the form of life cycle cost analysis. Because of the capability in providing high resolution spatial information from a distance away, commercializing scanning laser technologies poses different issues than those faced by other structural monitoring science and technologies including *issues of privacy and security, issues about redundancy in image information* and *legality and technology relevancy issues related to commercialization*. With this in mind, the technical barriers would be the issue of retrieving *useful and legitimate* CRS data from multi-variant imaging and data sources that can be best used in enhancing bridge infrastructure management. The potential of generating massive data in the order of petabytes also challenges the computing sciences and pushing the technology envelope towards unprecedented scale.

These are not easy issues to address, which have to be answered by all bridge managers, government and stakeholders. However, the societal benefits are tremendous: for example, there is the potential of conducting temporal effect studies. By temporal, we mean more frequent inspections/visits than current federally mandated two-year cycles, which would allow reliable imaging of the transformation of damage through time. With the digital, high resolution imaging techniques, it may also be possible to identify critical problems prior to disastrous failure, such as what happened to the Minnesota I-35 Bridge. It is also possible to report real time tracking of excessive truck loading and environmental loadings.

Finally, bridge management decisions are complex processes that must consider bridge conditions as well as bridge utility, available funding and impact of rehabilitation techniques. Bridge managers need a tool to enable "total management" decision-making. This consideration is critical to the development of RS technologies for structural health monitoring, since it helps us understand how remote sensing data is associated with the bridge utilities. However, such considerations also represent a need to revisit the definitions and operations of existing bridge management systems.

6. Conclusion

This chapter introduces two types of scanning laser technologies: 1) static range finding laser and 2) dynamic measurement laser. Several bridge monitoring applications of scanning laser technologies have been described including damage detection, underclearance measurements, load deflection measurements, post-event validations, damage detection, modal behavior identification and cable tension measurements. These examples of scanning laser application provide realistic scenarios for industry-wide implementation of scanning laser techniques for bridge monitoring and both technologies are ready to be commercialized. However, it is also pointed out that such applications can significantly influence conventional bridge inspection and management by forcing the integration of geoscience technologies and advanced image capture and processing techniques. However, much work needs to be done to ascertain critical issues such as data security, privacy and the management of massively generated data are first addressed.

7. References

Adams, R.D., Cawley, P., Pye, C.J. & Stone, B.J. (1978). A Vibration Technique for Nondestructively Assessing the Integrity of Structures. *Journal Mechanical Engineering Science*, Vol. 20, No. 2, pp.93-100.

Aktan, A.E., Farhey, D.N., Helmicki, A.J., Brown, D.L., Hunt, V.J., Lee, K.L. & Levi, A. (1997). Structural Identification for Condition Assessment: Experimental Arts. *ASCE Journal of Structural Engineering, Vol.* 123, No. 12, pp.1674-1685.

Al-Turk, E. & Uddin, W. (1999). Infrastructure Inventory and Condition Assessment Using Airborne Laser Terrain Mapping and Digital Photography, *Transportation Research Records*, TRB, Vol. 1690, pp. 121-125.

Bian, H., Bai, L., Wang, X., Liu, W., Chen, S.E. & Wang, S. (2011). Effective LiDAR Damage Detection: Comparing Two Detection Algorithms, *Structural Engineers*. Vo. 27, No. 137, pp.327-333.

Boehler, W. and Marbs, A. (2002). 3D Scanning Instruments, Proceedings of *CIPA, Heritage Documentation – International Workshop on Scanning for Cultural Heritage Recording*, Corfu, Greece.

Cawley, P. and Adams, R.D. (1979). The Location of Defects in Structures from Measurements of Natural Frequencies, *Journal of Strain Analysis*, Vol. 14, No. 2, pp.49-57.

Charlotte City Council (2011). *City of Charlotte Zoning Ordinance*, City of Charlotte.

Chen, S.E. & Petro, P. (2005). Nondestructive Bridge Cable Tension Assessment Using Laser Vibrometry, *Experimental Techniques*, Vol. 29, No. 2, pp. 29-32.

Chen, S.E., Rice, C., Boyle, C. & Hauser, E. (2011). Small Format Aerial Photography for Highway Bridge Monitoring, *ASCE Journal of Performance of Constructed Facilities*, Vol. 25, No. 2, pp. 105-112.

Chen, S.E., Petro, S., Venkatappa, S., GangaRao, H. and Moody, J. (2002). Modal Testing of an AVLB Bridge, *Experimental Techniques*, Vol. 26, No. 6, pp.43-46.

Chen, S.E., Venkatappa, S., Petro, S.H. & GangaRao, H. (2000). A Novel Damage Detection Technique Using Scanning Laser Vibrometry and A Strain Energy Distribution Method, *Materials Evaluation*, Vol. 58, No. 12, pp.1389-1394.

Cho, N. (1994). *Preproduction Qualification Test (PPQT) of the Armored vehicle Launch Bridge (AVLB) – MLC 70 Upgrade Program*, U.S. Army Tank-Automotive Command Final Report # CSTA-7567.

Choi, S. & Stubbs, N. (1997). Nondestructive Damage Detection Algorithm for 2-D Plates, *Smart Systems for Bridges, Structures, and Highways, Proceeding of SPIE*, Vol. 3043, pp.193-204.

Curless, B. & Levoy, M. (1996). A Volumetric Method for Building Complex Models from Range Images, *Proceedings of SIGGRAPH*, pp.303-312.

DiPasquale, E., Ju, J.W., Atilla, A. & Cakmak, A.S.(1990). Relation Between Global Damage Indices and Local Stiffness Degradation. *ASCE Journal of Structural Engineering*, Vol. 116, No. 5, pp.1440-1456.

Drain, L.E., *The Laser Doppler Technique*, Wiley-Interscience, Sussex, England, 1980.

Fritsch, D. & Kilian, J.C. (1994). Filtering and Calibration of Laser-Scanner Measurements, *Proceedings of ISPRS Commission III Symposium*, SPIE, Vol. 2357, pp. 227-234.

Fuchs, P.A., Washer, G.A., Chase, S.B. & Moore, M. (2004a). Application of laser-based instrumentation for highway bridges. *ASCE Journal of Bridge Engineering, Vol.* 9, No. 6., pp.541-549.

Fuchs, P.A., Washer, G.A., Chase, S.B. & Moore, M. (2004b) Laser-based instrumentation for bridge load testing, *ASCE Journal of Performance of Constructed Facilities*. Vol. 18, No. 4, pp.213-219.

Girardeau-Montaut, D. & Roux, M. (2005). Change detection on points cloud data acquired with a ground laser scanner. *Proceedings of Workshop Laser scanning 2005*, Enschede, the Netherlands.

Hausner, G.W., Bergman, L.A., Caughey, T.K., Chassiakos, A.G., Claus, R.O., Masri, S.F., Skelton, R.E., Soong, T.T., Spencer, B.F. & Yao, J.T.P. (1997). Structural Control: Past, present, and Future, *ASCE Journal of Engineering Mechanics*, Vol. 123, No. 9, pp.897-958.

Hearn, G. & Testa, R.B. (1991). Modal Analysis for Damage Detection in Structures. *ASCE Journal of Structural Engineering,* Vol. 117, No. 10, pp.3042-3063.

Jelalian, A.V. *Laser Radar Systems*. Artech House, Boston, London, 1992.

Junge, B. (1994). Experiences with Scanning Laser Vibrometry in Automotive Industries, *Proceedings of SPIE,* Vol. 2358, pp.377-382.

Kim, Y.R., Hummer, J.E., Gabr, M., Johnston, D., Underwood, B.S., Findley, D.J. & Cunningham, C.M. (2008). *Asset Management Inventory and Data Collection,* Final Report, FHWA/NC/2008-15.

Lefevre, R. J. (2000). Radar bridge clearance sensor. *Proceedings of IEEE 2000 International Radar Conference,* Alexandria, VA, pp. 660-665.

Leonard, J.W. (1988), *Tension structures – behavior and analysis*, McGraw-Hill, New York, NY.

Lichti, D.D. & Gordon, S.J. (2004). Error propagation in directly georeferenced terrestrial laser scanner point clouds for cultural heritage recording, *Proceedings of WSA2 Modeling and Visualization*, Athens, Greece.

Liu, M., Frangopol, D.M. & Kim, S. (2009). Bridge Safety Evaluation Based on Monitored Live Load Effects, *ASCE Journal of Bridge Engineering,* Vol. 14, No. 4, pp. 257-269.

Liu, W.Q., Chen, S.E. & Hauser, E. (2010a). LiDAR-Based Bridge Structure Defect Detection, *Experimental Techniques,* DOI: 10.1111/j.1747-1567.2010.00644.x.

Liu, W.Q., Chen, S.E. & Hauser, E. (2011). Bridge Clearance Evaluation Based on Terrestrial LiDAR Scan, *ASCE Journal of Performance of Constructed Facilities,* doi:10.1061/(ASCE)CF.1943-5509.0000208.

Liu, W.Q., Chen, S.E., Boyajian, D. &Hauser, E. (2010b) Application of 3D LiDAR Scan of Bridge Under Static Load Testing, *Materials Evaluation,* Vol. 68, No.12, pp.1359-1367.

Nessler, G. & Lovelace, D. (1997). *Final Report on Acquisition of Frequency Response Data on an Armored Vehicle Launched Bridge (AVLB).*

Oliver, D.E. (1995). Scanning Laser Vibrometer for Dynamic Deflection Shape Characterization of Aerospace Structures, *Proceedings, SPIE,* Vol. 2472, pp.12-22.

Pandey, A.K., Biswas, M. & Samman, M.M. (1991). Damage Detection from Changes in Curvature Mode Shapes, *Journal of Sound and Vibration,* Vol. 145, No. 2, pp.321-332.

Park, S. & Stubbs, N. (1995). Bridge Diagnostics via Vibration Monitoring, *Proceedings of SPIE,* Vol. 2719, pp. 36-45.

Pieraccini, M. & Parrini, F. et al., (2007). Static and dynamic testing of bridges through microwave interferometry. *NDT&E International,* Vol. 40, pp. 208-214.

Rueger, J.M., *Electronic Distance Measurements: An Introduction,* Springer-Verlag, New York, 1990.

Shinozuka, M. & Rejajaie, S.A. (2000). Analysis of Remotely Sensed Pre- and Post-Disaster Images for Damage Detection, *Smart Structures and Materials, Proceedings of SPIE,* Vol. 3988, pp. 307-318.

Smith, D.E., Zuber, M.T., Neumann, G.A. & Lernoine, F.G. (1997). Topography of the Moon from the Clementine LiDAR, *Journal of Geophysical Research,*Vol.102, No. E1, pp. 1591-1611.

Watson, C., Chen, S.E., Bian, H. & Hauser, E. (2011). 3D Terrestrial LiDAR for Operational Bridge Clearance Measurements, *ASCE Journal of Performance of Constructed Facilities*, doi:10.1061/(ASCE)CF.1943-5509.0000277.

6

The Support of Geomatics in Glacier Monitoring: The Contribution of Terrestrial Laser Scanner

Danilo Godone[1] and Franco Godone[2]
[1]*Turin University, Faculty of Agriculture, Deiafa Department / NATRISK, Research Centre on Natural Risks in Mountain and Hilly Environments*
[2]*National Research Council, Research Institute for Geo-Hydrological Protection*
Italy

1. Introduction

Worldwide glacier monitoring was initiated more than a century ago and it is now integrated into global climate-related observation systems. Glacier mass changes have been clearly recognised as high-confidence climate indicators with respect to early detection strategies of greenhouse effects on climate (Hoelze et al. 2003).

Combinations of in-situ observations with remotely sensed data, traditional measurements with new technologies, are an integrated and multi-level strategy applied to the worldwide glacier monitoring within major climatic zones by the international institution of the Global Terrestrial Network for Glaciers (GTN-G) and Global Terrestrial Observing System (GTOS) (Haeberli et al., 2005).

Many examples of successful applications of space-borne synthetic aperture radar interferometry (InSAR) are available in the scientific literature of the last decades applied to detection of ground surface deformations (induced by seismic or volcanic activities) and for monitoring of glacier motion (Wasowski & Gostelow, 1999; Massonet & Feigl, 1998)

The applications of Ground Remote Monitoring techniques (Ground Based - Synthetic Aperture Radar (GB-SAR) and Terrestrial Laser Scanner (TLS) for extraction of topography and flow properties of glaciers are less frequent.

As reported by Sailer et al. (2005), the observational parameters include "accurate topographic data of the observed target and its surroundings, as well as of the terrain that may be affected down-streams of a glacier by floods or avalanches. Depending on the size and accessibility of the area to be observed and on the repeat interval, satellite-borne, airborne or ground based imaging sensors or scanners (InSAR, photogrammetry, GB-SAR, terrestrial laser scanner - TLS) are the preferable observational tools".

Even in the above mentioned work, three types of basic observational requirements for each of the glacial hazard type are mentioned:

- base maps (extent of glaciers, glacial lakes, surface morphology);
- topographic data (DEM);
- surface motion (of glacier and/or moraine).

This chapter shows the methodology and application results of the Terrestrial Laser Scanner for the extraction of topographic data and surface motion at the Belvedere Glacier, in Italian Alps.

The monitoring campaigns have been performed in Summer/Autumn 2006 and 2007 and they consist on measurements of the central/lower part of the glacial body, of the glacier snout and measurement of a landslide (the Locce landslides), affecting the terminal moraine of the Locce Glacier, tributary on the right side of the Belvedere Glacier.

2. The Belvedere glacier

The Belvedere Glacier (WGI code I4L01211009) is located in the Anzasca valley, Macugnaga (45°58' N, 7°58' E), in the north-eastern sector of Piemonte (Italy).

It is a rock covered glacier (Mazza, 1998) situated at the base of the gigantic Monte Rosa east face. According to the Glaciorisk database (GRIDABASE http://www.nimbus.it/glaciorisk/gridabasemainmenu.asp) the glacier elevation ranges from a maximum of 4520 m a.s.l. to a minimum of 1760 m a.s.l., its length is about 6 km and its maximum width is up to 500 m, with an average surface of 5.58 km^2. It's oriented at 45° north with an average slope of 10°. The glacier main body is connected, in its upper part, with the Northern Locce, Signal and Monte Rosa Glaciers and it ends with a forked snout.

The glacier has been subject of scientific studies and expeditions from the end of the 18th century, when its magnitude has been firstly described and highlighted (De Saussure, 1779 - 1796).

At the beginning of the present century - since 1999, according to Mazza (2003) - the glacier has undergone a remarkable increase in horizontal displacement speed and surface elevation. According to the displacement rate measurement, carried out by remote sensing techniques, the speed of the glacier has increased from 32 – 43 m per year, recorded in the last 5 years of the 20th century to 92 – 112 m per year (02/09/1999 – 06/09/2001); in the autumn of the same year the speed had still increased to its maximum of 100 – 200 m per year (Kaab et al., 2005).

According to the definition of Haeberli et al (2002) the Belvedere glacier has been subject to a "surge-type movement", currently a unique phenomenon in the Alps; conversely Mazza (2003) have explained the phenomena with the kinematic waves theory due to the reduced horizontal displacement speed, the lack of the formation of a terminal moraine and the absence of the typical block-flow movement; in fact, due to the high displacement speed, the glacier is unable to keep a viscous flow and breaks into small ice blocks. Moreover, the kinematic wave theory is supported by previous events, though characterised by reduced magnitudes, occurred between 1984 and 1985 and at the end of 1992 (Mazza, 2003).

The surface elevation has shown a similar trend, with an increase in ice thickness of up to 20 m reaching the level of the Little Ice Age moraine (Mortara et al., 2003), with the consequent appearance of large crevasses on the whole glacier. Due to the described phenomena the formation of a new moraine has been observed. Contemporarily, in the upper part, the glacier surface has shown an opposite state, by lowering its elevation and in the originated depression a supraglacial lake has been observed. The lake has been identified as the "*Effimero Lake*" (Figure 1).

The lake has undergone different fillings until reaching its maximum volume of $3 \cdot 10^6$ m^3 and a 57 m depth with evident risk of GLOF (Glacial Lake Outburst Flood) or *jökulhlaup* due to termokarst processes or rockfall from Monte Rosa East Face. During the most critical phase the lake level has been rising with a rate of 1 m/day so an overflow event was also evaluated (Mercalli et al., 2002; Mortara et al., 2003; Mortara and Mercalli, 2002; Tamburini and Mortara, 2004).

Fig. 1. The "Effimero lake", in the left of the image the "Locce lake" (Photo G. Mortara)

In the past, similar events have been observed in Belvedere glacial basin. Due to intense precipitation, the consequent outbursts of a water pocket inside the glacier have provoked different collapses in the lateral moraines and the consequent trigger of debris flows (1868, 1896, 1904, 1922) as reported by Haeberli et al (2002).

Since 1990 (Fischer et al., 2006; Fischer et al, 2011) the Monte Rosa east face has also been involved in several episodes of ice and rock avalanches, due to the permafrost degradation in high mountain slopes caused by the increase of 0° C isotherm altitude (Beninston, 2003; Carrasco et al., 2005).

Permafrost and ice play a key role in high mountains slope stability, the presence of ice in discontinuities improves the bond between rocks by its adhesion effect. Melting results in the loss of this factor and in the increase of flowing water, increasing the pressure in joints and, consequently, lowering the shear strength (Davies et al., 2001). Unfortunately such events have been observed in the rest of the Alps in the last years (Deline et al., 2002; Deline et al., 2004a; Giani et al., 2001; Godone et al., 2007).

In the last years two main events have been observed: on 25 August 2005 a massive ice avalanche has fallen down from the Monte Rosa east face with a total volume of up to $1.1 \cdot 10^6$ m^3; on 21 April 2007 a similar event has happened (Figure 2), with nearly halved volume involved (~500000 m^3) (Cat Berro et al., 2008). In every occasion, the debris has covered the upper sector of Belvedere and due to its mixed composition, rocks and ice, can be considered as a contribution in glacier mass balance (Federici et al., 2008).

Fig. 2. The 2007 ice avalanche deposit (photo G. Mortara)

Both phenomena, the changes in surface dynamics and the supraglacial lake, have triggered a civil protection procedure in order to reduce risk related to the explained events for the inhabitants of Macugnaga (Mercalli et al., 2002; Mortara et al., 2003; Mortara and Mercalli, 1999). Fortunately the lake has ended, by itself, the emergency state by two endoglacial outbursts, due to the improvement in subglacial drainage system, finished without consequences (Tamburini and Mortara, 2004; Tamburini et al., 2003). The two episodes occurred on 19 and 20 June 2003 with a discharge of, respectively, 14 and 9 m^3/s; among them a marked discharge reduction has been observed, probably due to a temporary obstruction in the drainage system.

3. Materials and methods

The traditional measurements carried out using GPS and ablatometric stakes provide only punctual evaluations of the phenomena. In glacier monitoring an extensive approach could be preferred in order to obtain a complete description of the surveyed object. According to this statement, different Terrestrial Laser Scanner campaigns have been planned at different sites of the glacier surface, in order to evaluate the most interesting geodynamical phenomena during their evolution. The measurements includes the central body of the glacier, the snouts and a landslides affecting the frontal moraine of the Lago delle Locce and the right side of the Belvedere Glacier in its upper part (Figure 3).

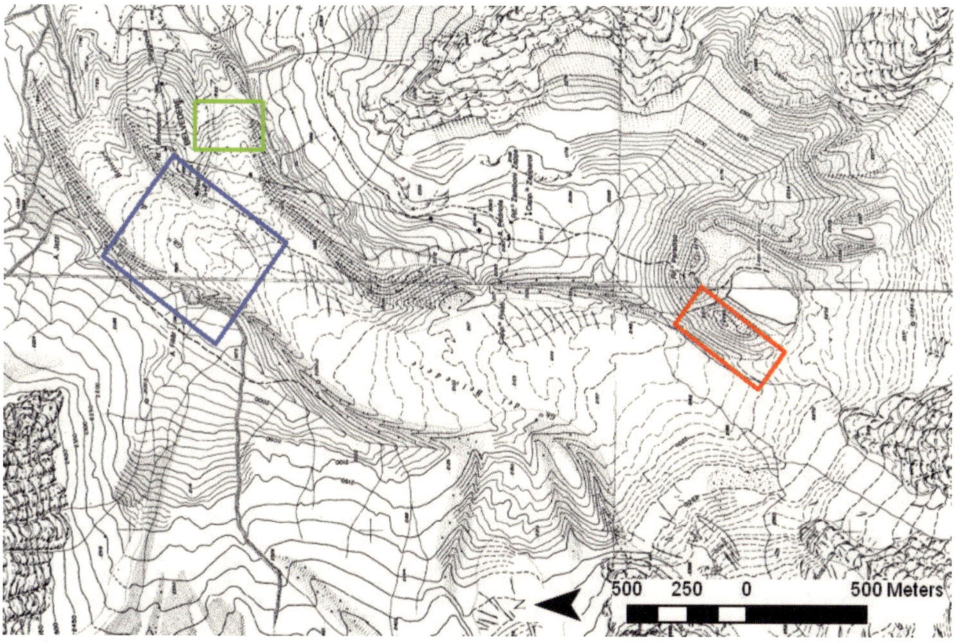

Fig. 3. Terrestrial Laser Scanner survey sites: right snout (green square), Locce landslide (red square) and glacier surface (blues square). Regional Technical Map (original scale 1:10000) in background

Fig. 4. The Optech ILRIS-3D laser scanner during a field masurements (photo P. Federici)

The Laser Scanner work as a range finder, projecting thousands of laser pulses towards the survey target and through the reflected beams is capable of reconstructing its geometry; moreover, by measuring the object from different points of views it's possible to complete the description avoiding shadow effects, induced by object features.

The instrument employed in the following experiment was the Optech ILRIS-3D™ (Figure 3). The laser is characterised by a quite long range, according to target reflectivity, and it assures safe operational conditions according to laser safety standards, as reported in the following table (Table 1).

Dynamic scanning range	3 m - 1500 m to an 80% target
	3 m - 800 m to an 20% target
	3 m - 350 m to an 4% target
Data sampling rate (actual measurement rate)	2500 points per second
Beam divergence	0.00974°
Minimum spot step (x and y axis)	0.00115°
Raw range accuracy	7 mm @ 100 m
Raw positional accuracy	8 mm @ 100 m
Laser wavelength	1500 nm
Laser class (IEC 600825-1)	Class 1
Digital camera	Integrated digital camera 6 Mpixel (CMOS sensor)
Scanner field of view (ILRIS-3D)	40° x 40°

Table 1. ILRIS-3D technical characteristics (www.optech.ca/i3dtechoverview-ilris.htm)

The laser is built in a robust metal case, allowing its employment in fieldwork conditions, moreover, thank to a special backpack the instrument is fully portable (Figure 5). The complex weight is up to 13 kg and with the aid of a second operator carrying the battery pack and the tripod is fully operational without further equipments.

Fig. 5. ILRIS-3D mounted on tripod and the special backpack employed in instrument transport (left) and ILRIS-3D interface control - red squares represent surveyed areas, the green one is the area under measurement (right)

At the beginning of the survey, the instrument acquires a picture of the object with its incorporated digital camera, and displays it on the controller screen. This enables the operator to decide the area, or the areas – e.g. the objects and targets - of the picture to be scanned by simply drawing rectangular selection windows and by specifying acquisition parameters for each one.

The aim of the whole experiment was the comprehension of glacier dynamics in time, so a multitemporal approach has been adopted when planning the surveys. In order to maximize the productivity and avoid risks for the operators, the employment of targets placed on the object has been strongly limited. One advantage of the laser scanner is to work without contact with the surveyed objects, so allowing to measure dangerous or inaccessible places, too. This peculiarity has already been employed in landslide and volcanoes monitoring (Hunter et al., 2003; Oppikofer et al., 2008).

For each area under observation, two scan sessions have been executed with the aim of measuring changes in glacier features; during the third test on the central body of the glacier also measurements of some targets have been used in order to make a comparison with ablation stakes measurement accomplished with GPS, thus obtaining co-registered data.

The acquisition strategy has been planned with the aim of measuring an overlapping sector among adjacent scans and to include stable areas external to the survey object. Overlapping parts have been used while collecting tie points (Figure 6) to join scans into a single point cloud and stable areas have been employed for the alignment of multitemporal scans, as reference areas.

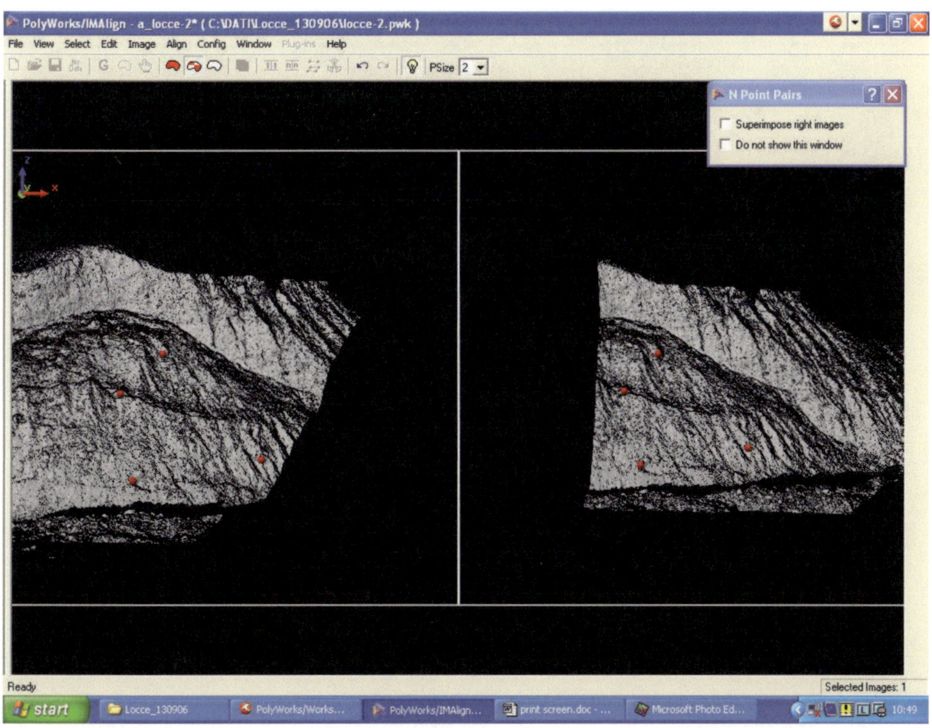

Fig. 6. Scans co-registration though common feature collimation (red points)

Data processing have been performed using Innovmetric Polyworks™, the software is subdivided in modules specifically designed for the different work phases. The ImAlign module is dedicated to the alignment of different scans into a unique reference system, in the described experiment this goal has been achieved by homologous feature collimation; using an interactive closest point algorithm (Beinat, 2006; Besl and Mc Kay, 1992; Chen and Medioni, 1992), the matching has been achieved with good results in alignment accuracy by employing only few couples of selected points.

The point cloud is not really modified by the operation, as the software compiles a rototranslation matrix used in locating the original data in the three dimensional reference system. By comparing multitemporal matrix parameters a computation of object displacement can be achieved, too.

The module allows also performing manual object removal and data cleaning in order to obtain an essential point cloud without minor details such as trees or buildings. By selecting different sectors of a point cloud the software allows to separate stable and unstable areas with the purpose of using them, respectively, in multitemporal alignment and comparisons (Figure 7).

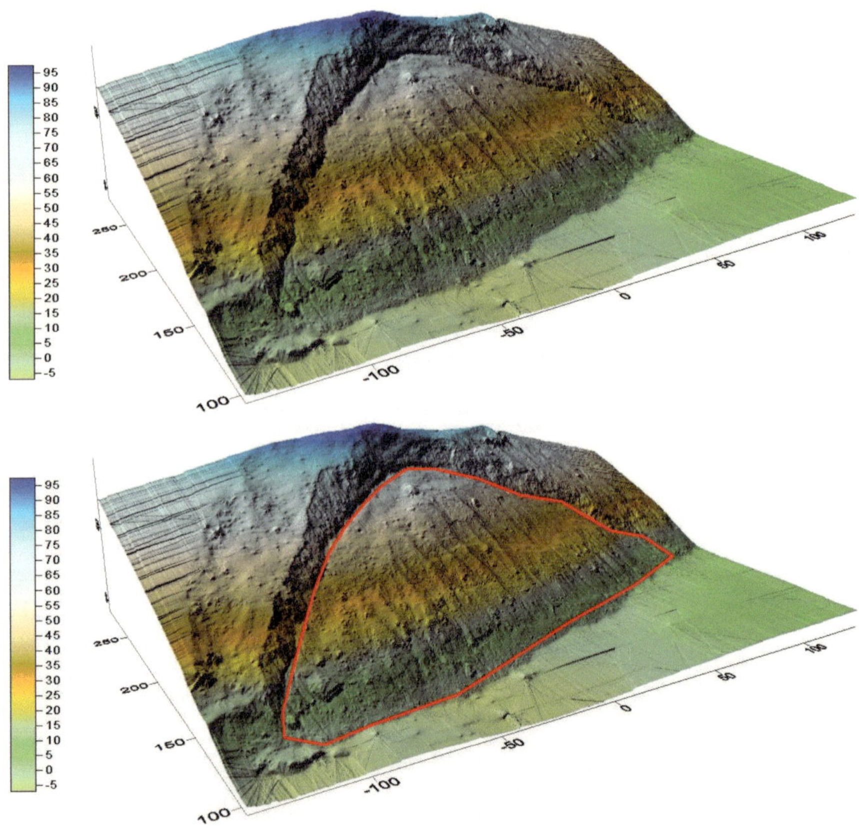

Fig. 7. Triangulated surface of the Locce landslide and selection on the moving/stable areas (red line) for the alignment and movement determination.

The software processes 3D data according to their reference system so when a complex or huge amount of data needs to be aligned, the operator may select only the usable sectors and perform the alignment; then by applying the rototranslation matrix to the whole dataset, the software computes the alignment parameters achieving the final results and reducing computational time and efforts.

At the end of the alignment procedure, the software opens the second module, ImMerge, where the point cloud is transformed into triangulated surfaces. Surfaces are computed according to the Delaunay triangulation (De Smith et al, 2007) thus generating triangular networks (Lambers et al, 2007). The meshing process is necessary as the comparison algorithm need to compute differences between two continuous surfaces and not on point clouds; the second option is however feasible but computation time is quite longer and the success in the process conclusion is not guaranteed.

The computed surface may be edited in the following module, ImEdit, where volume calculation and cross section extraction may be executed, too.

On the other hand the surface comparisons are carried out in the ImInspect module. This one is dedicated to the inspection of meshes and the analysis of differences between them according to predefined directions e.g. along one axis, along the shortest distance or along a user specified vector.

The inspection could also be carried out manually, by vector plotting on homologous features in two different, multitemporal, meshes. In this module the mesh georeferencing is also achievable by assigning coordinates to point picked on the surface; the software has a series of functions, that help the operator in finding the centre of the target by interpolation algorithms, as usually target coordinates are referred to the target centre.

3.1 Moraine survey

The Locce Landslide has been measured according to the same approach adopted in the snout survey, explained in details in the next paragraph; the first survey has been carried out on 01/08/2006 and the second one on the 13/09/2006. Also in this experiment the object has been measured by various scan positions.

Each scan session has been accomplished with a 7 cm resolution obtaining up to 9 million of 3D points (see again figure 7).

Fig. 8. Panoramic image of Belvedere right snout

3.2 Terminus survey

The right snout of the Belvedere Glacier (Figure 8) has been surveyed in two separated sessions on the 02/08/2006 and on the 14/10/2006. Each session (Figure 9) has been subdivided in different scan positions, in order to completely measure the glacier snout and to include, in sessions, the mountainside and the Little Ice Age moraine, with the purpose of employing their feature as reference areas.

As showed in the figure 9, reporting the instrument interface, the scans have been executed with a 20 cm resolution collecting up to 11 millions of points in the first session and 14 millions in the second one.

Data has been downloaded in a folder and then processed in the previously explained mode. The comparison between the two scan sessions has been carried out by the assessment of differences, between surfaces along the z axis, in order to quantify the ablation, in the snout zone. Moreover, by manual object recognition a few boulders have been determined in both surfaces and their locations have been linked by a vector, in order to analyze the main snout displacement direction.

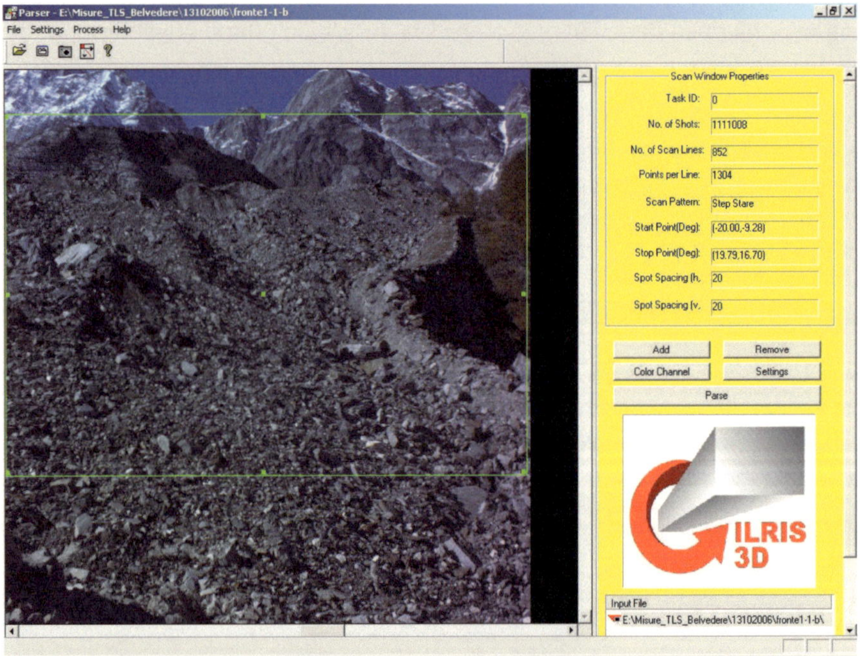

Fig. 9. ILRIS-3D interface during snout survey

3.3 Glacier surface survey

The survey of glacier surface has required a more complex procedure as the georeferencing of the final result was necessary.

During a survey campaign five targets have been placed on the glacier and their position has been measured employing the differential GPS, as in stake positioning (Figure 10). The coincidence of the two measurements has been repeated during the second laser survey in order to have the same time span covered by two kinds of survey methods.

Target coordinates have been processed as well as the one belonging to stakes and stored in an ASCII file in order to be used in surface georeferencing.

The laser measurement has then been executed, during the survey different resolutions have been chosen when scanning the glacier (7 cm resolution) or targets (1 cm resolution). According to these parameters, during the first session the glacier surface has been measured by 14 millions of 3D points and by 19 millions in the second session.

The data obtained from the first scan has been aligned in one overall point cloud and then georeferenced by the employment of targets coordinates. The triangular meshes obtained from the two scan sessions have then been co-registered on reference areas i.e. the lateral moraines.

The surface has then been compared along the z axis in order to evaluate the ablation and manually processed in order to find boulders or other homologous features on both sessions.

Boulders have been processed by computing their rototranslation matrix (from session 1 to session 2), with the purpose of obtaining vectors parameters and evaluating glacier displacement.

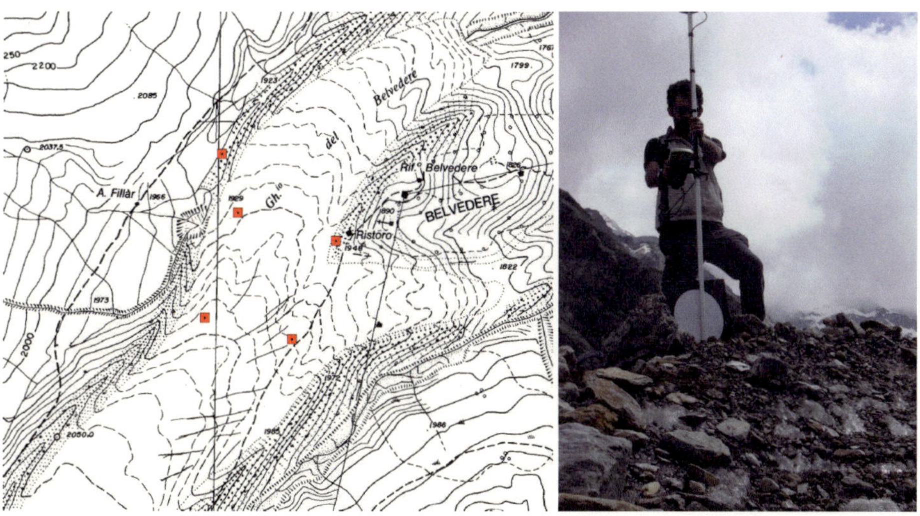

Fig. 10. Targets location (Regional Technical Map, original scale 1:10000, in background) and survey (Photo P. Federici)

3.3.1 Comparison with ablatometric stakes

Thank to the availability of georeferenced laser scans, an independent check of the results of multi-temporal point cloud comparisons, could have been carried out with punctual ablation stakes surveys carried out in the same monitoring campaigns (Godone et al., 2010). In early June 2006 ablation stakes have been installed on the Belvedere Glacier surface and stakes measurements have been repeated regularly during 2006 and 2007 ablation seasons (figure 9). Stakes were introduced in ice up to a depth of 8 meters, after perforation of ice by means of a stream driller, as shown in figure 11, and periodically checked with differential GPS measurements. At the same time the ablation rate was measured at each point.

This procedure has been employed in order to evaluate laser scanning reliability in glacier monitoring in comparison with a well-established technique.

Ablation and displacement data obtained from both methodologies have been statistically compared, by Student's *t* test, in order to evaluate results differences and comparability, and moreover to highlight critical aspects.

Fig. 11. Installation of an ablation stake using a stream driller and D-GPS measurement (Photo P. Federici)

4. Results and discussions

The laser scanner approach allows to extend the survey and analysis to an entire object, or a portion of it, and not to narrow only on discrete locations as in traditional surveys, e.g. GPS, topographical surveys.

Moreover the analysis tools available in data processing software allow the accomplishment of several investigations and testing with the purpose of extending the comprehension of the analysed phenomena.

The accuracy obtained in pre-processing (scans alignment) allowed to execute the comparisons. In the last experiment it has been computed as decimetric and centimetric in the other two tests

4.1 Moraine landslide survey

The measurement of the Locce landslide allowed, although not directly, the understanding of the global dynamic of the glacier. The survey approach was the same of the other two cases with multiple and multitemporal (01/08/06 and 13/09/06) scans.

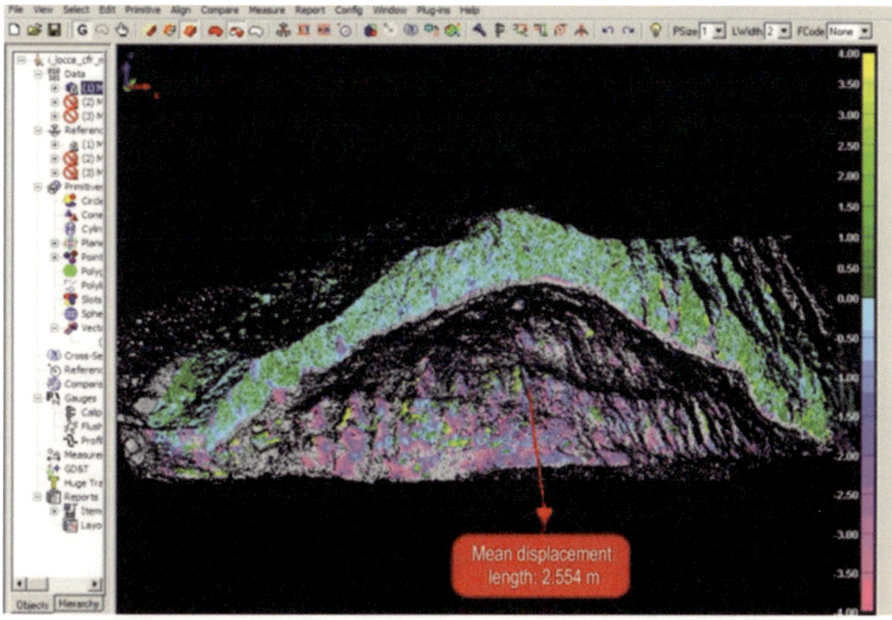

Fig. 12. Map of landslide collapse computed by subtracting the two multitemporal surfaces along the average displacement vector

Fig. 13. Measure of the total displacement of the landslide from its edge

The surface comparison gave an average measure of landslide movement of over 2 m (Figure 12) with an average rate of 0.05 m/day and in addition, through manual measurement, the total vertical shift of the landslide mass has been estimated in 36.62 m since the trigger of the event (Figure 13).

In this experiment the accuracy reached in the alignment phase was centimetric.

4.2 Terminus survey

The right snout has been measured on the 02/08/06 and on the 13/10/06 with multiple scans, point clouds obtained from the measurement have than been separately merged into a unique surface and then compared, after the alignment according to fixed features. The alignment processes have been accomplished obtaining centimetric accuracy allowing the execution of the next measurements.

The comparison has been carried out in two ways, automatic and manual. Firstly the data comparison has been performed by assessing the differences between the two surfaces along the z axis (Figure 14), by evaluating an estimate of the front ablation of about 4 m and then by manually plotting vectors linking the same feature represented in both point clouds, usually large boulders, in order to estimate the displacement (Figure 15). At the end of the process up to six vectors have been measured (Table 2), obtaining an average displacement of 6.93 m and a rate of 0.09 m/day.

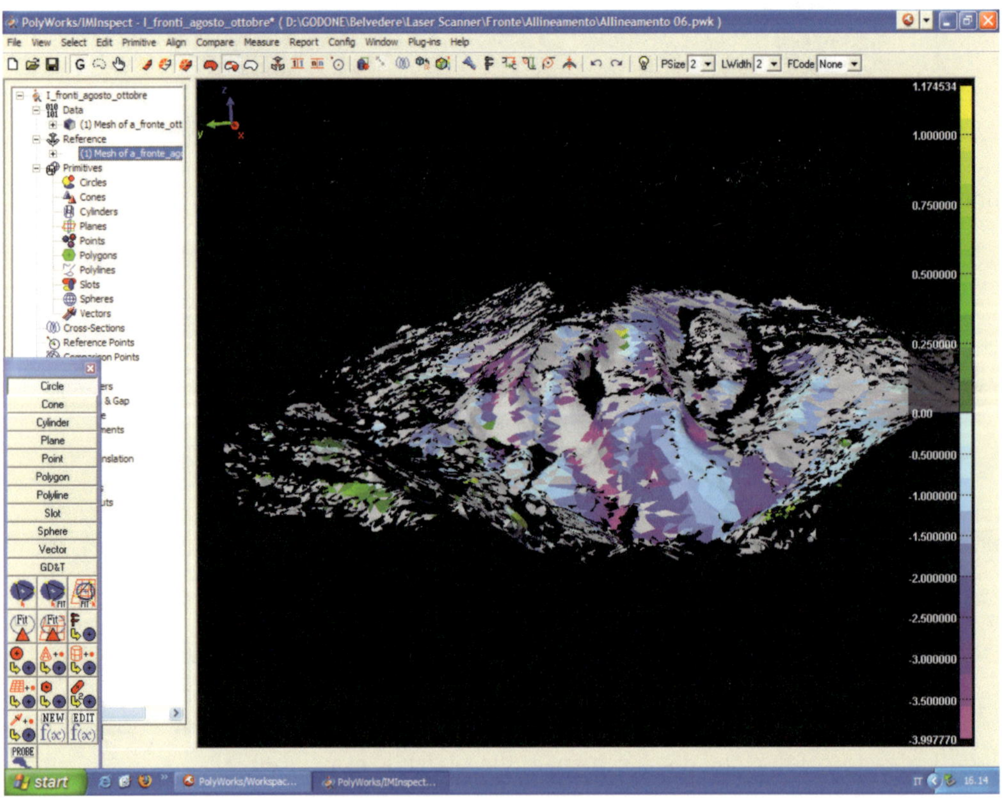

Fig. 14. Map of snout ablation computed by subtracting the two multitemporal triangulated surfaces along z axis

Vector	x	y	z	rx	ry	rz	Length
1	112.76	54.55	53.88	-0.89	0.02	-0.45	11.03
2	163.03	85.88	67.58	-0.93	0.04	-0.37	5.73
3	162.13	37.35	79.50	-0.54	-0.79	-0.30	8.06
4	93.01	-6.80	39.29	-0.71	-0.44	-0.55	6.05
5	73.65	1.53	33.70	-0.87	-0.22	-0.44	4.92
6	59.15	17.04	30.14	0.00	-0.75	-0.66	2.40

Table 2. Vectors measured from natural benchmarks (vectors 1, 2, 4 and 5 have been employed in snout displacement computation)

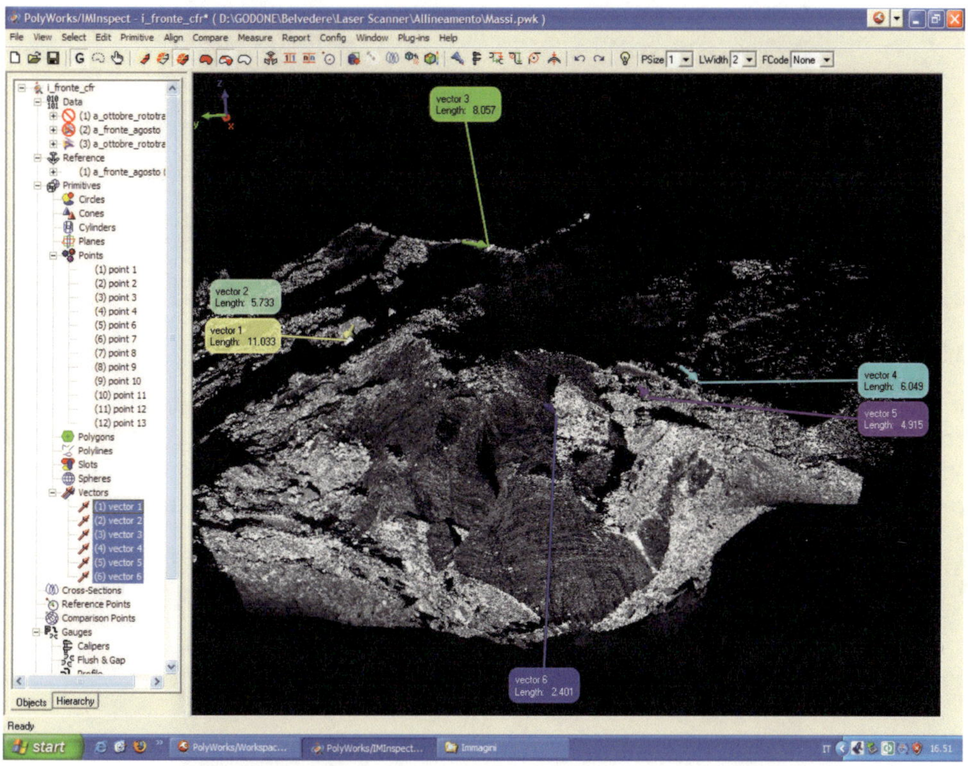

Fig. 15. Vectors plotted on the glacier 3D point cloud

4.3 Glacier surface survey

The multitemporal survey (29/06/2007 - 01/09/2007) of the Belvedere surface has provided a global overview of glacier dynamic, and despite the traditional approach, i.e. stake measurement, results are widespread to the whole surface investigated and not only related to the position of every single stake.

4.3.1 Data alignment and calibration

The georeferencing of the first surface, through target alignment, has assured enough accuracy in order to continue the comparison process. The two multitemporal surfaces have been aligned using the "natural benchmark" approach; in other words employing natural features such as boulders and rocks on glacier cover, with an accuracy of ± 0.2 m (Figure 16).

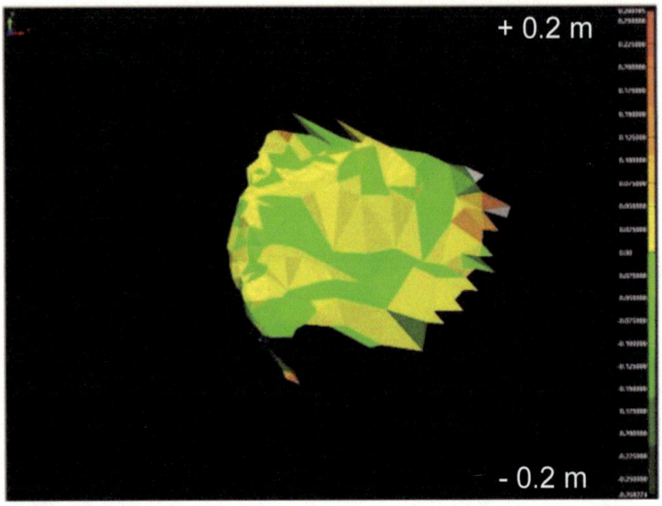

Fig. 16. Residuals of the alignment of the two surfaces computed on natural benchmarks

4.3.2 Displacement

The analysis executed on "natural benchmarks" on the surface of the glacier has allowed identifying up to 40 boulders, usable for displacement vectors computing (Figure 17, Figure 18). The calculation carried out on the rototranslation matrices has provided a displacement measurement ranging from 0.436 to 7.615 m, with an average rate of 0.059 m per day

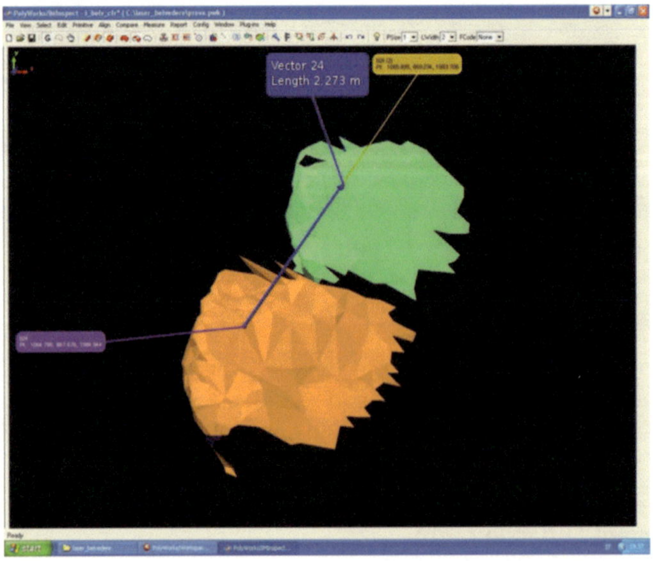

Fig. 17. Measurement of displacement vector on two multitemporal positions of the same boulder

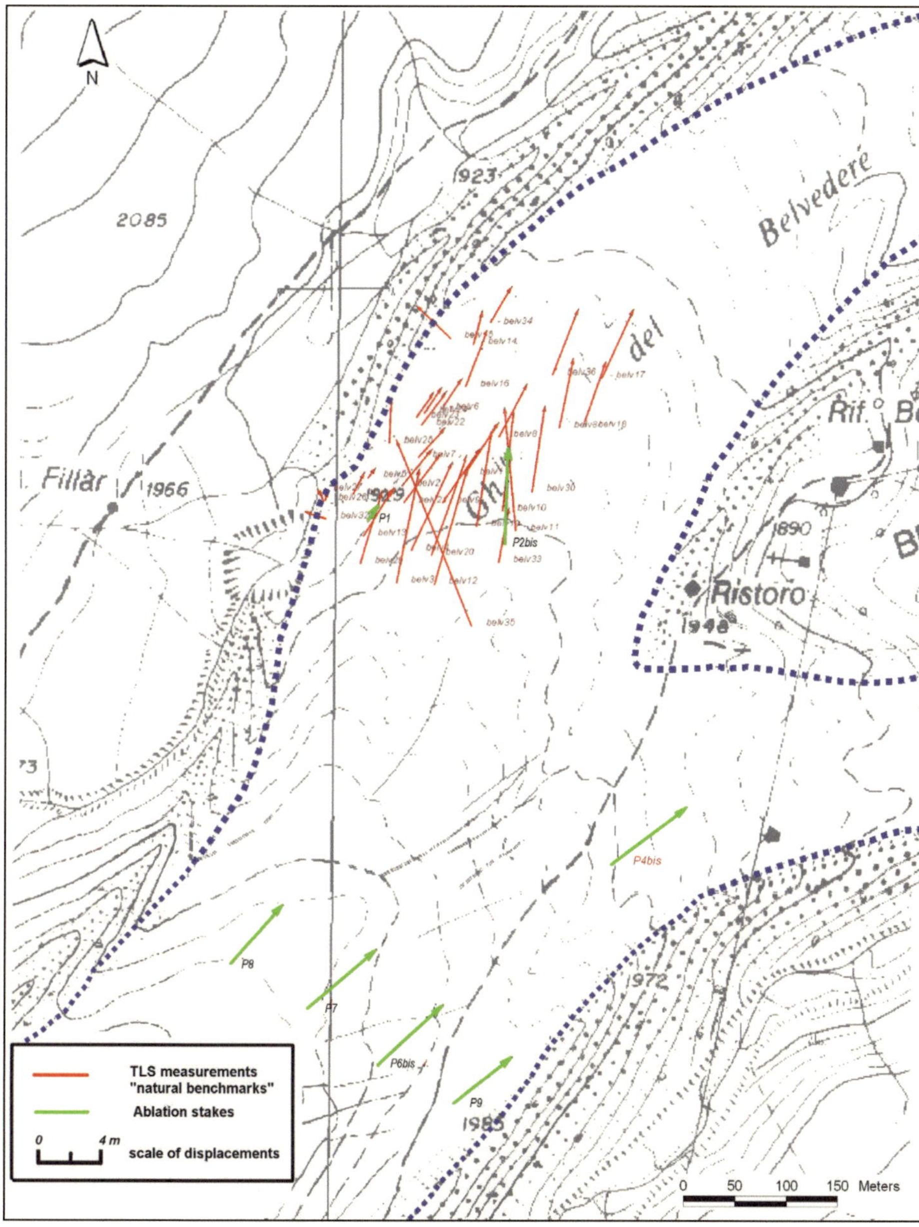

Fig. 18. Displacement measured with TLS on "natural benchmarks" (in red) and from ablation stakes positions (green)

4.3.3 Ablation

The difference of surfaces' elevation, computed along the z axis, has resulted as the measure of the ablation of the investigated portion of the glacier. The ablation ranges from a minimum value of 2 m to a maximum of 8 m (Figure 19). The average ablation rate is 0.048 m/day; the rate measured with traditional method in the same period - 65 days - is 0.028 m/day.

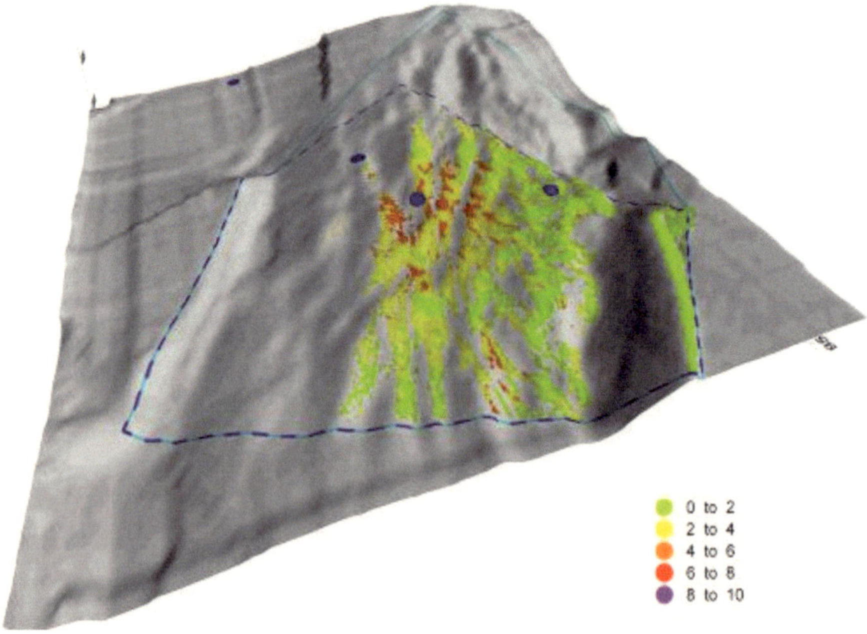

Fig. 19. Ablation map computed by subtracting two multitemporal surfaces

5. Discussion

The terrestrial laser scanner measurement has been highly experimental, as currently is hard to find extensive experiments on glacier carried out with this technology, e.g. the Miage lake (Mont Blanc Massif) survey (Deline et al., 2004b).

Compared with GPS, total stations and other field instruments, the laser is more difficult to handle and carry during the way to measuring station but the high speed survey and final results anyway encourage its employment.

The opportunity of completely reconstruct the surveyed object is highly valuable in glaciology, as it allows to carry out measurement otherwise impossible or highly dangerous to operators. The resolution and the achievable accuracy (Figure 20), let to perform, not only global but also detailed measurements and analysis and moreover through a multitemporal approach change detection investigations are possible, with meaningful results.

Considering the glacier or its portion, the survey has always been subdivided in two or more measuring sessions; in the post processing phase the scan were aligned according to homologous feature common to two o more scans. In the first and the second case showed, feature measured, and employed for the alignment, has been selected in areas outside the analysis sector, and considered stable, such as the mountainside surrounding the survey target. This method is useful when the georeferencing of the object is not mandatory and a relative analysis is the objective. When needed, as in the last experiment, a set of target placed on the glacier and georeferenced has been used in order to align, and put in map coordinates, the scans. It's necessary to state that, sometimes, the second methodology is not achievable; due to the impossibility to reach the survey object for safety reasons, as in the second and, particularly, in the first experiment.

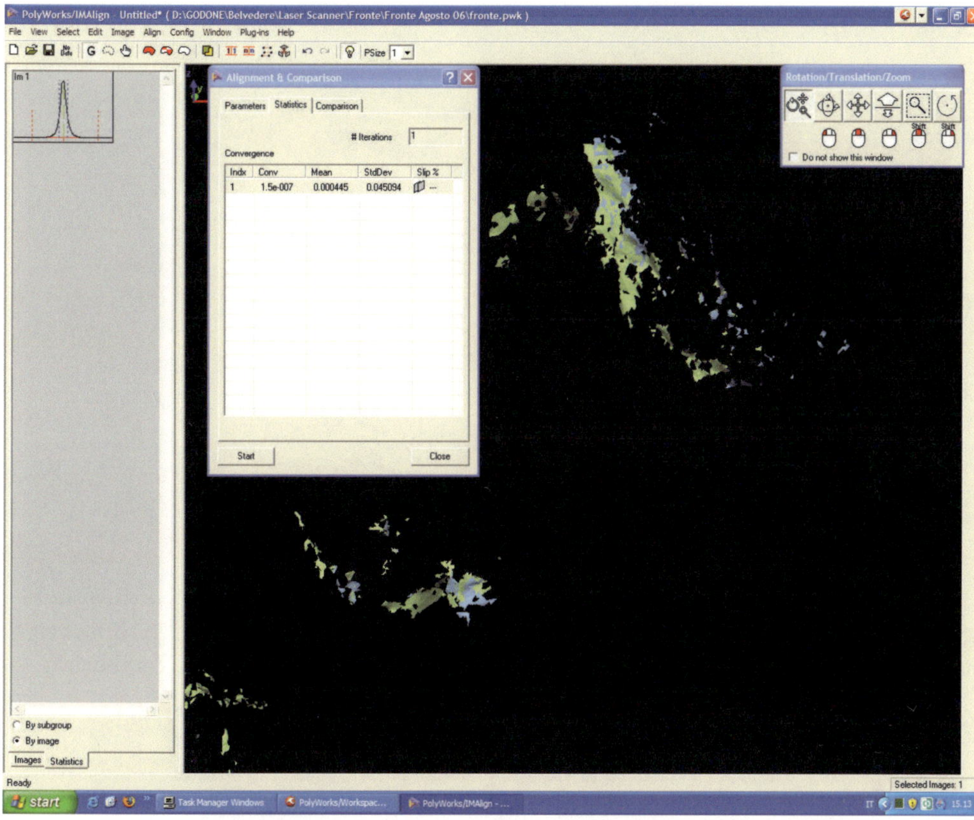

Fig. 20. Alignment accuracy assessment

5.1 Moraine survey

The survey of the Locce landslide has two goals, to monitor the process of the landslide in time and to give contributing factors to Belvedere dynamic comprehension.

As the following experiment, the measurements have been carried out with the change detection approach. The results obtained from the first analysis show the landslide trend to move towards the glacier surface (Figure 21). This fact may be partially explained by the pressure of the uphill lake, called "Lago delle Locce" that exerts a pressure on the surrounding moraine, including the landslide sector.

On the other side, however, there is the glacier surface and according to the recent dynamics, has been characterized by a sudden mass movement down valley, without the contribution of upper glaciers and without particularly abundant snowfall in the previous seasons. This flow has caused an increase of surface elevation in the lower sector of the glacier, and a consequential decrease in the upper part, including the Locce moraine sector (Haeberli et al., 2002; Kamb et al., 1985; Mazza, 2003; Miller, 1971).

Fig. 21. Evolution of the landslide from 01/05/2005 (upper left) to 26/06/2005 (upper right), year 2006 (lower left) and 2007 (lower right) (Photo courtesy: L. Schranz, A. Tamburini)

Fig. 22. Upper edge of the Locce landslide with evidences (red line) of further collapses (Photo G. Mortara)

The evidences, of these dynamic, are evident also on the top of the Locce moraine, where the main ridge is still collapsing and a rotational movement is clearly recognizable (Figure 22).

The terrestrial laser scanner has allowed studying the phenomenon as it prevented the operator to stay close to the object during measurement and provided a complete reconstruction in order to perform every kind of analysis requested, in a virtual environment. Due to the need of relative evaluations, the scans have been aligned on common features in stable areas of the survey, in order to avoid target placement in dangerous places. Notwithstanding the dimension of the landslide, every survey has been completed in one day, with multiple scans, assuring an accurate and detailed description of the site.

The analysis has been carried out both automatically and manually, highlighting the flexibility of the methodology when studying such events.

The magnitude of the phenomenon deserve further studies and deepening, as the Locce sector has been already involved in several lake outbursts in '70 (Mortara and Mercalli, 2002) and the current trend of the landslide is not leading to a safe and stable situation, if considering Belvedere glacier reduction, too.

5.2 Terminus survey

Measurement of glacier front, usually, are based on repeated surveys (often along predefined bearings) from a reference point placed at a certain distance form the glacier, in order to measure fluctuations in time (Bonardi et al., 2006). Obviously, this approach tends to excessively simplify the complexity of the front, with a single or discrete representation, however this approach is highly convenient as it requires cheap instrumentations. It is really fast to perform, but results are only referred to few point measured on the glacier. On the other hand, the employment of laser scanner requires skilled operators and the availability of such equipment, but the results are nearly impressive both under visual and geometrical point of view.

With terrestrial laser scanner, not only displacement evaluation have been executed, also ablation data are available after the scans alignment, by differentiating the obtained surface, a global evaluation is feasible; another advantage of this technique. In this case is more appropriate to refer the measurement to the snout rather than to the simple front.

The entire measurement has been completed in two session of one day duration, including the reaching of the site. As explained above, the snout has been measured from three different positions in order to scan the entire ice cliff and the surrounding mountainsides. The post processing phase has been subdivided in three phases; at the beginning the scans have been aligned, reconstructing the front in the two epochs, then the two surfaces have been aligned thank to external features and then, finally, the multitemporal analysis has been carried out. The two surfaces have then been compared globally along the z axis with the aim of obtaining an estimate of the ablation in the front sector. These data, compared with those measured in the, traditional, stake approach, in the same period, have shown similar values ($p = 0.380$) confirming the reliability of the method.

The ablation estimation has been carried out in a completely automatic way; on the contrary the measure of glacier displacement needs to be performed manually, as it is based on object recognition on both surfaces. In order to measure the displacement on the two surfaces, features common to the two multitemporal scan have been searched. When a rock or a boulder was recognized, in both scans, was then used to manually generate a displacement vector, linking the two objects in the two scans.

Moreover the vector obtained needs to be checked in order to exclude those characterized by anomalous directions, probably due to falling of rocks and not strictly related to glacier displacement

Due to the 2.5 month period, only 6 vectors were traced and, among them, only 4 were then used in order to compute the displacement speed. The selection has been carried out based on vector direction in comparison with the glacier main flow direction. The methodology has allowed measuring glacier surface displacement rate, without risks for the survey operators, but due to the time span between the two surveys the number of available features on the glacier was nearly scarce, in order to measure the displacement in a more detailed way, measurement should be repeated at an interval never greater than one month. The proof of this advice is in the result of the first analysis; in the global evaluation of the front ablation, positive values have been found in the bedrock close to the ice cliff, this can be explained by the accumulation of boulders, fallen, from the surface due to the high displacement speed, as shown in the next figure (Figure 23), highlighted by white ellipses.

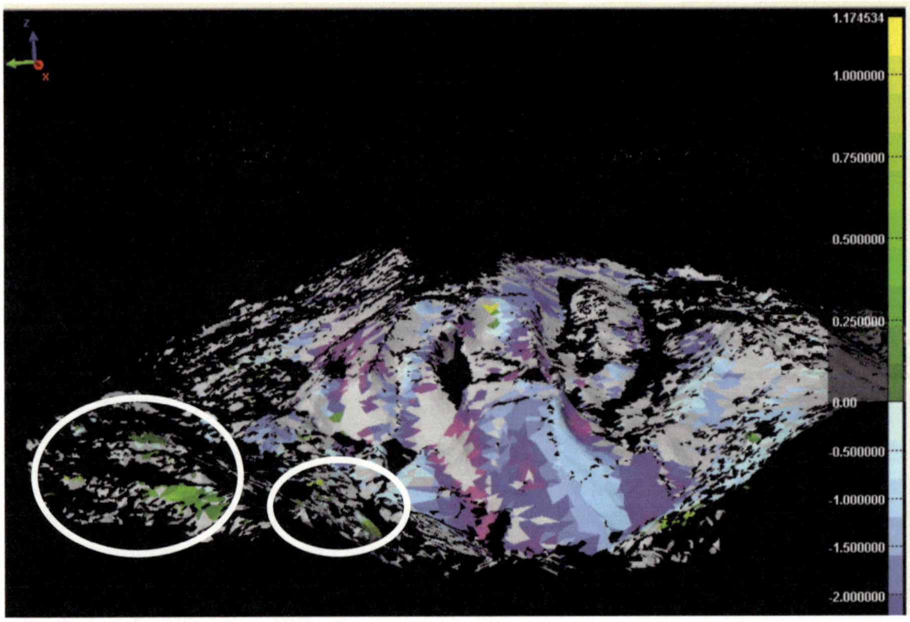

Fig. 23. Snout ablation map, accumulation of rocks, fallen from the glacier surface, are highlighted by the white ellipses in the lower left part of the image

5.3 Glacier surface survey

The survey of the surface has been the most complex field activity in the whole experiment. From the logistic point of view, the laser instrument, equipped with tripod and batteries had to be transported to a raised position, over the glacier surface, and due to the location of the measuring station the entire equipment has been moved only by instruments operators.

The survey has also been georeferenced, in order to analyze glacier displacement according to its bearing and to compare it with stakes measures. In order to accomplish the georeferencing, 4 targets have been placed on the glacier during the first survey and their positions have been measured with differential GPS. In order to optimize the stake and target survey, one GPS campaign has been performed with both aims.

The employment of targets comprises additional tasks to achieve, in fact every single target need to be measured in a separate scan session; this detail suggest that targets should be used only when strictly needed as they require quite a lot of operational time, to spend in measurement of their position, in their scanning and computational time in data processing.

5.3.1 Data alignment and calibration

Every measurement has been characterized by multiple scans in order to survey the largest part of the glacier, each scan has been then aligned thank to targets.

The alignment, between the two surveys, has been performed easily as in previous experiments. In this one, the "natural benchmark" approach has been used in a twofold way; during the alignment phase, as already described in the two previous tests, the two scans have been aligned on features tracked down in glacier surroundings; and during the analysis phase as described below.

5.3.2 Displacement

The glacier displacement has been measured according to the new approach, already tested on Liligo Glacier, Karakoram, Pakistan (Diolaiuti et al., 2003) and on Belvedere snout, which considers as reference points boulder and rocks placed on glacier surface. By recognizing the relative positions of these "natural benchmark" a vector computation is possible, allowing measuring local glacier displacements, limited only by the number of recognizable features. In spite of the scarce results obtained on the right snout of the glacier, in this experiment up to 40 features, and consequently, displacement vectors have been computed with a highly detailed analysis of the phenomenon (Figure 18). Displacement data, compared with those measured by GPS at stakes positions are comparable with no significant differences ($p = 0.245$) and similar variances ($p = 0.323$), as shown in the same figure.

5.3.3 Ablation

Laser data offers also the chance to measure the glacier ablation in the scanned area. In this case the automatic comparison of the two surfaces along z axis, as already performed on snout experiment, have proven its effectiveness providing a fast and accurate measurement extended to the entire scanned area. The comparison between laser data and stake measurement has provided an unexpected result: the two datasets have shown significant differences in variances ($p = 4.700 \cdot 10^{-3}$) and highly significant differences in their values ($p = 7.000 \cdot 10^{-4}$), according to *t* test. The analysis have been repeated, by comparing laser data with the glacier elevation losses, measured with GPS at stake locations; in this second test the two datasets have shown non significant differences ($p = 0.751$). This different test results suggest the importance of the correct interpretation of the laser survey results, due the high complexity of a glacier system, several factor (e.g. debris cover or bare ice…) should be taken into account before identifying the correct meaning of a new technique results.

6. Conclusions

The cryosphere is characterized by undoubtedly remarkable phenomena with meaningful magnitudes. Hazards related to these events may endanger human life and settlements, as already stated by different authors (Huggel, 2004; Rott et al., 2005). The monitoring of these phenomena should assure enough accuracy as they have to describe complex phenomena and their potential hazards. Contemporarily, they should be characterised by a handy employment and management due to the difficult environment of application.

The geomatic methodologies employed in the monitoring and analysis of cryospheric phenomena has provided a satisfactory response to the briefly listed requirements.

GPS and Terrestrial Laser Scanner have confirmed their effectiveness in the research. The ease in their field employment has allowed the execution of fast and reliable surveys.

The Terrestrial Laser Scanner, thank to its high acquisition frequency, provides a complete description of the glacier surface in short time. The output point cloud is immediately ready to process and allows the extraction of geometrical features of the object.

As a remote technique, the accessibility of the survey object is not needed so measurement of unsafe areas is feasible (Biasion et al., 2005). When the object georeferencing is not requested also the target positioning is not necessary, point cloud processing and comparison may be executed in a relative approach, with no loss in the final accuracy. The proposed "natural benchmark" approach has proved its applicability introducing a new method in the processing of data.

In both cases the accuracies obtained by the two techniques has assured their applicability to cryosphere monitoring reaching the requirement specified for this kind of measurements (Rott et al., 2005).

The fast execution of measurement has allowed completing a survey session in few hours enabling their repetition during the season. A high frequency monitoring is highly recommendable in these events in order to achieve the maximum number of information on the phenomenon dynamics. Moreover, due to the phenomena complexity, multidisciplinary approaches are recommendable (Zublin et al, 2008) in order to achieve the maximum detail in the survey object reconstruction and consequent analyses.

The measurements are still in progress as the explained phenomena deserve a continual monitoring to deepen the comprehension of their dynamics.

Other techniques are developing, like the GB-SAR (Ground Based Synthetic Aperture Radar), in cryosphere monitoring and the integration with the other methodologies is highly recommended in order to obtain data from different sources, integrate and cross validate them. This multi disciplinary approach should lead to the definition of the optimal method in the monitoring of these phenomena.

7. Acknowledgment

Lamberto Schranz, alpine guide, who assisted and cooperated during field surveys.

Andrea Tamburini (IMAGEO SrL), Giovanni Mortara (Italian Glaciological Committee) and Paolo Federici (RSE SpA – Research on Energy Systems) for their valuable cooperation during field works, data analyses and for their helpful comments in the preparation of this chapter.

The project has been partially funded by "Comitato Scientifico Centrale del Club Alpino Italiano".

8. References

Beinat, A., 2006. Tecniche di Registrazione. In: F. Crosilla and S. Dequal (Editors), Laser Scanning Terrestre. Collana di Geodesia e Cartografia. International Centre for Mechanical Sciences,, Udine, pp. 39 - 53.

Beninston, M., 2003. Climatic Changes in Mountain Regions: a Review of Possible Impacts. Climatic Changes, 59: 5 - 31.

Besl, P.J., Mc Kay, N.D., 1992. A Method for Registration of 3D Shapes. IEEE Transactions on Pattern Analysis and Machine Intelligence, 14(2): 239 - 256.

Biasion, A., Bornaz, L., Rinaudo, F., 2005. Laser Scanning Applications on Disaster Management. In: P. Van Oosterom, S. Zlatanova and E.M. Fendel (Editors), Geo-information for Disaster Management. Earth and Environmental Science. Springer, Berlin, pp. 19 - 33.

Bonardi, L., Catasta, G., Righetti, F., D'Adda, S., 2006. Manuale di Osservazioni Glaciologiche. I Quaderni del Servizio Glaciologico Lombardo, 5. Servizio Glaciologico Lombardo, Milano, 47 pp.

Carrasco, J.F., Cassassa, G., Quimtana, J., 2005. Changes of the 0 C Isotherm and the Equilibrium Line Altitude in Central Chile During the Last Quarter of the Twentieth Century. Hydrological Sciences Journal, 50: 933 - 948.

Chen, Y., Medioni, G.G., 1992. Objec Modeling by Registration of Multiple Range Images. Images and Vision Computing, 10(3): 145 - 155.

Davies, M.C.R., Hamza, O., Harris, C., 2001. The Effect of Rise in Mean Annual Temperature on the Stability of Rock Slopes Containing Ice-Filled Discontinuities. Permafrost and Periglacial Processes, 12: 137 - 144.

De Smith, M.J., Goodchild, M.F., Longley, P.A., 2007. Geospatial Analysis: A Comprehensive Guide to Principles, Techniques and Software Tools, 414 pp.

Deline, P., Chiarle, M., Mortara, G., 2002. The Frontal Ice Avalanche of Frebouge Glacier (Mont Blanc Massif, Valley of Aosta, NW Italy) on 18 September 2002. Geografia Fisica e Dinamica Quaternaria, 25: 101 - 104.

Deline, P., Chiarle, M., Mortara, G., 2004a. the July 2003 Frebouge Debris Flow (Mont Blanc Massif, Valley of Aosta, Italy): Water Pocket Outburst Flood and ice Avalanche Damming. Geografia Fisica e Dinamica Quaternaria, 27: 107 - 111.

Deline, P., Diolaiuti, G., Kirkbride, M.P., Mortara, G., Pavan, M., Smiraglia, C., Tamburini, A., 2004b. Drainage of Ice-Contact Miage Lake (Mont Blanc Massif, Italy) in September 2004. Geografia Fisica e Dinamica Quaternaria, 27: 113 - 119.

De Saussure, H.B., 1779 - 1796. Voyages dans les Alpes, 4, Neuchatel.

Diolaiuti, G., Pecci, M., Smiraglia, C., 2003. Liligo Glacier (Karakoram): reconstruction of the recent history of a surge-type glacier. Annals of Glaciology, 36: 168 - 172.

Federici, P., Luzi, G., Noferini, L., Mecatti, D., Macaluso, G., Tamburini, A., Martelli, D., 2008. Report on Use of Ground Remote Monitoring (GB-SAR and TLS) for Extraction of Topography and Flow Properties of Glaciers. Deliverable D8.1, Project Galahad, EC FP6 project no. 018409, www.galahad.eu.

Fischer, L., Kaab, A., Huggel, C., Noetzli, J., 2006. geology, Glacier Retreat and Permafrost Degradation as Controlling Factors of Slope Instabilities in a High-mountain Rock Wall: the Monte Rosa East Face. Natural Hazards and Hearth System Sciences, 6: 761 - 772.

Fischer L.,Eisenbeiss H., Kaab A., Huggel C., Haeberli W.,2011 Monitoring Topographic Changes in a Periglacial High-mountain Face using High-resolution DTMs, Monte Rosa East Face, Italian Alps Permafrost and Periglacial Processes, 22 (2), pp. 140–152, April/June 2011

Giani, G.P., Silvano, S., Zanon, G., 2001. Avalanche of 18 January 1997 on Brenva glacier, Mont Blanc Group, Western Italian Alps: an unusual process of formation. Annals of Glaciology, 32: 333 - 338.

Godone, F., Godone, D., Tamburini, A., Mortara, G., 2007. La valanga di roccia della Cima Thurwieser (SO): determinazione del volume con tecniche di fotogrammetria digitale., 11a Conferenza Nazionale ASITA, Centro Congressi Lingotto, Torino 6 - 9/11/2007.

Godone D., Godone F., Tamburini A., 2010. Belvedere glacier monitoring: a multidisciplinary approach, GI4DM 2010 Conference "Geomatics for Crisis Management", 2 - 4/02/2010, Torino, Italy.

Haeberli, W., Kaab, A., Paul, F., Chiarle, M., Mortara, G., Mazza, A., Deline, P., Richardson, S., 2002. A Surge-type Movement at Ghiacciaio del Belvedere and a Developing Slope Instability in the East Face of Monte Rosa, Macugnaga, Italian Alps. Norsk Geografisk Tidsskrift - Norwegian Journal fo Geography, 56: 104 - 111.

Haeberli, W., Noetzli, J., Zemp, M., Baumann, S., Frauenfelder, R., Hoelzle M., 2005. Glacier Mass Balance Bulletin No. 8 (2002-2003). Compiled by the World Glacier Monitoring Service (WGMS). IUGG (CCS)/UNEP/UNESCO.

Hoelzle M., Haeberli W., Dischl, M., Peschke, W., 2003. Secular glacier mass balances derived from cumulative glacier length changes. Global and Planetary Change, Vol. 36, Issue 4, May 2003, pp. 295-306.

Huggel, C., 2004. Assessment of Glacial Hazards Based on Remote Sensing and GIS Modeling, Matematisch-Naturwissenschftlichen Fakultat, Zurich, 75 pp.

Hunter, G., Pinkerton, H., Airey, R., Calvari, S., 2003. The Application of a Long-range Laser Scanner for Monitoring Volcanic Activity on Mount Etna. Journal of Volcanology and Geothermal Research, 123: 203 - 210.

Kaab, A., Huggel, C., Fischer, L., Guex, S., Paul, F., Roer, I., Salzmann, N., Schlaeffi, S., Schmutz, K., Schneider, D., Strozzi, T., Weidmann, Y., 2005. Remote Sensing of Glacier- and Permafrost-related Hazards in High Mountains an Overview. Natural Hazards and Hearth System Sciences, 5: 527 - 554.

Kamb, B., Raymond, C.F., Harrison, W.D., Engelhardt, H., Echelmeyer, K.A., Humprey, N., Brugman, M.M., Pfeffer, T., 1985. Glacier Surge Mechanism: 1982-1983 Surge of Variegated Glacier. Science, New Series, 227(4686): 469 - 479.

Lambers, K., Eisenbeiss, H., Sauerbier, M., Kupferschmidt, D., Gaisecker, T., Sotoodeh, S., Hanusch, T., 2007. Combining photogrammetry and laser scanning for the recording and modelling of the Late Intermediate Period site of Pinchango Alto, Palpa, Peru, Journal of Archaeological Science, 34(10):1702 - 1712.

Massonet, D., Feigl, K.L. ,1998. Radar interferometry and its application to changes in the Earth's surface. Rev. Geophys., 36, 441-500.

Mazza, A., 1998. Evolution and Dynamics of Ghiacciaio Nord delle Locce (Valle Anzasca, Western Alps) from 1854 to the Present. Geografia Fisica e Dinamica Quaternaria, 21: 233 - 243.

Mazza, A., 2003. La Teoria delle Onde Cinematiche: Possibile Applicazione al Ghiacciaio del Belvedere (Valle Anzasca, Alpi italiane). Ipotesi Preliminari. Terra Glacialis, 6: 23 - 33.

Miller, M.M., 1971. glaciers and Glaciology. Encyclopedia of Science & Technology. Mcgraw-Hill, pp. 218 - 229.

Mortara, G., Chiarle, M., Tamburini, A., 2003. The Emergency Caused by the "Effimero" Lake on the Belvedere Glacier (Macugnaga, Monte Rosa Group, Italian Alps). In: D. Richard and M. Gay (Editors), Glaciorisk EVG1 2000 00512 Deliverable.

Oppikofer, T., Jaboydeoff, M., Blikra, L.H., Derron, M., 2008. Characterization and Monitoring of the Aknes Rockslide using Terrestrial Laser Scanning. In: J. Locat, D. Perret, D. Turmel, D. Demers and S. Leroueil (Editors), 4th Canadian Conference on Geohazards: From Causes to Management. Presse de l'Université, Laval, Québec, pp. 211 - 218.

Rott, H., Nagler, T., 2005. Observational requirements for improving monitoring and forecasting - Glaciers. In: R. Sailer (Editor), Technical Report on User Requirements for Improved Monitoring of Landslide, Glaciers and Avalanche Hazards and on Recommendations for Methodological and Technical Development in the Project. GALAHAD Deliverable, pp. 28 - 30.

Sailer, R, Schaffhauser, A., Fromm R., Jörg P., Herrera G., Bardasano L., Ponce de León D., Rott, H., Nagler, T., 2005. Technical report on user requirements for improved monitoring of landslides, glaciers and avalanche hazards and on recommendations for methodological and technical developments in the project. Deliverable D1-1. Project Galahad:, EC FP6 project no. 018409, www.galahad.eu.

Somigliana, C., 1917. Primi Rilievi del Ghiacciaio di Macugnaga. Rivista Club Alpino Italiano, 36(3 - 4): 65 - 67.

Tamburini, A., Mortara, G., 2004. The Case of the "Effimero" Lake at Monte Rosa (Italian Western Alps): Studies, Field Surveys, Monitoring. In: F. Maraga and M. Arratano (Editors), Progress in Surface and Subsurface Water Studies at Plot and Small basin Scale. Technical Documents in Hydrology. Unesco - International Hydrology Programme, Turin, pp. 179 - 184.

Tamburini, A., Mortara, G., Belotti, M., Federici, P., 2003. The emergency caused by the "Short-lived Lake" of the Belvedere Glacier in the summer 2002 (Macugnaga,

Monte Rosa, Italy). Studies, survey techniques and main results. Terra Glacialis, 6: 37 - 54.

Wasowski, J., Gostelow P., 1999. Engineering geology landslide investigations and SAR interferometry. Proc. of FRINGE'99, Liege, Belgium, November 1999, http://esrin.esa.it/fringe99.

Züblin M., Fischer L., Eisenbeissa H., 2008 COMBINING PHOTOGRAMMETRY AND LASER SCANNING FOR DEM GENERATION IN STEEP HIGH-MOUNTAIN AREAS The International Archives of the Photogrammetry, Remote Sensing and Spatial Information Sciences. Vol. XXXVII. Part B6b. Beijing 2008

7

Integrated Reverse Modeling Techniques for the Survey of Complex Shapes in Industrial Design

Michele Russo
INDACO Dept., Politecnico di Milano
Italy

1. Introduction

In general a survey can be considered as a synthetic and open instrument of knowledge, applied to investigate the constitutive information of a real object. In particular from the geometrical point of view, the survey methodology becomes an activity finalized to improve the comprehension of a product, supporting the project activity during a design process.

Commonly two different survey approaches allow to acquire the geometrical information of a physical object in the space: the first is based on the traditional survey practices, founded on the direct measure of a model surface, and the production of sections or orthogonal projections. The second one refers to the digital 3D survey techniques, that permit to acquire also complex free-form surfaces in a short time, generating high density point clouds.

In the last 20 years scientific and technological improvements have allowed to refine the 3D acquisition instruments that, are today suitable to acquire the different geometrical and material characteristics of a physical model. In the Cultural Heritage field for example, the introduction of tridimensional acquisition techniques has represented an innovative solution to survey real objects with a non-contact approach, representing both complex architecture and single artifacts. The same has happened in the Industrial Design field, in which the products present a level of geometrical and material complexity unsuited for a direct survey.

In this field the introduction of an innovative medium for the interpretation of reality has led to a time compression in conceiving, design and development of the product. In fact the possibility of translating a physical object to a digital mould with 3D acquisition instruments allows to improve the use of digital models inside the product cycle, foreseeing the final design results and verifying in advance the quality and functionality of the product with 3D digital simulation. Morphological or aesthetical inspections on the digital model represent in this way a real opportunity to get better the whole industrial process, transferring easily the physical variation on the digital project.

The readaptation of the well known existing 3D survey methodologies (Bernardini & Rushmeir, 2002; Beraldin, 2004; Guidi et al., 2010) to the different needs highlighted in the

Industrial Design field has led to research different reliable processes suited for this field. An initial outcome of these studies (VV.AA., 2006) was represented by a first codification of the passage from real to digital products in Industrial Design, applying mono-instrumental 3D acquisition approach on different case-studies, with the aim to reach digital representations useful to support the project. This codified pipeline was named "Reverse Modeling", indicating the opposite modeling approach that don't start from an idea but from the physical object. At the end different alternative approaches were defined, in order to overcome some bottlenecks present inside that process, improving the performances of the 3D acquisition instruments.

The aim of this chapter is to analyze the 3D scanning survey approach applied inside a project of Industrial Design. Starting from the role of the digital acquisition techniques in the Design field, pro and cons of the Reverse Modeling approach are discussed. Than a relation between the complexity of a physical Design product and the 3D survey approach has been defined, in order to suggest a selection of instruments for the best 3D survey of Industrial Design objects. Besides this, the principal limits of 3D active systems are described, from one side highlighting the importance of the instrument characterization, from the other justifying the integration of different 3D acquisition instruments to overcome the bottlenecks of the mono-instrumental approach. At the end some particular case studies are presented, in order to show the different methodologies that have led to the integration or sensor fusion, comparing with the same results obtained with a "traditional Reverse Modeling approach". Conclusions at the end will close the chapter.

2. The role of 3D laser technology in industrial design field

The role of 3D survey inside a project is univocally defined by the different typology of starting data used at the beginning of the project. For this reason, the Reverse Modeling pipeline can be allocated inside a process of an Industrial Design production in accordance with three different scenarios.

The first one consists in a traditional approach, based on the progressive refinements of sketches and maquettes, in which the manual creativity of the Designer is especially exploited. In this case the Reverse Modeling allows to recreate a digital mould of the physical model, extracting the technical drawings useful for the project variations.

In the second scenario the application of CAID instruments (Computer Aided Industrial Design) in the first conceiving step allows to reach immediately the digital definition of the project. Anyway also in this approach the creation of a physical prototype is considered an essential passage. This condition demonstrates that the virtual reality can't substitute the reality itself and implies the use of Reverse Modeling techniques as an obliged passage to update the original (virtual) project with the (physical) modifications.

The last scenario happens when the digital representation of a final industrial product doesn't exist and the reactivation of its production is required (Fig. 1). Either a "restyle process", due to exterior modifications, or a "redesign process", due to functional variations, requires to work on a digital shape very close to the real product. For this translation the Reverse Modeling represents the only reliable way.

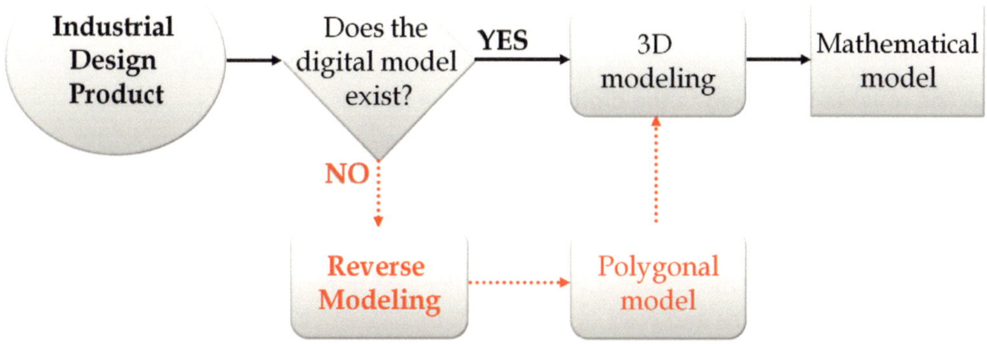

Fig. 1. Transformation process from an existing product to a digital one

In all these cases the use of the 3D laser scanning technologies depends directly to the technological level of the industrial process, that in the last years has been relevant but not evenly distributed. In fact the use of different productive processes in the Industrial Design panorama represents an ambiguity for the presence of mixed representations like traditional drawings, the merge of digital and traditional productive ones, or the exclusive use of advanced digital instruments. The variation of these different systems of representation and production changes radically the role of the tridimensional survey.

2.1 The reverse modeling approach in a non-digital productive process

A "traditional" process of Design consists in an initial passage of conceiving, followed by the refining of the project through different iterative processes, in which the physical model represents the best way to understand the development of the project. Every step corresponds to the production of a different physical model: from the first maquette (a raw model that represents the initial concept) made by hand to better refined models, up to detailed final products.

In this way the physical model represents the best instrument for verifying the different project phases, providing more useful feedbacks both on the global project and the single details than a (digital or manual) bi-dimensional or tridimensional representation. As essential part of the project evolution, the physical model becomes a masterpiece on which the Designer works directly, using his craftsman abilities.

"The contents of the virtual space hold particular characteristics. At first there is a feeble, mediate presence of gravity, which produces an unstable perceptive situation. The essential perceptive structure that characterizes a real situation, defined by the pervasive presence of the gravity attraction, lacks in the virtual scene. In virtual space the vision has an absolute primacy, while the real space represents itself a perceptive system that involves the other four senses." (Maldonado, 1997).

The use of 3D laser scanner acquisition starts when it's necessary to transfer with high accuracy the shape variations of the physical model on the digital one. Normally this doesn't happen if the maquette variations belong to the conceiving phase or if the Designer tries to optimize the costs of the project. On the contrary, if the physical alterations produced are decisive for the formal characterization of the project, the best way to avoid the loss of geometrical information is represented by the application of 3D laser scanner inside a Reverse Modeling process.

2.2 The reverse modeling in the digital productive process

The introduction of digital techniques inside the Industrial Design production has modified the representation and communication channels of a standard project, changing the relationship between real and unreal word. This transformation leads towards a stream of digital data and technological innovations that introduce the traditional representation of productive processes in the virtual environment. For example, nowadays it's almost difficult to be compliant with time and costs optimization, managing every step of a Design process only using hand drawings. On the contrary, with a digital system a better global control of the project is possible, thanks to the simulation capacities, multi-view representations and real-time visualization, using the digital model as "project repository" of data that can be extracted in every moment of the Design process.

In industrial production based principally on digital technologies, the role of the Reverse Modeling is minor because the digitalization process leads to a lower use of physical maquette. Beside this, a constant research in the optimization of costs and time-production (Time-to-Market) is pursued, bounding further the role of Reverse Modeling.

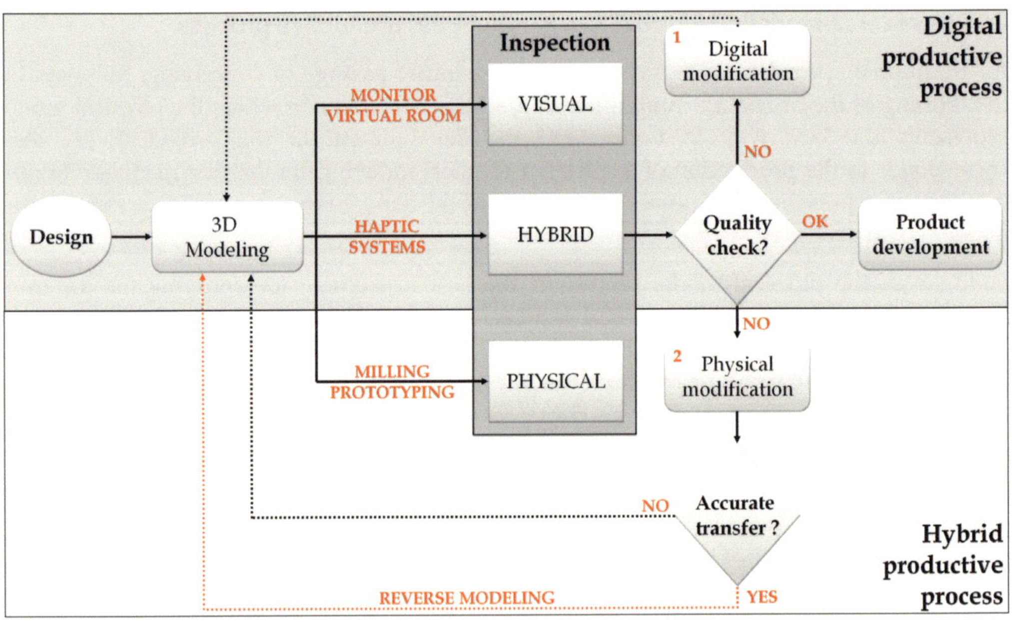

Fig. 2. Digital or hybrid production cycle in the Design field

In this scenario the production of physical models remains an essential step only when high level of Design productions are required. An example is given by the automotive or boat fields, in which digital advanced technology and manual manufacturing coexist, taking both advantages of the relevant investments that are normally devoted to these Design processes. In these fields the physical models modifications have to be digitally translated with high accuracy. The use of 3D acquisition and modeling techniques allows to reach this goal, minimizing the level of geometrical data loss.

On the contrary, in the creation of low level of Design objects often only one physical model is realized to test all the production aspects. In this direction an exception is represented by the ergonomic design products, for which different and accurate physical analysis are necessary to reach an acceptable evaluation on the quality of the project.

So, excepting for some particular products or processes, the Reverse Modeling approach is today applied only if the variations of a physical model have to be translated with high accuracy on a digital one. (Fig. 2).

2.3 A possible solution: The digital methodologies crossbreeding

In a Industrial Design process that tries to privilege higher optimized solutions, the Reverse Modeling and Laser Scanning techniques assume a different role. This can be central when the high Design quality and the physical inspections on the final product are needed or when it's necessary to transfer the variations from the physical to the digital model.

These evaluations have to be done also in relation with the close future, when will probably happen both the fall in the costs of the 3D acquisition instruments and the exponential increase of the instrument performances. In addition, the growing researches towards a progressive simplification of the 3D data process represent an evident sign on the possible future application of 3D instruments. This simplification should lead to two different results: from one side a more frequent use of 3D digital data, from the other the spread of "black box" systems, characterized by a low control of the software, typical (d)evolution of the commercial digital system (like O.S. Windows). But in the future the 3D acquisition approach should assume a relevant role, with an higher accessibility in cost and competences, enlarging its applicative field.

An unfavorable aspect remains the man-hours spent to manage the global process, from the 3D acquisition to the generation of a polygonal or mathematical model. For this limit a solution should be represented by the integration between modeling and visualizing systems that could lead to an optimization of the Reverse Modeling process according to the "Time-to-Market" requests.

An assumption has to be done, in order to identify a different role of the 3D acquisition survey inside a standard production cycle based on an hybrid digital technology: the inspections on a physical model and on its digital visualization could give both a contribute to the global knowledge of the product that otherwise would present lacks.

So in an optimized process these steps have to be considered. In particular the first result of a Reverse Modeling process is a polygonal model that can be analyzed directly through the simulation instruments present in modeling or visualizing systems, avoiding the time-consuming reconstruction of a mathematical model that is usually defined in this step of the process. (VV.AA., 2006).

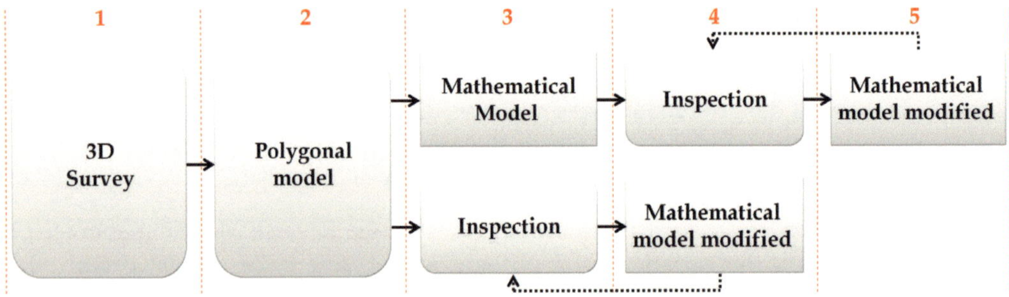

Fig. 3. Two different processes in which the polygonal model assumes a different role

In this way a fast but global quality evaluation of the model is possible, thanks to the physical and visual inspection given by the integration of digital technologies. The modeling phase can start from the 3D acquired surface, choosing the distance between the polygonal model and the mathematical reconstruction in relation with the inspection results obtained on the polygonal model and the purpose of the project (Fig. 3). Following this experimental scheme, a compression of the modeling time is consequent, adopting a production methodology more close to a Designer mental process. In addition the direct advantage in the use of polygonal model, instead of mathematical one, for the inspection and simulation step, produces a consistent time reduction of the whole process, however dependents to the level of complexity of the physical model.

It would be desirable for the future to reach an improvement of the digital systems, that should allow a reliable translation of polygonal complex model to a mathematical one. This transition nowadays represents one of the clearer bottlenecks of the entire Reverse Modeling process and its simplification should radically change its role, improving the integration with the process of the Industrial Design products.

3. The complexity idea in 3D survey

In general the complexity can be represented by a structured system, consisting in a lot of parts that interact mutually with a relationship of autonomy-dependence. In the past some mathematicians have tried to understand this idea, analyzing for example the relation between beauty, order and complexity. Even if they reached the conclusion that Art and Creativity can't be expressed by a formula, the role of the complexity has surprised many artists, deeply moved by the Fractal, the non-Euclidean geometry and the Chaos Theory (VV.AA., 1985).

So which is the idea of complexity in the Industrial Design field? How can be analyzed?

This argument can be faced starting from the following citation: "The complexity of a (natural or artificial) system can be considered a property that depends of the particular representation available in that moment for that system, described by one or more codex (language). So the complexity is intended in the codex and not in the object nature."[1]

As consequence of this thinking, the analysis of an object complexity can consist in the search of an interpretation codex. In the Architectural field a codified language, resulted from a grammar and syntax evolution, allows to understand, or at least to guess, a monument, passing through the subdivision of a building in its essential parts, the comprehension of the coexistence rules and the translation of its geometric principles. (De Luca et al., 2007). On the contrary, in the Industrial Design field is feebler the presence of an interpretation codex that allows to combine shapes with its specific formal or constructive meaning.

In this sense the 3D survey can be considered an interesting instrument of knowledge that allows to extract geometrical information from the real object and represents them in a more intelligible way, suggesting a digital codex of interpretation of the system. So the complexity in the Industrial Design field can be represented by the level of difficulty in the object survey or, in other words, by the instrument capacity of translating the geometrical complexity of the product in a coherent new digital shape[2].

This kind of research can't point to the absolute definition of the level of complexity, because nowadays an objective rule that allows to calculate that level doesn't exist (Brunn et al., 1998). But a "relative" evaluation on the product complexity can be reached through the identification of some factors, bounded to the geometrical or material characteristics and environmental conditions, which lead to an higher or lower global complexity of the physical object survey.

3.1 Geometrical characteristics

The geometrical characteristics can be principally defined as the general dimensions, the spatial distribution of the object in the space and the ratio between the principal dimension and the smallest particular on the external surface. In the Industrial Design field a great variation of products in terms of dimension is often clear and represents a problem only if associated with the request of particular survey precision, another critical and recurrent aspect in this field.

The volume can be defined by the three dimensions along the principal axes. The spatial distribution represents another critical factor of the Reverse Modeling process, due to the possible lack in 3D data alignment compensation of range maps.

At the end, the ratio between the diagonal of the envelope polyhedron and the dimension of the smallest particular give, in relation with the details distribution, an immediate feedback of the dimensional dynamics of the object. Some other factors, like thickness or shadows areas were not considered in this experimental evaluation for their complex quantification or because they are typically related to particular typology of products.

---

[1]Jean-Louis Le Moigne, Théorie du système général - Théorie de la modélisation.

[2]Some algorithms have been implemented to answer to this request, al least partially, defining a group of processes and strategies to identify models in relation with the level of complexity.

3.2 Material characteristics

The critical aspects regard the material itself, the surface treatment and the color, that can condition in a deep manner the optical response of the light beam.

The material answer depends by the nature of the material itself, diffusive or specular, even if the real behavior is often hybrid and between these two possibilities. Depending to the major or minor presence of one of these effects, materials can be differentiated in opaque, translucid or transparent. For example transparent materials with absence of reflected light waves represent a glary case of non-consistent material for the application of 3D active systems. Another example is referred to the porous materials, in which the reflection of the incident light doesn't happen on the external surface, as should be, but inside the object, producing an altered response of the signal and an error in the metrical measurement.

The surface treatment can modify radically the optical response, because it represents the first material layer encountered by the light wave. For example it's clear the difference between a non-treated glass and a ground glass. These treatments are divided in matte, eggshell, semi-gloss, gloss and reflecting. An high level of reflectivity creates deep problems to the 3D acquisition for its strong specular behaviour.

At the end the color is a decisive factor. Starting from the assumption that a better light response is related with the electromagnetic spectrum of the light wave, the deviation from this one involves a worse light feedback. Both black and white colors represent the worst or the best conditions for a 3D acquisition

3.3 Environmental conditions

Some of the most relevant external conditions are the brightness of the environment, the spaces for the instruments handling and the time availability, which can determine a decrease or increase of the internal critical factors of the survey. To simplify this point, three different work environments are considered: favourable, intermediate and adverse, in relation with the major or minor presence of critical external factors.

3.4 A relative evaluation of survey complexity

For all these aspects a linear or logarithmic scale is suggested (Fig. 4), representing on a single axis the different variables due to the major or minor digital survey complexity of products.

Case Study	**LOC (Volume)**
Sugar *bowl*	1.8
Car *chassis*	2.2
Car *maquette*	2.4
Ship *hull*	6.5

Table 1. List of the case studies and relative level of survey complexity (LOC) calculated from the tridimensional star diagram

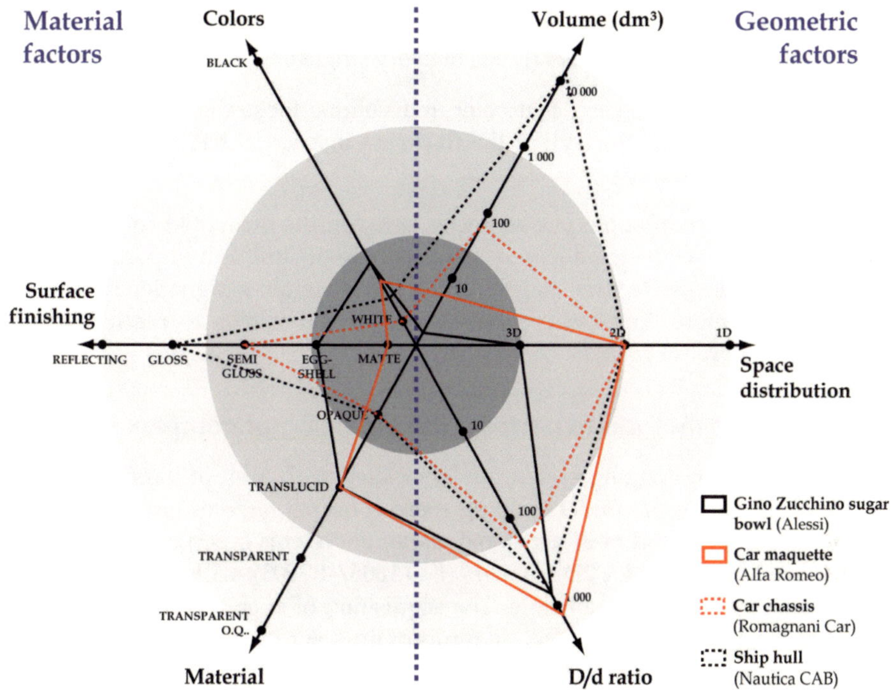

Fig. 4. Star diagram of the complexity factors of four different case studies.

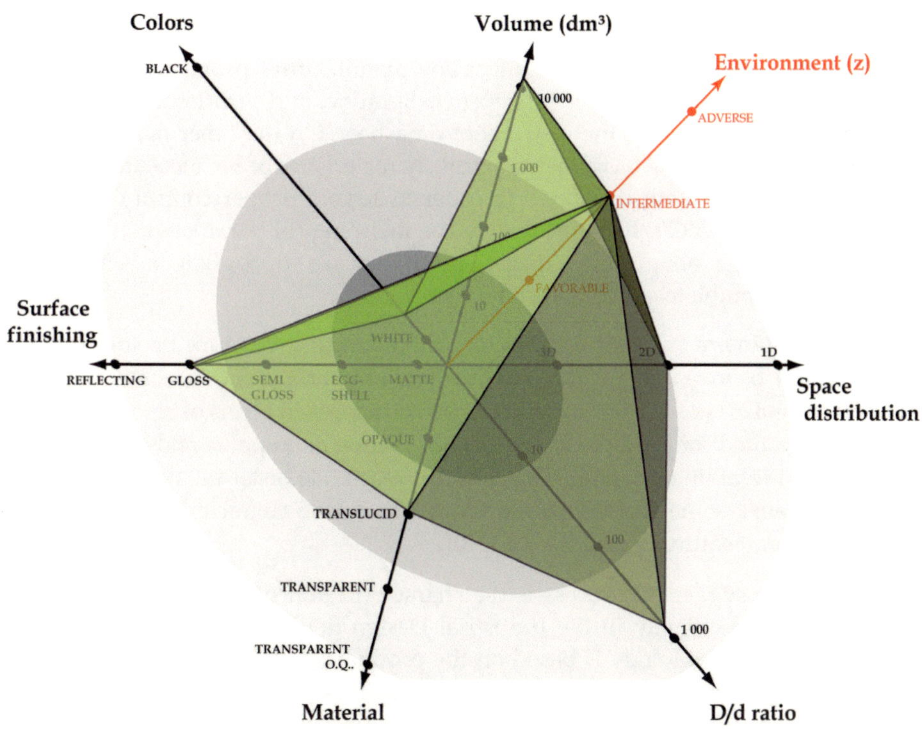

Fig. 5. Volumetric diagram of the complexity factors of the ship hull

In relation with the volume evaluation (Fig. 5), an order of products is suggested, giving an idea of the level of survey complexity (LOC) of every single object (Table 1).

From this experimentation is clear that color and volume factors have weighted upon the complexity evaluation. Also the lack of the thickness inspection has not allowed to define correctly the LOC of car chassis.

At the end, the presence of some incoherencies between the evaluated complexity and the real problems faced in surveying depends principally to the uniform importance ascribed to the whole factors; a deeper factors analysis and a better weight assignation should lead to a more reliable evaluation. Anyway the suggested method allows to reach a relative but coherent idea of the product survey complexity.

4. Characterization method to perform the digitizing of complex shapes

The performances of an imaging system are associated with a lot of parameters according with the typology of sensor. In 2D imaging these concepts have originated standards for photographic equipment and bi-dimensional scanning systems based on specific targets for images (ISO 12233:2000) and for 2D scanners (ISO 16067-1:2003), with procedures that allow to calculate different system parameters. The acquisition of such targets allows to estimate for example the overall resolution level (in terms of lines per millimetre) taking into account both the optical factors and the digital ones.

On the contrary, no standard test protocols currently exist for evaluating the performances of 3D imaging systems. In addition the testing procedures adopted by the 3D imaging industry are not univocally defined and the official technical features for end-user are not often easily comparable to each other. In fact any manufacturer provides few metrological information generally obtained after the system calibration, and certifies the correctness and completeness of parameters on the instrument datasheet. On the other hand the final user has to know in advance some basic metrological characteristics of his measurement system - like field of view or working distance - in order to have a proper control over its survey project (Beraldin et al., 2007). If unfortunately the metrological parameters of a scanner are different from what the manufacturer declares, or if they are appropriate but change in time, the user may have completely wrong results.

The use of a 3D camera without a defined data checking procedure by the metrological point of view may be in general a "jump in the dark". In particular this lack is evident when an active 3D system is chosen for its particular performances in terms of resolution, accuracy and uncertainly, suited for example for the 3D acquisition of complex and refined Industrial Design products (Beraldin & Gaiani, 2005). For this reason, in order to take full advantage of 3D imaging systems, some studies are developed to define characterization processes for assessing 3D camera features (Guidi et al., 2010).

The purpose of the suggested process is the characterization of some Triangulation-based 3D scanners, used specifically in the Industrial Design field for the particular acquisition performances. The methodology is based on the production of 3D images with specifically developed 3D targets from which parameters of resolution, accuracy and uncertainly can be extracted. These experiments represent a preliminary approach for defining a process of range camera characterization. Future work for generating 3D version of other 2D ISO targets has to be done, in order to enhance the number of quantitative data attainable with simple test objects and procedures, but this system seems to give already usable results for verifying instruments performances in standard laboratory activities.

4.1 Target objects

Some target objects recreate in 3D space the 2D characterization condition, because they define slanted edge analysis (Burns, 1999, 2000) or generate a z-behaviour in the point cloud with abrupt jumps on z, and a progressively reduced size on the horizontal plane.

The first object is a "set of steps" (Fig. 6a), a set of coaxial cylinders carved form made with a single block of iron, with diameters varying linearly from 100 mm to 10 mm at steps of 10 mm, and height varying non linearly from 15 micrometers up to 7.68 millimetres, with a doubling of the step size for each transition (i.e. 15, 30, 60, 120, 240, 480, 960, 1920, 3840, 7680 micrometers).

The second target is a parallelepiped block (Fig. 6b), made in rectified iron, with a white flat varnish to avoid reflection. It has dimension 100x60x270 mm and it was used to investigate scanners' behaviour at edges.

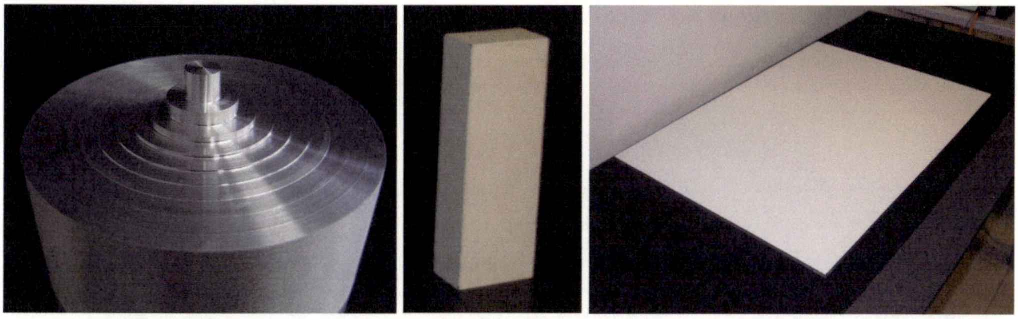

(a) (b) (c)

Fig. 6. Target object: (a) Set of step; (b) Parallelepiped block; (c) Glass plate

The last target is a glass plate of 700x528x11 mm (Fig. 6c). The particular manufacture process of this material allows to obtain a plate characterized by peak deviation inside a range of few microns, so proper to be used as target test. One side of the plane was painted with matt white flat varnish obtained with a furnace treatment, avoiding the clearness problem of the glass, obtaining a perfectly smooth and optically cooperative surface.

To determine how far the manufactured objects deviates from their supposed ideal geometrical form, with the aim to get reliable test objects, every target have been measured with a CMM measurement system[3], which give the reference parameters and dimensions for the experimental analysis. In addition all objects have been recorded once in the same defined condition with different 3D range camera, providing a reliable method to estimate the instruments' performance under practical application conditions. Each measurement on these objects was made at a temperature around 20 °C.

[3]CNC milling machine (Biglia CNC B301)

4.2 Characterization of the 3D range camera

The described targets allowed to extract for each range map the best parameters of resolution, accuracy or uncertainly.

Principally two different resolution aspects were examined: limitation in z-resolution due to the opto-geometric configuration (distance-to-baseline ratio) and horizontal resolution capability mainly due to optical performances.

In the first case the "set of steps" object allowed to test the z resolution. This process was executed maintaining the set of the scanner axis aligned with the object one, positioning the camera at different distances for each acquisition. The selection of circular sets of 3D points corresponding to the different steps has been done according to their distance from the object axis. The influence of uncertainty and accuracy on the optical resolution of each analyzed device imposed some visual conditions that determined if a set of 3D points was distinguishable from the adjacent ones. When one set of points, which should lie on the same plane, was not distinguishable and selectable from those belonging to the adjacent step, was classified as an undistinguishable step (Fig. 7). The first distinguishable step was defined as the z resolution limit. Such capability has been detected by fitting a plane on the set of points nominally belonging to the same plane, and evaluating the average distance of that set of points from the fitted primitive.

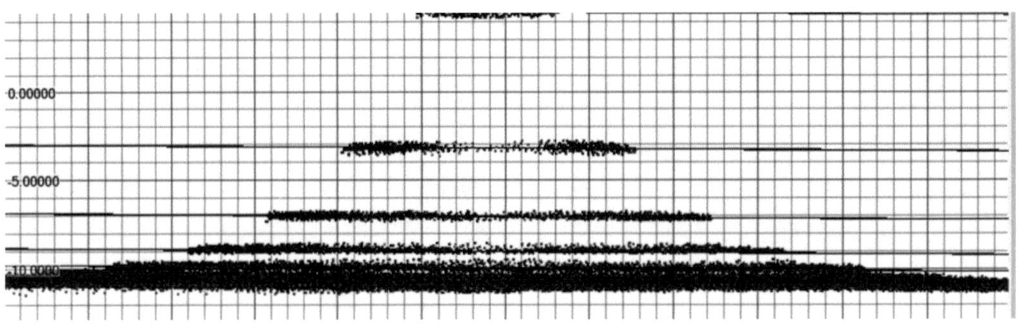

Fig. 7. Point cloud originated by the 3D scanning of the "set of steps" object

On the contrary the xy resolution has been estimated by acquiring a sharp edge of the metallic parallelepiped. Differently from the work by Goesele (Goesle et al., 2003), a vertical z jump was used for testing the equipment rather than the triangular shape of its corner. This implies that in our case the laser scanner was oriented nearly orthogonally respect to one of the parallelepiped faces, and the recorded range map was converted into a monochromatic image. The edge of the range map has been transformed in a B/W image by coding the distance (z) as a gray level. On the resulting image the ISO slanted edge analysis has been applied. Resolution capability of the optical system from image with slanted edge can be found by evaluating the SFR (Spatial Frequency Response) behaviour, also called MTF (Modulation Transfer Function) (Russo et al, 2009). The image generated from the range map had a 1:1 correspondence between 3D points and image pixels, with no re-sampling. In order to have the best possible correspondence between device performances and image indexes, any post-processing on range maps, such as smoothing, cleaning, etc., was also avoided.

At the end accuracy and uncertainly performances were tested and verified with the plate test object that minimizes the least square deviation from the acquired data, carrying out a theoretical data set. For the accuracy analysis, the absolute accuracy parameter was distinguished in this approach from the relative one: this latter is considered as the deviation of a set of 3D points from the shape they should describe (rather than from their actual position in space) divided by the diagonal of the range map. Typical example is the 3D scan of a plane, where all the points should be coplanar, but for a number of reasons they are displaced from such theoretical behaviour. The following experimental formula was applied:

$$A_{LR} = \frac{\Delta z_{\max} + abs(\Delta z_{\min})}{d} \quad (1)$$

where Δz_{max} and Δz_{min} are the maximum and minimum deviation of the actual range data from the fitted primitive expressed in micrometers and d is the diagonal of the framed area expressed in millimetres. This parameter represented in practical terms the average accuracy error along z (in micrometers), for each millimetre explored by the range map along the xy plane (Guidi et al., 2010).

Fig. 8. The different phases of parameters extraction through the comparison between range map and best fitting plane.

The uncertainly parameter, described by a random error distribution, can be estimated by calculating the distance of each measured point from its theoretical value, which is given in this situation by the point on the corresponding fitted primitive. The standard deviation extracted from the statistical distribution of this sequence of distances gives an evaluation of the instrument uncertainty (Fig. 8).

5. Limitations in the mono-instrumental approaches

Some application limits of 3D laser scanner are now discussed, integrating the analysis on the correct use of these instruments.

A first general assumption regards the presence of different working principles inside the world of 3D non-contact camera, that allow to survey dissimilar geometrical entities. In this sense a range camera permits to obtain a fast 3D acquisition of free-form surfaces but it is limited by a grid sampling step, that normally prevents the geometrical acquisition of edges of the object. On the contrary a photogrammetric or topographic approach lets the acquisition of few points, defining simple break-lines of the object or more complex curves in relation with the number of surveyed points. In addition systems for processing digital photogrammetric images have been presented recently, introducing the possibility to generate dense point clouds from stereo-matching procedure (Remondino, 2006, 2011).

Probably in the next years the photogrammetric approach would become suitable for every application, but nowadays some aspects of the software that supply these procedures have to be improved, in order to reach the same quality of 3D laser scanning output. So, despite the technological and methodological advances shown in the photogrammetric and photo-modeling fields in terms of instrument performances, 3D active systems are considered yet the better instruments to acquire 3D complex surfaces.

Starting from this assumption, the relation between a survey instrument and the level of complexity of an Industrial Design product becomes essential to understand the mono-instrumental limits.

A shape can be considered "simple" from the surveying point of view when it is composed by homogeneous geometries that can be acquired only by one instrument. If a product can be simplified through cubes, parallelepipeds, cylinders, spheres or in general any shape that can be described by mathematical primitive geometry, during the 3D survey it's necessary to acquire only the reconstructive reference points. These measures can be acquired manually or, if necessary, applying topographic or photogrammetric instruments.

But in the Industrial Design field the products are often characterized by free-form surfaces, that can be surveyed in a coherent way by 3D active systems that produce dense point clouds with a fixed sampling step. So even if the product is composed only by free-form surfaces can be considered "simple", because an active system allows to acquire all the shape without using other survey instruments. From the 3D acquisition point of view the complexity concept is introduced only if the 3D acquisition of the physical product requires the integration of different survey instruments.

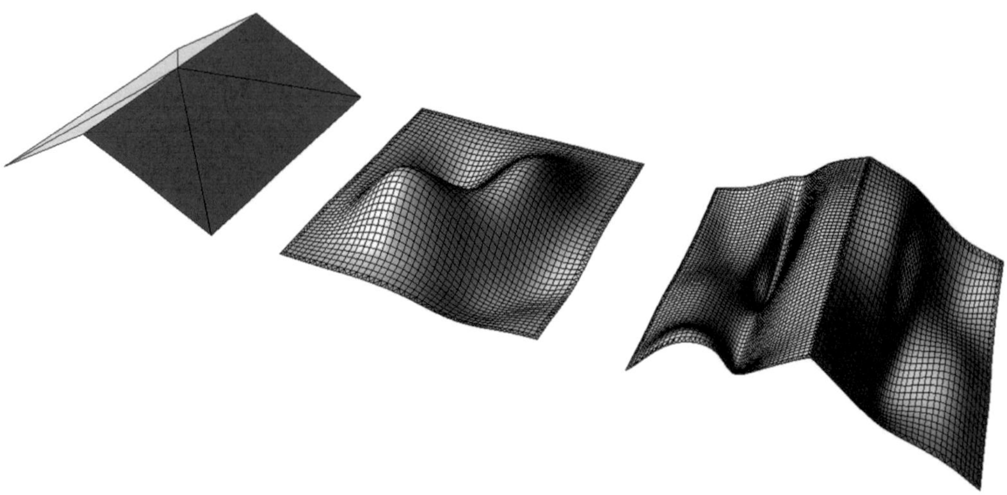

Fig. 9. Sequence of model: the grey models present an uniform geometry, while the red ones show a complexity condition of survey for the merge of linear and free-form geometry

A complexity condition can be verified if geometric primitives and free-form surfaces coexist in the same product. In this situation an integration of different instruments has to be considered, otherwise some parts of the shape have to be deduced later through a massive data post-processing (Fig. 9).

Following these concepts, most part of the Industrial Design products belong to the complex shapes, a condition that can interferes with correct representation of the object and the the 3D knowledge investigation, consequently with a clear comprehension of the global geometry.

6. State of the Art in the sensor integration and sensor fusion

In some situations the integration between different instruments is necessary to reach an acceptable survey (Beraldin et al., 2002). The integrated methodologies suggested in the last ten years were oriented principally to the generation of digital models that could represent correctly the complex geometrical variations present inside an object.

Starting from a background analysis in this field in the last decade, three principal purposes for integration data can be identified:

- to increase object information; a first 3D acquisition is progressively enriched with other 3D data in order to improve its interpretation (Levoy et al. 2000; Bernardini et al., 2002; Guidi et al., 2002);
- to improve uncertainly, accuracy and resolution control on the digital model; this allow to avoid frequent local errors of alignment, rectifying in the same time the geometrical characteristics that represent error sources (Blais et al., 2000; Borg & Cannataci, 2002; Guidi et al., 2004, 2005);
- to verify the accuracy level of the whole model; this control is essential especially if the instruments are applied separately and represents an important data check to guarantee an high level of survey quality (Guidi et al., 2003; Georgopoulos at al., 2004; Hans de Roos, 2004; Ohdake & Chikatsu, 2005).

A scientific background analysis on sensor integration and fusion requires a complex and articulate research starting from the nineties until today (Beraldin, 2004), in which the use of digital photogrammetry and 3D laser scanning is raised in a lot of different applications.

In general the possibility to exploit completely the digital capacities represents the principal advantage of the integration (El-Hakim, 2000; El-Hakim et al., 2002), that maximize the instrumental efficiency and minimize or overcome the single technological limits. In this sense, different authors described pro and cons of the active and passive optical systems applied in different fields, focusing both on the quickness of a photogrammetric survey approach and the huge amount of data produced by a 3D laser scanner, highlighting that scanning and prost-processing time is partially compensated with the completeness of the 3D data acquired (Böehler & Marbs, 2004). The accuracy of the final model produced by an integrated approach and the reduction of time in acquisition or data process represent other different aspects that have to be considered (Velios & Harrison, 2001).. In addition the relationship between the dimension of the surveyed object and the time of the acquisition and modeling process is analyzed, highlighting that the photogrammetric method is a scale-invariant approach and the increase in object dimension corresponds to an exponential increase of the post-processing time. So the object dimensions affect in particular the 3D active techniques (Guidi et al., 2008). But also the complexity aspects have to be considered. In this sense a comparison between the use of active and passive camera justifies from one side the primacy of the latter for the level of accuracy reached with medium and wide objects, from the other a more precise data for the survey of little products by 3D laser scanner (Remondino et al., 2005).

In conclusion the background analysis points out the close complementarity between active and passive systems, both for the global control of the 3D acquisition and the high accuracy performances, either necessary when the surveyed object presents a strong dimensional dynamic.

Until few years ago the difference in the working principles of the 3D non-contact instruments, defined as "instrumental gap", caused by the great difficulty to cover the important range of acquisition distance of 3 to 10 meters by Triangulation-based or Time of Flight (TOF) 3D scanner (Guidi et al., 2010). Nowadays this lack is weaker for the massive introduction in the last years of Amplitude Modulation TOF technology that is able to cover this useful range both for Architectonic and Industrial Design purposes. The introduction of these new instruments resolves partially the distance between Triangulation-based and TOF 3D scanner, and the combination of these two technologies can still lead to an evident improvement of instrument performances and an enlargement of the applicative fields for the survey system (Fig. 10). In addition, the integration between active and passive camera can bring further advantages in terms of global metrological control, increasing significantly the scale variation that can be covered with a 3D survey.

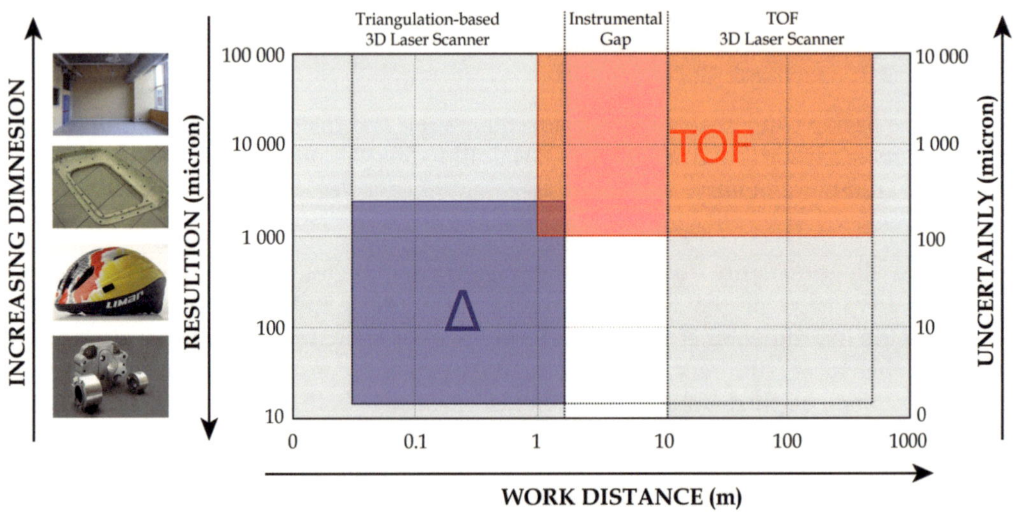

Fig. 10. Diagram on the applicative field of Triangulation-based and TOF 3D laser scanners, with the instrumental gap and the integration capacity (broken line) of the two systems

As said before, if significant geometric variations characterize an object and it is not possible to survey the whole surface with the application of one instrument, the product can be defined complex. In this scenario, coming back to the Industrial Design field, the principal aim of the instrument integration is to acquire the highest number of geometric information of a complex object. This approach minimizes the errors present in a single measure and in the global model, improving at the same time the instrument performances and data reliability. At the end can be defined a 3D acquisition campaign coherent with the product complexity.

7. Case studies

In this paragraph some experimental processes applied on Industrial Design case studies are presented, highlighting the bottlenecks in a Reverse Modeling process (Gaiani et al., 2006) and suggesting how can be solved through the instrument integration. In particular two different approaches are described: multi-resolution and sensor fusion.

7.1 The multi-resolution approach

The "multi-resolution" or "multi-res" term specifies the application of two or more different sampling steps during the 3D acquire, in order to survey correctly the entire surface of the physical object. This approach is necessary when an Industrial product presents a non uniform geometry, a significant variation of dimensions and an exceeding level of complexity. The multiple sampling steps can be obtained by one or more instruments, modifying the internal scan settings during the 3D acquisition. In both these cases, two 3D acquisition campaigns have to be planned: the first with the aim of surveying the object wireframe skeleton (edges), the second for the whole surface skin. This double analysis can face some troubles in the physical models that present an high level of geometrical complexity; in that cases the limits in the use of a single instrument are more evident.

A possible solution for the survey of these kinds of objects is represented by the dynamic variation of the 3D acquisition sampling step in relation with the geometric characteristics of the product. The analyzed models for this experimental step are the following two:

- Gino Zucchino sugar bowl (by Alessi Factory)
- Maquette of a car prototype (by Alfa Romeo Group)

7.1.1 Gino Zucchino case study

This little object of 140x90x80 mm (Fig. 11a) is designed by Guido Venturini and has been produced in Italy since 1993 with such a wide success to be clearly fixed in the people's memory. For this reason, the principal aim of this survey project was to produce a digital model very close to the real product, in order to allow its redesign (VV. AA., 2006).

The object presents different survey limits due on one hand to the sequence of simple surfaces and complex details, on the other to the non-Lambert light response caused by the slightly porous and smooth translucent material.

For these reasons, in the survey project a multi-resolution sampling step was considered, maintaining coherence between the complexity of the product and the instrument performances in term of resolution and spot size. In addition the material has obliged to low the laser power, minimizing the errors due to an incorrect light reduction still reduced inside a light-controlled laboratory.

(a) (b)

Fig. 11. (a) Physical model; (b) final polygonal model of the sugar bowl

The first standard approach considered to adopt one Triangulation-based 3D laser scanner SG100 (ShapeGrabber) at the maximum resolution step for the whole surface.

On the contrary, in the second multi-resolution approach the acquisition was carried on with the same instrument and survey set, modifying dynamically the resolution from 0.1 mm to 0.25 mm for the external surface of the product, while the closing joints were scanned at the highest resolution. In this way was defined a coherent polygonal model with the geometrical complexity of the physical object (Fig. 11b).

Starting from the comparison between a fixed resolution approach and a dynamic one, the latter has allowed to create a model that can be easily managed and edited in few time (Fig. 12). Moreover both the methodologies preserve the geometrical characteristics, but the mono-resolution model presents a surplus of 3D data that have conditioned the length of the post-processing step.

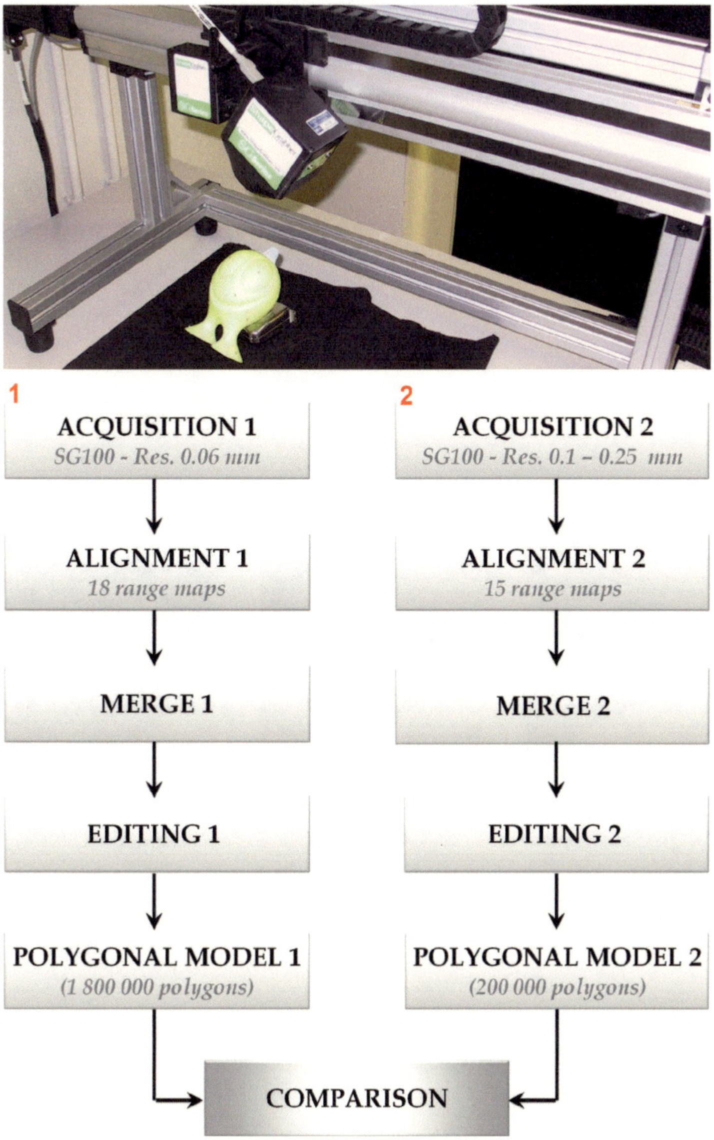

Fig. 12. Block diagram of the Reverse Modeling processes of the sugar bowl

7.1.2 Car *maquette*

In this case study a minor geometrical complexity of the physical object is replaced by a wider dimension that introduces new bottlenecks in the Reverse Modeling process.

The maquette analyzed represents a prototype car in scale 1:10 defined by a resin material and an overall dimension of 420x180x90 mm (Fig. 13a). The principal aim of this experiment is to improve a Reverse Modeling process applied in this condition, trying to reach a field of view optimization and a global control in the alignment of many range maps with ICP (Iterative Closest Point) algorithm.

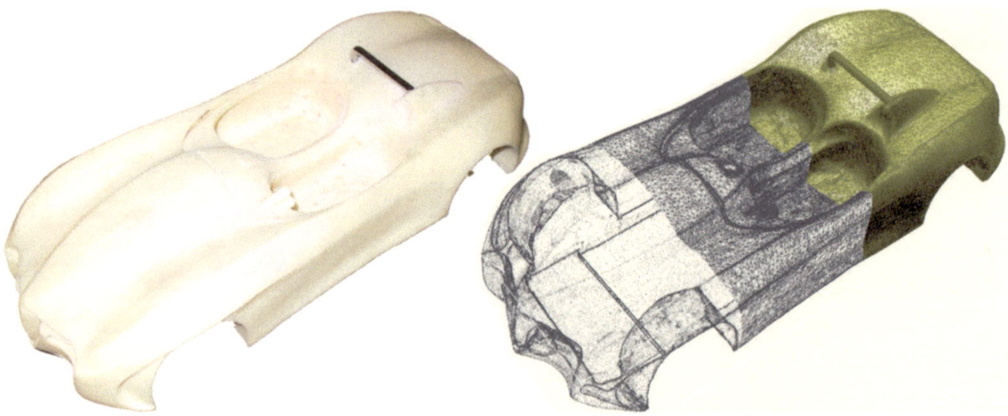

(a) (b)

Fig. 13. (a) Physical model; (b) final polygonal model of the car maquette

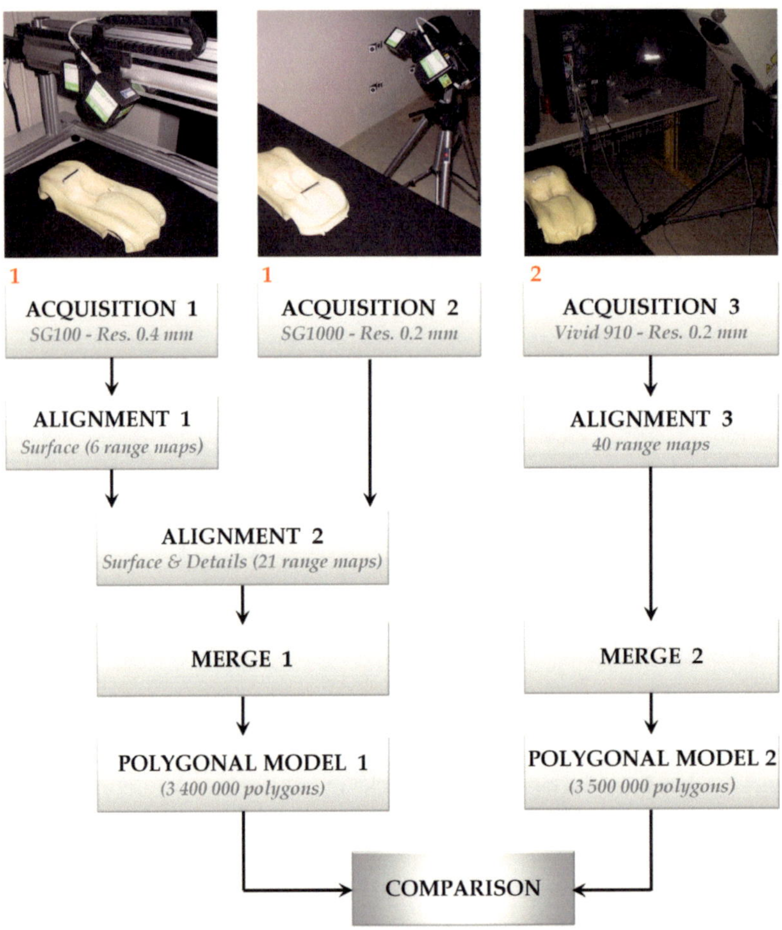

Fig. 14. Block diagram of the Reverse Modeling processes of the car maquette

In this experiment the use of a single Triangulation-based 3D laser scanner Vivid 910 (Minolta) was compared with the application of two different 3D laser scanners, suited for the geometrical characteristics of the physical model: SG100 and SG1000 (ShapeGrabber). The whole 3D survey was conducted in a controlled-light laboratory.

In the multi-resolution approach two different survey sessions were realized. During the first one, wide range maps were acquired through a SG100 rail scanner, defining a base skeleton of 3D data. In the second step, detailed and more complex parts were acquired with the SG1000 rotational scanner and aligned to the first group of global scans, reaching an optimized polygonal model (Fig. 13b).

In the mono-instrumental experiment a Vivid 910 with TELE lens (25 mm) was used to cover the whole surface with a resolution similar to the integrated approach (Fig. 14). At the end of the experiment, the multi-resolution methodology results more flexible in the variation of scan area and sampling step that can be chosen in relation with the geometrical characteristics of the product, allowing a better quality and polygon distribution of the final model.

7.2 The sensor fusion approach

A "sensor fusion approach" represents instead the integration between different 3D acquisition techniques, based on the use of survey instruments characterized by different working principles. This methodologies can be applied to solve some particular limits in the Reverse Modeling process, like the different levels of details in the 3D acquisition, the alignment uncertainly and the global accuracy of the model. The Triangulation-based or TOF 3D laser scanners and the photogrammetric approach allow to reach different results in terms of data density, accuracy, sampling step, working distance, field of work, time of acquisition, data process etc. In addition, these instruments are characterized by relevant differences in terms of costs and management.

For all these reasons, the sensor fusion approach tries to exploit the best performances of the different survey systems, in order to obtain results otherwise not reachable by the use of a single technology. The case studies analyzed are two:

- Chassis of a car door (by Romagnani Car)
- Hull of a ship mould (by Nautica CAB)

7.2.1 Door chassis

This case study presents a 900x500x20 mm dimension and a material of polyester resin and glass fibers with a white-opaque surface characterization that produce a slight effect of light retro-diffusion (Fig. 15a). It has a 2 cm small thickness and a complex geometric distribution of wide free-form surface with many details. For these, the 3D acquisition and the alignment process represent the principal bottlenecks, due principally to the error propagation of particular shape and the complex positioning in the same reference system of the two sides.

The sensor fusion approach in this particular case has integrated two different Triangulation-based 3D laser scanner, Vivid 910 (Minolta) and a SG1000 (ShapeGrabber) with a photogrammetric system. This second one is composed by a EOS 10D camera (Canon) and a group of rectified planes of 15x21 cm dimension, on which a set of 11/12 targets with 2 cm diameter, recognizable both from 3D laser scanners and photo-camera, were attached. At the end a "certified space" front-face mirror (Leica) of 24x24 cm was employed, allowing at the same time the acquisition of two 3D scans, one direct and one reflected, thanks to the refection of the external surface without errors in the measurement. The whole 3D survey was conducted in a controlled light laboratory.

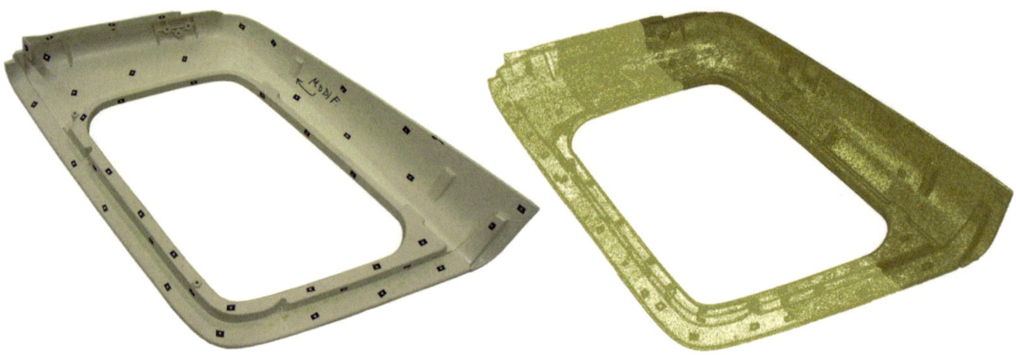

(a) (b)

Fig. 15. (a) Physical model; (b) final polygonal model of the door chassis

Two different approaches were followed, the first characterized by the use of one Vivid 910 with its set of lens, obtaining a big amount of range maps for the whole surface. In this case the thickness problem was faced acquiring in addition a 3D reference object positioned inside the products. The entity of global error obtained in the ICP alignment step avoided the definition of a final polygonal model (Fig. 15b).

Instead the second approach employed all the instruments described above: different range maps were acquired with SG1000, including also the three rectified planes, exploiting the more suitable condition of the rotational scanner to minimize the alignment errors. The thickness problems were solved acquiring a set of scans in the four corners of the object with the front-face mirror, generating some particular double range maps that avoided the interpenetration of the two sides. The two sets of 3D data acquired for each side of the product were aligned at the scan reference corner, defining the final shape (Fig. 16).

At this point, the metrical quality reached in the second process was verified, comparing the distances between targets acquired in the final polygonal model (by Matlab procedure) with the same ones surveyed by the photogrammetric system (Guidi et al., 2003).

The existing metric differences between these two sets of measures demonstrated the validity of the second survey approach and the better global accuracy of the photogrammetric system. The utility of the sensor fusion approach for an inspection of the quality level of a complex product is than verified (VV.AA., 2006).

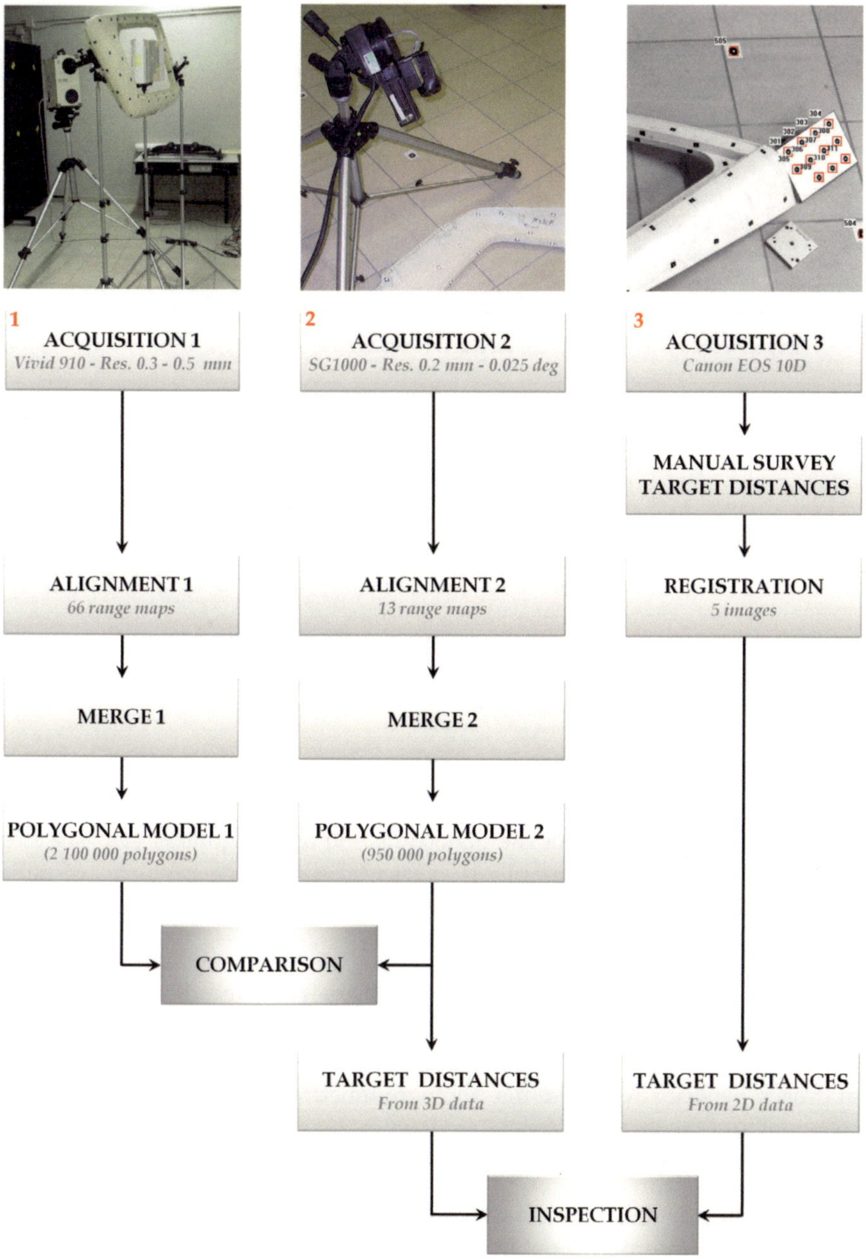

Fig. 16. Block diagram of the Reverse Modeling processes of the door chassis

7.2.2 Ship hull

This case study analyzes the application of a sensor fusion approach for Design products of big dimensions, the hull survey of a rubber boat of 6400x2200x820 mm, characterized by a predominant 2D geometry defined with a plastic material reinforced by incorporated fiberglass and a red smooth and shining surface (Fig. 17a).

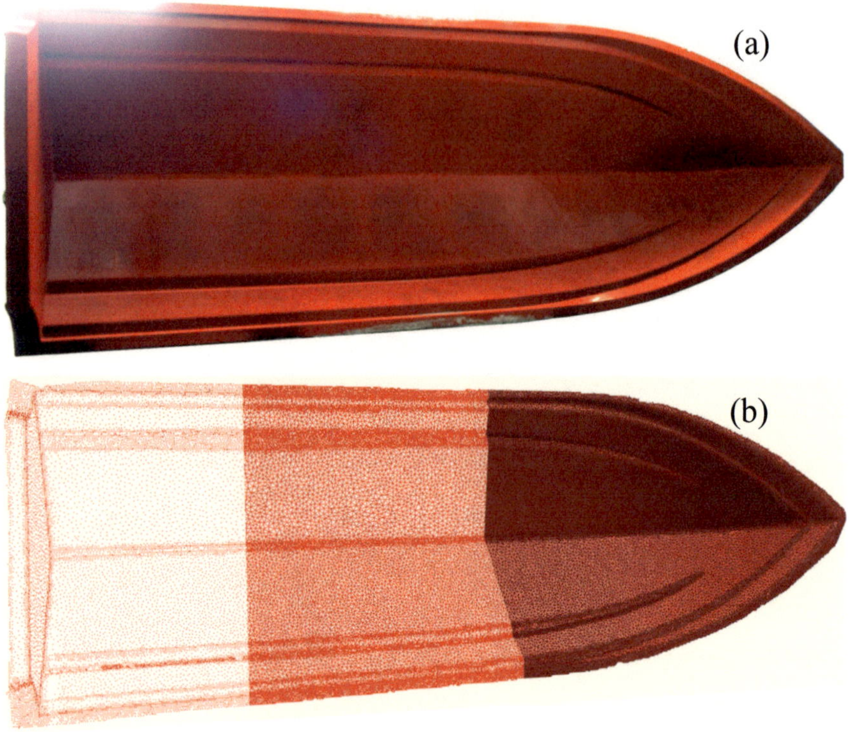

Fig. 17. (a) Physical model; (b) final polygonal model of the ship hull

In this situation, the complexity of the 3D acquisition and alignment step are stressed in comparison with the previous example. In particular the sliding problems of range maps and the control of the global accuracy have to be solved through the application of sensor fusion approach during the acquisition and alignment step.

In this experiment three different approaches were analyzed and discussed. The firsts two consider the application of one single Triangulation-based 3D laser scanner SG1000 (ShapeGrabber) and a Laser Radar CW LR200 (Leica), while the third employs a sensor fusion approach between SG1000 and a photogrammetric system composed by a EOS 10D camera (Canon) and the target systems described in the previous paragraph. The whole 3D survey was conducted in a floodlit industrial ambient.

The first mono-instrumental approach was studied minimizing the number of scan positions and angular rotation, in order to optimize the scanning time and the acquisition of points external to the object. The range maps obtained by the Triangular-based sensor were than aligned with ICP, showing a good "apparent" value of standard deviation. A polygonal model from these scans was than created.

On the contrary, the second mono-instrumental approach used only one position to acquire the entire surface of the hull, defining one single point cloud of the object that was triangulated. For the high metrological characteristics of the Laser Radar, the polygonal model obtained was used as "gold standard" to compare it with the other two (Fig. 18).

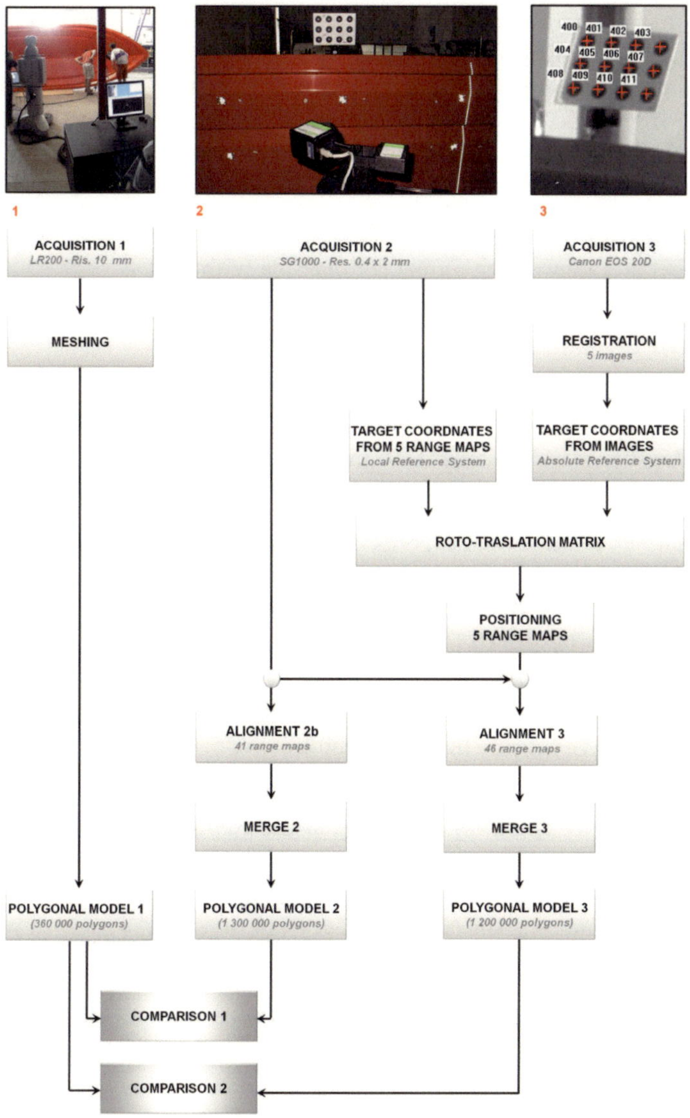

Fig. 18. Block diagram of the Reverse Modeling processes of the boat hull

At the end a sensor fusion approach was employed to improve the single instrument performances (Guidi et al., 2004). Five different rectified plates were positioned around the hull, while other single targets were attached around the object. During the survey step both the hull and the plates were acquired with the 3D scanner. The definition of the 3D coordinates of the targets from the range maps and the photogrammetric system allowed to position in the global reference system the five 3D scans with a level of accuracy defined by the photogrammetric approach. These 3D data connections were used as references to align the other 3D data, avoiding the problems due to the sliding range maps. The standard deviation values obtained in this step were comparable with the other ones reached with the mono-instrumental approach. At last the final polygonal model was defined (Fig. 17b).

The comparison between the three polygonal models and the sections extracted demonstrates the quality of the sensor fusion approach, that allows to reach a metrological performance comparable with a very expensive Laser Radar, lowering the initial 75.6 mm variation between the gold standard and the mono-instrumental approach (major than 1% of the total length) until 5.6 mm, minor than one part of thousand (Guidi et al., 2005).

In conclusion the sensor fusion approach allows to improve a single survey system, overcoming the ICP limits, due to the 2D shape of the object and the smooth surface, obtaining a global accuracy of the model comparable with very expensive and complex 3D survey systems.

8. Conclusion

At the end of the experimental phase, some conclusions about the use of 3D laser scanner techniques for the survey of Industrial Design shapes can be outlined.

From the process point of view, the 3D survey represents an interesting knowledge approach that allows to understand better a physical product, suggesting a different but clear codex of representation of its shape through the definition of a coherent digital mould of the real object. For this capacity, the application of 3D laser scanner in the Industrial Design process can supply useful information for the Design project, in particular for the possibility of translate in an accurate way a physical maquette or a finished product in a digital model. The role of this step depends essentially from the level of technology and professional skills involved inside the industrial process.

From the object point of view, a Design product can be considered "simple" or "complex" in relation with its geometrical and material properties that supply a first relative idea of the survey complexity of the physical object. The complexity generates some bottlenecks inside the Reverse Modeling process, due to the 3D acquisition limits of the survey instrument, the errors in range map alignment, the lacks in the 3D data and the accuracy of the final model.

Both the 3D acquisition approaches, multi-resolution and sensor fusion, allow to overcome some of these bottlenecks. The multi-resolution approach represents the first step in which the integration can be applied to solve a complex situation of survey, while the sensor fusion method is the second one. The multi-resolution methodology use one or more 3D acquisition instruments in relation with the sampling steps that are suited for the morphological characteristics of the product. In this sense, an experimental ratio can be considered to recognize when a dynamic resolution approach has to be applied: the ratio between the diagonal of the envelope polyhedron (D) and the dimension of the smallest measurable particular of the object (d):

$$D/d \geq 1000$$

Below this value, a standard Reverse Modeling approach should be suited to obtain a correct survey of the product, anyway considering the other parameters of complexity. The increase of this value corresponds to a higher necessity to adopt a multi-resolution approach, initially to optimize the geometrical characteristics of the digital polygonal model, preserving the morphological and material complexity and optimizing the whole process of 3D survey.

The experimental phase on the survey of complex industrial objects verifies that the standard Reverse Modeling process can't allow to reach good results "in all the situations", on the contrary the multi-resolution approach permits to optimize the acquisition and construction phases.

The sensor fusion approach is necessary to reach good metrological results in complex survey situations, in which the matter or dimensional characteristics of objects create huge problems in the 3D acquisition process. Its application allows to enlarge the use of 3D sensor in Industrial Design applications, overcoming the problems due to mono-instrument methods and improving the 3D sensor performances, that lead in addition to an optimization of the whole Reverse Modeling process (Fig. 19).

In this scenario the possibility to translate a physical object to a digital one would probably grow up in the future, foreseeing a stronger role of the Reverse Modeling inside the Industrial Design process.

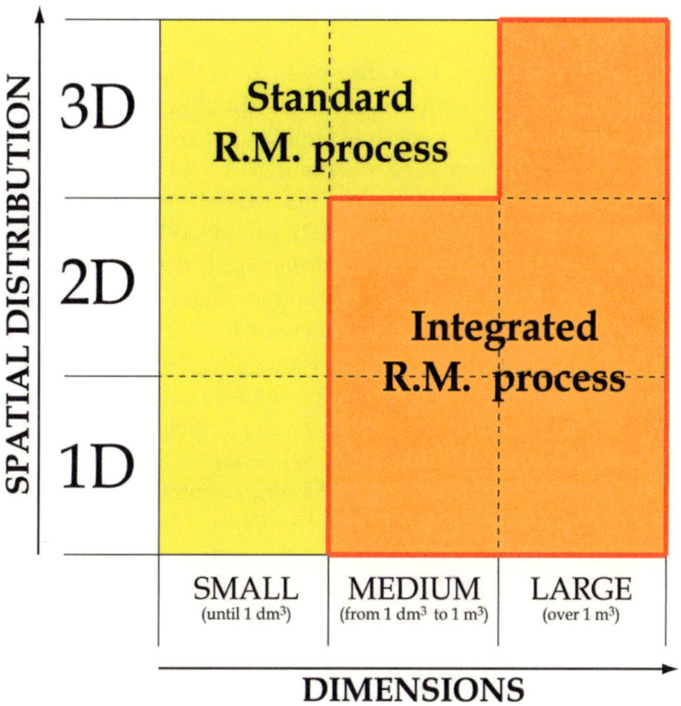

Fig. 19. Synthesis diagram on the Reverse Modeling applied in Industrial Design field.

9. Acknowledgment

The author desires principally to acknowledge Prof. Gabriele Guidi (Mechanical Dept., Politecnico di Milano) for the scientific and humane support, a mile-stone for the personal academic growth of the author. A special thank also to Laura Micoli for the important experiences shared on the experimental case studies and the help in the chapter revision.

In addition the author would like to acknowledge for the scientific contributes given by Prof. Monica Bordegoni and Ing. Grazia Magrassi (Mechanical Dept., Politecnico di Milano).

At the end a final thank to Prof. Marco Gaiani, scientific local leader of the national projects PRIN02-04 and PRIN04-06 during that part of the experimental case studies were analyzed.

10. References

Beraldin, J.-A., Picard, M., El-Hakim, S.F., Godin, G., Valzano, V., Bandiera, A. & Latouche, C. (2002). Virtualizing a Byzantine Crypt by Combining High-resolution Textures with Laser Scanner 3D Data, *Proceedings of VSMM 2002*, pp. 3-14, ISBN 89-952475-1-7 93000, Gyeongju, Korea, September 22-27, 2002

Beraldin, J.-A. (2004). Integration of Laser Scanning and Close Range Photogrammetry – the Last Decade and Beyond, *Proceedings of the XXth ISPRS Congress, Commission VII*, pp. 972-983, Istanbul, Turkey, July 12-23, 2004

Beraldin, J.-A. & Gaiani, M. (2005). Evaluating the Performance of Close Range 3D Active Vision Systems for Industrial Design Applications, *Proceedings of Electronic Imaging, Videometrics IX (SPIE)*, pp. 162-173, ISBN 9780819456380, San Josè, California, USA, January 16-20, 2005

Beraldin, J.-A., Blais, F., El-Hakim, S., Cournoyer, L. & Picard, M. (2007). Traceable 3D Imaging Metrology: Evaluation of 3D Digitizing Techniques in a Dedicated Metrology Laboratory, *Proceedings of the 8th Conference on Optical 3-D Measurement Techniques (Optical3D)*, Zurich, Switzerland, July 9-12 , 2007

Bernardini, F. & Rushmeier, H. (2002). The 3D Model Acquisition Pipeline. *Computer Graphics forum*, Vol. 21, No. 2, (June, 2002), pp. 149-171, ISSN 1467-8659

Bernardini, F., Rushmeier, H., Martin, I.M., Mittleman, J. & Taubin, G. (2002). Building a digital model of Michelangelo's Florentine Pieta. *IEEE Computer Graphics and Application*, Vol. 22, No. 1, (January/February 2002), pp. 59–67, ISSN 0272-1716

Blais, F., Beraldin, J.-A. & El-Hakim, S.F. (2000). Range Error Analysis of an Integrated Time-of-Flight, Triangulation and Photogrammetry 3D Laser Scanning System, *Proceedings of Laser Radar Technology and Applications V (SPIE)*, pp 236-247, ISBN 9780819436610, Orlando, Florida, USA, September 5, 2000

Böhler, W. & Marbs, A. (2004). 3D Scanning and Photogrammetry for Heritage Recording: A Comparison, *Proceedings of the 12th International Conference on Geoinformatics*, pp. 291-298, ISBN 91-974948-1-X, Gävle, Sweden, June 7-9, 2004

Borg, C.E. & Cannataci, J.A. (2002). Thealasermetry: a hybrid approach to documentation of sites and artefacts, *Proceedings of the Commission V Symposium Close-Range Imaging, Long-Range Vision, XXXIV part 5*, pp. 93–104, Corfu, Greece, September 2-6, 2002

Brunn, A., Gulch, E., Lang, F. & Forstner, W. (1998). A hybrid concept for 3D building acquisition. *ISPRS Journal of Photogrammetry and Remote Sensing*, Vol. 53, No. 2, (April, 1998), pp. 119–129, ISSN 0924-2716

Burns, P. & Don, W. (1999). Using Slanted-Edge Analysis for Color Registration Measurement, *Proceedings of PICS: Image Processing, Image Quality, Image Capture, Systems Conference*, pp. 51-53, ISBN 0-89208-215-1, Savannah, Georgia, USA, April 1999

Burns, P.D. (2000). Slanted-Edge MTF for Digital Camera and Scanner Analysis, *Proceedings of PICS: Image Processing, Image Quality, Image Capture, Systems Conference*, ISBN 0-89208-227-5, Portland, Oregon, USA, March 2000

De Luca, L., Véron, P. & Florenzano M. (2007). A generic formalism for the semantic modeling and representation of architectural elements. *The Visual Computer*. Vol. 23, No. 3, (March, 2007), pp. 181-205, ISSN 1432-2315

De Roos, H. (2004). The Digital Sculpture Project Applying 3D Scanning Techniques for the Morphological Comparison of Sculptures. *Computer and Information Science*, Vol. 9 No. 2, (n.d., 2004), pp. 85, ISSN 1401-9841

El-Hakim, S.F., Beraldin, J.-A. & Blais, F. (1995). A comparative evaluation of the performance of Passive and Active 3-D Vision Systems, *Proceedings of St. Petersburg-Great Lakes Conference on Digital Photogrammetry and Remote Sensing (SPIE)*, pp. 14-25, ISBN 9780819420190, St. Petersburg, Russia, June 25-30, 1995.

El-Hakim, S.F. (2000). 3D Modeling of Complex environments, *Proceedings of Electronic Imaging, Videometrics and Optical Methods for 3D Shape Measurement (SPIE)*, pp. 162-173, ISBN 9780819439871, San Josè, California, USA, January 21-26, 2000

El-Hakim, S.F., Beraldin, J.-A. & Picard, M. (2002). Detailed 3D reconstruction of monuments using multiple techniques, *Proceedings of the Commission V Symposium Close-Range Imaging, Long-Range Vision, XXXIV part 5*, pp. 13-18, Corfu, Greece, September 2-6, 2002

Gaiani, M., Micoli, L.L. & Russo, M. (2006). Fasi, attività e metodo della ricerca. Casi di studio, In: *Metodologie innovative integrate per il rilevamento dell'architettura e dell'ambiente*, Docci M. editor, pp. 167-179, Gangemi Publisher, ISBN 8849207786, Rome

Georgopoulos, A., Tsakiri, M., Ioannidis, C. & Kakli, A. (2004). Large scale orthophotography using dtm from terrestrial laser scanning, *Proceedings of the XXth ISPRS Congress, Commission V*, pp. 467-472, Istanbul, Turkey, July 12-23, 2004

Goesele, M., Fuchs, C. & Seidel, H-P. (2003) Accuracy of 3D Range Scanners by Measurement of the Slanted Edge Modulation Transfer Function, *Proceedings of Fourth International Conference on 3-D Digital Imaging and Modeling (3DIM)*, pp. 37, ISBN 0-7695-1991-1, Banff, Canada, October 6-10, 2003

Guidi, G., Tucci, G., Beraldin, J.-A., Ciofi, S., Ostuni, D., Costantini, F. & El-Hakim, S.F. (2002). Multiscale Archaeological Survey Based on the Integration of 3D Scanning and Photogrammetry, *Proceedings of the Commission V Symposium Close-Range Imaging, Long-Range Vision, XXXIV part 5*, pp.58-64, Corfu, Greece, September 2-6, 2002

Guidi, G., Beraldin, J.-A., Ciofi, S. & Atzeni, C. (2003). Fusion of range camera and photogrammetry: a systematic procedure for improving 3D models metric accuracy. *IEEE Transactions on Systems Man and Cybernetics Part B-Cybernetics*, Vol. 33, No. 4, (n.d., 2003), pp. 667-676, ISSN 1083-4419

Guidi, G., Beraldin, J.-A. & Atzeni, C. (2004) High accuracy 3D modeling of Cultural Heritage: the digitizing of Donatello's "Maddalena". *IEEE Transactions on Image Processing*, Vol. 13, No. 3, (n.d., 2004), pp 370-380, ISSN 1057-7149

Guidi, G., Micoli, L.L. & Russo, M. (2005). Boat's hull modeling with low cost triangulation scanners, *Proceedings of the Videometrics VIII in Electronic Imaging (SPIE)*, pp. 28- 39, ISBN 9780819456380, San Josè, California, USA, January 17-20, 2005

Guidi, G, Frischer, B., Lucenti, I., Donno, J. & Russo, M. (2008). Virtualizing ancient Imperial Rome: from Gismondi's physical model to a new virtual reality application. *Special Issue on Digital Matter and Intangible Heritage (IJDCET)*, Vol. 1, No. 2/3, (n.d., 2008), pp. 240-252, ISSN 1753-5212

Guidi, G., Russo, M., Magrassi, G. & Bordegoni, M. (2010). Performance evaluation of triangulation based range sensors. *Special Issue in Instrumentation, Signal Treatment and Uncertainty Estimation (Sensors)*, Vol. 10, No. 8, (July, 2010), pp. 7192-7215, ISSN 1424-8220

Guidi, G., Russo, M. & Beraldin, J.-A. (2010). *Acquisizione e modellazione poligonale,* McGraw Hill Publisher, ISBN 9788838665318, Milan

ISO 12233:2000. (February, 2011). Photography – Electronic still picture cameras – Resolution measurements, In: *ISO Standards By TC42 Photography,* September 2011, Available from: < http://www.iso.org/iso/iso_catalogue/catalogue_tc.htm>

ISO 16067-1:2003. (March, 2009). Photography - Spatial resolution measurements of electronic scanners for photographic images - Part 1: Scanners for reflective media, In: *ISO Standards By TC42 Photography,* September 2011, Available from: <http://www.iso.org/iso/iso_catalogue/catalogue_tc.htm>

Levoy, M., Pulli, K., Curless, B., Rusinkiewicz, S., Koller, D., Pereira, L., Ginzton, M., Anderson, S., Davis, J., Ginsberg, J., Shade, J. & Fulk, D. (2000). The Digital Michelangelo Project: 3D Scanning of Large Statues, *Proceedings of the 27th annual conference on Computer graphics and interactive techniques (SIGGRAPH),* pp. 131-144, ISBN 1581132085, New Orleans, Louisiana, USA, July 23-28, 2000

Maldonado, T. (1997). *Critica della ragione informatica,* Feltrinelli Publisher, ISBN 8807102218, Milano

Ohdake, T. & Chikatsu, H. (2005). 3D modelling of high relief sculpture using image-based integrated measurement system, *Proceedings of the ISPRS Working Group V/4 Workshop (3D-ARCH),* Venice, Italy, August 22-24, 2005

Remondino, F., Guarnieri, A. & Vettore, A. (2005). 3D modelling of close-range objects: photogrammetry or laser scanning?, *Proceedings of the Videometrics VII in Electronic Imaging (SPIE),* pp. 216–225, ISBN 9780819456380, San Josè, California, USA, January 17-20, 2005

Remondino F. & El-Hakim, S. (2006). Image-based 3D Modelling: A Review. *The Photogrammetric Record,* Vol. 21, No. 115, (September, 2006), pp. 269–291, ISSN 1477-9730

Remondino, F. (2011). Heritage Recording and 3D Modeling with Photogrammetry and 3D Scanning. *Special Issue Remote Sensing in Natural and Cultural Heritage (Remote sensing),* Vol. 3, No. 6, (December, 2010), pp. 1104-1138, ISSN 2072-4292

Russo, M., Magrassi, G., Guidi, G. & Bordegoni, M. (2009). Characterization and Evaluation of Range Cameras, *Proceedings of the 9th Conference on Optical 3-D Measurement Techniques (Optical3D),* pp. 149-158, ISBN 978-3-9501492-5-8,Vienna, Austria, July 1–3, 2009

Velios, A. & Harrison, J.P. (2001). Laser Scanning and digital close range photogrammetry for capturing 3D archeological objects: a comparison of quality and practicality, *Proceedings of CAA,* pp. 205-211, ISBN 1841712981, Visby, Gotland, Sweden, April 25-29, 2001

VV.AA. (1985). *La sfida delle complessità* (Bocchi G. & Ceruti M. editors), Feltrinelli Publisher, ISBN 9788842420729, Milan

VV.AA. (2006). *La rappresentazione riconfigurata* (M. Gaiani editor), Poli.Design Publisher, ISBN 88-87981-85-X, Milan

8

The Hidden World: Reality Through Laser Scanner Technologies – A Critical Approach to Documentation and Interpretation

Mercedes Farjas[1], J. Julio Zancajo[2] and Teresa Mostaza[2]
[1]Universidad Politécnica de Madrid
[2]Universidad de Salamanca
Spain

To my parents
Whatever hidden world you inhabit,
I miss you so much, I too want to be there.

1. Introduction

Our experience with laser scanner systems began in 2003, when the Leica company put some of their new devices at our disposal and gave us the opportunity to include them in our 3D modelling work (Figure 1). As we indicated in the press cutting (Farjas and Sardiña, 2003) this new equipment was able to model practically continuous surfaces without the need to discretise the data capture and ruled out subjectivity in choosing the singular points to represent the model.

Fig. 1. Laser scanner modelling of the Cibeles Fountain in Madrid by CYRAX 2500 Leica-Geosystem

These were surprising times in which we let ourselves be carried away not only by the spectacular results but also by the capacity for evolution of this new technology. This included longer-range devices, short-range precision, they could be controlled by a computer or left to work independently, they could use external or compact batteries, which made them independent of external power sources. Although it would be true to say we were highly impressed by this new digital tool, it should also be said that it was not all pure enjoyment; we also had our moments of suffering.

As researchers in the field of 3D modelling, we believe it is now time to pause and take stock of the system of automatic processes in which we now find ourselves working with this technology. Perhaps we should think about recovering the power to take decisions, to look back to the time of traditional survey methods and in this light to analyse the tools that technological advances have put into our hands. Again and again we find mention made in the conclusions of studies carried out with laser scanners that in order to surpass the present limits we need more powerful computers, faster information processing, more highly automated tasks, the intervention of the operator must be bypassed, etc. Our question is: do we really mean this, or are we putting these words into the mouths of the computers? Is this not what computers want, or would request if they could speak for themselves? Have computers already not practically taken over the data acquisition and information treatment processes? What role do we play in all of this? We have been reduced to merely obeying the orders given to us by the different programs, we labour at boring tasks and produce strange results that do not seem to be consistent with the day-to-day work of a scientist.

We consider that now is the time to think again about our real needs and objectives and for each researcher to adapt both processes and results to the context of his/her individual work. Now is the time to escape from the tyranny of 3D technology and to adopt a critical approach to laser scanner systems (Figure 2: data capture, treatment and visualisation). We are, so to speak, engineers and the new technologies are our raw materials, but we are slowly but surely turning ourselves into mere robots. 3D laser scanner technology has practically taken over the field and by itself directs entire projects, leaving us to carry out its orders from a pre-programmed protocol.

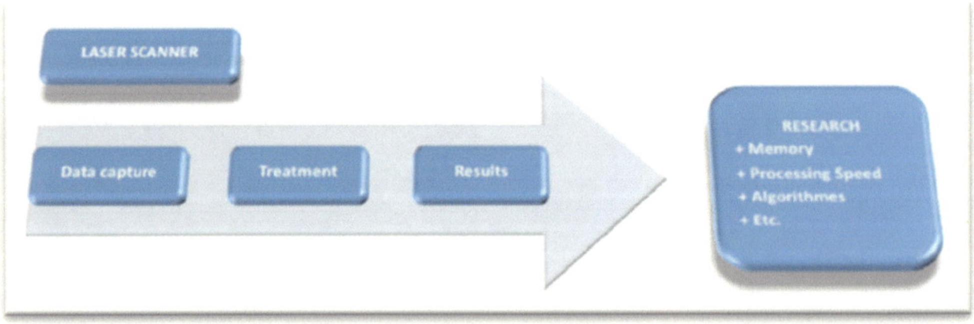

Fig. 2. Typical work scheme for laser scanner equipment.

We first suggested this idea in the workshops on "Graphic Documentation of the National Heritage- Present and Future" organised in Madrid by the *Instituto de Patrimonio Cultural de España* in November 2010 in a paper entitled *Cartografía en Patrimonio: la métrica en la documentación. ¿Una realidad pendiente?* (Farjas et alii, 2010 b). In this paper we compared a survey carried out by traditional methods of the 3D modelling of the Cloister of the Cathedral of Sigüenza by Silvia Peces Rata with another using laser scanner technology of the Escalada Monastery by María Expósito. The former used a Leica TCR 705 total station, which measures distances by means of visible laser beams without the need for a reflecting prism. A total of 1524 points were captured and subsequently edited by AutoCad 2005 to obtain the plans and elevations of the cloister on a scale of 1:200 from the point-cloud graphics file. A 3D model was finally produced as well as metric documents containing a description of the cloister. The general procedure used was similar to previous studies using the same equipment (Alonso et alii, 2002). The millions of points of the monastery captured in the laser scanner survey were stored in computer programs, but, as far as we are aware did not make any valuable contribution to the study itself.

Topography consists of the representation of a surface by means of points located in a system of coordinates. Developments in this field have been closely associated with technological advances, mainly in instrumentation. The appearance of computer programs, advances in applied electronics, the discovery of photogrammetry, global positioning systems and electromagnetic distance measurement have all played a part in the development process.

Terrestrial laser scanner equipment is one of the results of this evolution. This device carries out sweeps of the survey zone by laser beam to generate a 3D point cloud. The x, y and z coordinates of each point are stored in a reference system together with the reflected beam intensity and these provide us with a massive automated 3D metric data acquisition system. Data treatment is divided into different stages and includes work planning, locating work stations, the data capture itself, image taking, digital data treatment to eliminate unwanted points, joining the sweeps together, transforming coordinates, a sampling process to determine spatial resolution, the construction of a plane triangle network to define the object's geometry and the creation of the solid model. Finally, the model is textured from different digital images in order to give it a realistic appearance. Descriptions of the foregoing processes can be found in the proceedings of different congresses, such as: the International Conference CAA- Computer Applications in Archaeology, VAST- Symposium on Virtual Reality Archaeology and Cultural Heritage, VSMM- Conference on Virtual Systems and MultiMedia Dedicated to Digital Heritage and Arqueológica 2.0. Specific examples can be found in the authors' previous references (Farjas, 2007; Farjas and García-Lázaro, 2010).

In this chapter we provide examples of applications from a critical viewpoint in order to give ourselves the opportunity to question our achievements in both technical aspects and results. We have at the present time technologies, equipment and procedures that are used to a greater or lesser degree, but it is by no means clear to what extent they are used by the specialists in the field and even whether they are of any real assistance. We therefore propose going back to the general perspective of applying equipment, instruments and methods as appropriate. Our aim is that when what we really need is an image we will feel we are able to use a simple camera instead of the most advanced laser scanner system, or that when we need to film an object we can opt for traditional video footage instead of presenting badly textured hyper-realistic flights over that add nothing to the quality of the model as a great achievement. We should be able to freely choose whatever system of representation best suits our purpose and not feel obliged to follow an automated information processing chain, rejecting intermediate laser scanner process points and products for no reason at all, with the single-minded purpose of obtaining an empty 3D model.

As we have already mentioned, this chapter is divided into three sections. The first deals with an introduction to 3D laser scanner modelling methodology in a typical survey, indicating certain points that could be usefully subjected to analysis. The second describes traditional surveying methods and tries to determine to what extent they can be used to improve laser scanner modelling. Finally, the third section will take a look at future lines of research.

2. Laser scanner technology

2.1 General description

These topographical and map-making techniques have traditionally been linked to the representation of terrain in all kinds of states and situations and provide digital results in two or three dimensions. Data capture systems are in constant evolution; an important benchmark was the incorporation of 3D laser scanner systems. In Spain they appeared around 2003, when they started to be used experimentally in large-scale point cloud 3D modelling in archaeology and the national heritage.

A laser device emits a laser beam to obtain measurements. The word is an acronym of the description *Light Amplification by Stimulation Emission of Radiation*. The first laser was invented in 1960 by Theodore H. Maiman (1927-2007), a physicist from the University of Colorado, who had obtained a Master's Degree in Electrical Engineering and a Ph.D in Physics from the University of Stanford.

A 3D laser scanner captures the position of different points of an object with reference to its own position and gives each one a set of coordinates. Data captured are coordinates (x, y and z) and luminance (I), or RGB values if photographs are acquired simultaneously.

Terrestrial laser scanners can be classified according to their measuring system or their sweep system. There are three types of measuring systems: time of flight, phase shift and optical triangulation. Sweep systems can be divided into camera, panoramic and hybrid (Farjas et alii, 2010 a).

2.2 Measuring system

Time of flight

This type of scanner measures the time delay between the emission and return of a laser beam to define a vector that is completed with data from the orthogonal angles that define the point's position. The beam covers the entire study area and records a measurement for each point, according to the mesh definition selected by the user. This type of equipment is suitable for outside work and can be used to model objects at medium and long distances (up to a kilometre or more) with an accuracy to within one centimetre.

Phase shift

In this method, the distance between point and scanner is calculated from the phase shift between the emitted and received waves. A number n, which is the total number of complete wavelengths covered, is added to the measured shift. To determine the exact value of n, various wavelengths are emitted at different frequencies. Both orthogonal angles are also recorded in order to precisely locate the point.

These devices can be used for medium distances, usually less than a hectometre, in both interiors and exteriors and are accurate to within less than one centimetre.

Optical triangulation

The position of each point of the object is obtained by the principle of laser triangulation. A highly collimated laser beam is emitted while a camera receives the light reflected from the object. As the values of the base of the triangle formed by the light emitter and the camera and both angles are known, the position of each of the points can be obtained.

These devices are used for high-precision short-distance measurements and are accurate to within less than one millimetre.

2.3 Sweep systems

Camera

A laser beam is focused onto each of the points of the object by synchronised vertical and horizontal mirrors, so that the sweep creates a 60° scan window in both directions.

Panoramic

In these devices, the laser beam generally moves around the vertical axis, thus carrying out a horizontal movement. A vertical sweep is performed for each horizontal position, with the same movement as that in a total station. A 360° horizontal scan window and 300° vertical scan window can be obtained, limited only by the shadows produced by the shape of the device and its support.

Hybrid

The laser beam makes a horizontal sweep around the vertical axis, as in the panoramic method. It also makes a vertical movement through a mirror, as in the camera method, to record the points. A 360° horizontal and 60° vertical scan area can be obtained.

A laser scanner usually contains a data capture system together with a computerised control program. A camera is a required extra in optical triangulation but is also increasingly used with other systems to obtain model radiometric and colour information. GNSS receivers are now being incorporated into laser scanner technology as spatial positioning systems to provide geo-referencing.

When data acquisition is complete, the next stage is the processing of the information obtained. Most laser scanner devices include a data treatment and visualisation computer program designed to deal with the large number of points obtained from each sweep, which can be big enough to swamp traditional CAD systems. A 3D laser scanner project usually follows the work flow design shown in Figure 3, which begins with data acquisition, continues with data processing and finally provides usable 2 or 3D models in the form of point clouds.

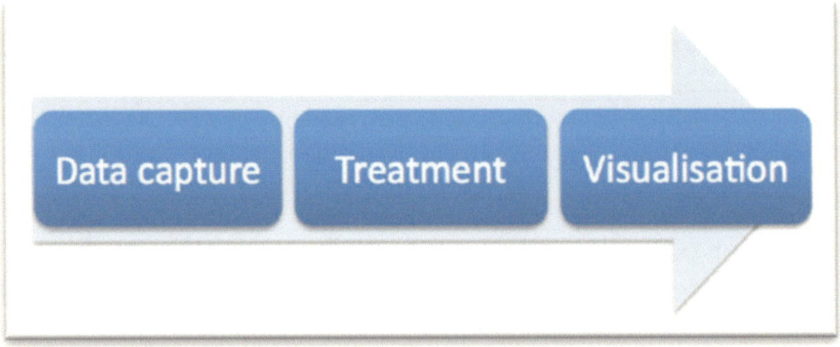

Fig. 3. Typical work flow of a laser scanner modelling project.

Close attention should be paid to three fundamental aspects in data acquisition: the number of work stations for the equipment and their locations, the point density registered by the device and the systems of coordinates that can be used. With regard to the first of these points, enough work stations should be planned to ensure complete digitalisation of the zone and minimising the number of gaps or zones in shadow. Another consideration is the way in which the different reference systems are to be linked. For stitching sweeps together (each of which is stored in the appropriate work station's instrument reference system) neighbouring sweeps must contain common points. Work station planning should include whether reference signals or auxiliary elements are to be used to join sweeps and if it is necessary to provide these points with coordinates for subsequent referencing of the model. If a camera or panoramic type scanner is to be used, the device's capture window should be taken into account during project design.

The point density requisites depend on the following parameters: the precision required by the work or project, the distance separating the device from the object, the resolution selected to carry out the work and capture time.

With respect to reference systems, it is important to remember that laser scanners register point coordinates in their own instrument reference systems and this will differ from station to station. It will therefore be necessary to reference all captures to a single global or local reference system, for which specifically designed target signals can be used to link, register or georeference different systems of coordinates from topographical measurements or from the specific signals provided by the device manufacturer, (flat) reflecting signals or (flat and curved) target signals that can be automatically recognised by the specific algorithms of the data editing program. If photogrammetric methods are to be used simultaneously with laser scanners (if the device is equipped with its own camera or images are obtained independently), the relationship between the camera's system of coordinates and that of the scanner should be considered.

After the data has been acquired it must be processed. Most scanners incorporate a data processing and viewing program in an appropriate format. In this respect, progress is being made in the direction of interoperability of devices and standardised formats are becoming available to make working with 3D point clouds more flexible. In April 2011, the American Society for Testing and Materials (ASTM) approved the E57 E2807 Data Exchange Standard, which is considered the first specification for the interchange of 3D data. The Association's webpage recently published a message from Gene V. Roe, chairman of the ASTM's E57.04 Committee; *"The new ASTM data exchange standard will allow the hardware vendors to export their data in a neutral, binary format that can be imported by the design software,"* and stresses, *"This has the potential to move the entire industry along the technology adoption curve by allowing the mass market access to laser scanned data."*

The 3D ASTM International E57 E2807 file interchange format has begun to filter through to commercial software. For example, since June 2011 it has been included in Trimble's RealWorks version 7.0. In this way 3D laser scanner professionals can use their design software of choice, depending on their applications and workflow.

Whatever the software selected, the general treatment of 3D scanner point clouds usually includes the following phases:

- Pre-editing of sweeps. If the sweep is too dense it can be subjected to a re-sampling or segmentation.
- Registering each point cloud in the chosen project reference system, generally either local or global.
- Elimination of unnecessary and erroneous points, as for example those obtained from the unexpected appearance of people or vehicles in the survey zone.
- 3D modelling.
- 3D segmentation of point clouds.
- Extracting geometries.
- 3D modelling of entities.
- Filling in "hollow" zones.
- Simplifying entities.

Most of these processes are carried out interactively, although research is now being carried out on the creation of automatic or semi-automatic or algorithms to optimise and simplify them. The opacity of the algorithms and the recent implementation of 3D laser scanner data acquisition technology mean that in certain cases they have to be performed manually using CAD tools.

3. 3D Laser scanner monologue: The methodology presents one of its applications and asks some questions about the process

In 2007 I was asked to make a precise 3D model of the façade of the Church of Santa Teresa in Ávila, Spain, for use in a possible restoration project (Prieto, 2008). The study zone considered for this model consisted of the façade of the church itself, the front of the adjacent building and the square in which both buildings were located. For this work I opted for a Trimble GS 200. What was this choice based on? Different technical criteria can be considered, but in practice the most important is price, so that normally a research team will make do with whatever equipment they already have, so that this question has really been decided beforehand. Different scanners have very different measuring characteristics and are thus designed for different applications but at the time of buying few in my special field are sufficiently familiar with the range of scanners to be able to distinguish between them. Their experience of the world of 3D is basically limited to the question of long, middle or short distance. In Ávila, data capture was possible with phase measuring equipment, so we used one of this type. Data were acquired from different positions and we obtained the complete model by joining the sweeps in a transformation process. The model of the façade was then geo-referenced in an official system of coordinates, using laser scanner capture points which had also been given coordinates in the official system by the radiation method with a Leica 705 total station. Geo-referencing models is worth doing, since they can then be related spatially, but you should not forget to indicate the precision obtained in the results, as this parameter could be an important criterion in the final model files. The complete survey process is described below.

3.1 Planning

Prior to the data capture process, the entire zone was studied in order to define the location of the stations. The number of stations should be kept to a minimum in order to get as few errors as possible when sweeps are joined together. We have to take into account the number of stations and joins that can be performed, how precision can be transferred from one sweep to another and how the joining of sweeps from different stations can affect the accuracy of the final model. Enough attention is still not given to the initial quantification of units of measurement and the control of rotations in this process.

The field equipment consisted of the following:

- Trimble GS 200 laser scanner.
- Carbon Fibre tripod.
- Accessories: Batteries, cables, generator, etc.
- Laptop with the PointScape 3.1 program.

In the laser scanner method, the common areas between stations must be surveyed in order to define common points in different sweeps and convert reference systems. I still have no precise overlap values, as these have never been experimentally defined or been considered in theoretical studies. In photogrammetry these values are indicated in the initial conditions to ensure that the quality of the results is acceptable.

When considering the location of the stations we also tried to keep the hidden zones of the object to a minimum. Some zones would of course be impossible to scan due to their height or the presence of nearby buildings. We finally decided to do the survey from a total of seven stations. This aspect must be given careful consideration in the planning stage, as in spite of the care we took during the field survey we continually found gaps in the takes during the editing process. Complementary methods can be used to analyse these shadows in the design phase, according to their relative importance in the project objectives. Possible alternative measuring methods can be established to supplement data acquisition or approximate surface laser scanner models can be implemented after editing by algorithms that are not very friendly to the reality they are required to represent. Is a visual inspection of the target zone enough to form the basis of these decisions?

The laser scanner registers the points of the model in the instrument's system of coordinates, which differs from station to station. To unite these points, as we have said before, well defined points are scanned in common zones. During editing, homologous points are identified in each point cloud to convert the coordinates of all the points into a single instrument system. The algorithms that integrate the sweeps into a common reference system are encoded and the operators do not normally know their mathematical formulation. We obtain the results and the fit precision parameters, but nothing else. Are these algorithms the most appropriate for the geometry of our points? Will it be necessary to use some kind of constraint in the fit that considers the location of the stations? In the future we may be able to analyse these questions, but at the moment the answers are hidden from us. In fact, these mathematical formulations and the way they are processed are the key differences between the different instrument manufacturers.

3.2 Laser scanner stations

Prior to data acquisition the scanner is placed at a point in the selected zone. In this study we did not wish to locate this point precisely as we intended to work with free stations. Are there any comments to make on this? The exact definition of the coordinates of the station could change the least squares adjustment of the observations, reducing the number of unknowns or degrees of uncertainty, or allowing new surfaces to be adapted to those already existing. Decisions can be taken here that could have a marked influence on the final precision obtained. At the present time scanners are not able to carry out forced centring and some do not even have plumb lines or other auxiliary systems. Let us leave these questions to future researchers, if they should think them worthy of study.

3.3 The data capture process

The device can be controlled by means of a laptop computer with Internet connection either through a control terminal or screen. In this study we used a laptop and PointScape 3.1. A window in the screen showed an image of the zone under study and indicated the area that could be scanned by the device from its present position at the chosen resolution. The precise scanning zone was then selected from the image in the window, after which the following parameters were set up:

- The first parameter that must be set is the scan resolution, i.e the distance between the registered points.
- The second is the approximate distance from the target. This value is vital if the device is to register the points at the desired resolution. From the average distance entered and the scan resolution the scanner computes the laser beam angle of inclination and programs the sweep.
- Finally, the number of times each point is measured must be entered. As in the case of measuring distance with a total station, each point is measured a certain number of times and the value of the observation distance is the average of the measurements performed.

In our project we varied measurement resolution due to the irregular outline of the church we were modelling. These values were established by bearing in mind that the principal objective was to map the façade of the Santa Teresa Church and that the other buildings were included merely to give the church spatial continuity. The façade was scanned with a resolution of approximately 1cm. Details of the façade, such as statues, shields, reliefs, etc. were taken at a resolution of 0.5cm, the other buildings at between 2 and 3cm with details between 0.5 and 1cm and the ground was given a resolution of 10cm. How was all this information indicated in the final model? How can the design precision affect decisions on the resolution of data acquisition? Is the target faithfully reproduced? Scanning is carried out automatically and all the points in the corresponding interval are captured. Surfaces are discretised at a certain interval that does not convey much information about the details, which could imply a *smoothing out* of the real object. It would appear that we need to design scans according to the surfaces and level of detail required, so as to eliminate improvisation in the field and ensure not only the resolution but also acquisition of the level of metric detail required by the project.

Table 1 shows the sweeps carried out by the laser scanner, with the number of sweeps performed from each station, the number of points included in each one, and total project points, which came to more than 15 million.

	Sweep 1	Sweep 2	Sweep 3	Sweep 4	Sweep 5	Sweep 6	Sweep 7	Total
Station 1	1,309,592	839,364	1,514,938	146,455				3,810,349
Station 2	316,761	586,130	762,631	178,965	794,604	1,689,909	492,897	4,821,897
Station 3	469,764	111,384	61,693					642,841
Station 4	553,404	274,535	84,954	121,729	205,363	299,610		1,539,595
Station 5	20,383	10,890	351,686	1,351,52	970,256	191,573		2,896,340
Station 6	1,028,361	26,211	30,095	61,035	38,105	26,529	29,612	1,239,948
Station 7	44,473	67,293	2,225	1,893				115,884
								15,066,854

Table 1. Points registered by station and total points.

After data capture and during the construction of the solid body a triangular mesh is obtained to define the surface geometry of the objects represented, using the Mesh Creation Tool to generate and edit the triangular mesh. Although a projection can be used in editing the mesh, in our case we decided against it. The editing options include eliminating vertices, triangles and rough edges and smoothing out the mesh itself.

To speed up the process, triangular mesh is initially generated one by one for individual part-elements, however, in this way problems may arise when joining meshes together (blank spaces and overlaps) and when placing photographs on the surface. We decided to try generating the triangular mesh with the entire surface in a single process and even though it took an hour and a half better results were obtained.

3.4 Data treatment

For this process we used Trimble's Realworks Survey 6.0 (Zazo et alii, 2011). The treatment was divided into the following steps:

1. Data cleaning
2. Joining sweeps
3. Geo-referencing
4. Data filtering
5. Generation and clean-up of the model
6. Texturing the model

3.4.1 Data cleaning

This step consists of removing foreign noise from the model (Figure 4), or the elimination of all points that do not strictly form part of the target object. These include, people, vehicles or animals that somehow intrude on the scene during data acquisition. Can other types of object come between the scanning device and the target? How can we control or quantify the accuracy of the measurements with respect to those of the actual object? What type of inaccuracies can be obtained in the details? Does a variable interfere by introducing into the model automatic processes which can only be detected at certain points? All these factors must be given a great deal of thought.

Fig. 4. Example of eliminating noise from a model.

3.4.2 Joining sweeps

The laser scanner uses a different reference system at each station. In order to give all the point clouds a common system of coordinates, we can choose between two options:

- Use aiming points such as cards or spheres arranged around the scanning zone.
- Use common scanning zones and locate common points between different stations.

Both options are solved in the same way by means of a 3D similarity transformation of 6 unknowns, 3 rotations and 3 translations. We chose the second option, since the model involved a building and common points could be easily identified in the different sweeps to an accuracy of around 1 cm. For this we used the *Cloud-Based Registration Tool,* which allowed three homologous points to be selected and after identification, showed the standard deviation of the fit on the computer screen. With the *Refine* option, a further fitting together of the point clouds can be performed, showing the standard deviation in each case together with an image of the model (Figure 5). With this program the process can be repeated as many times as required.

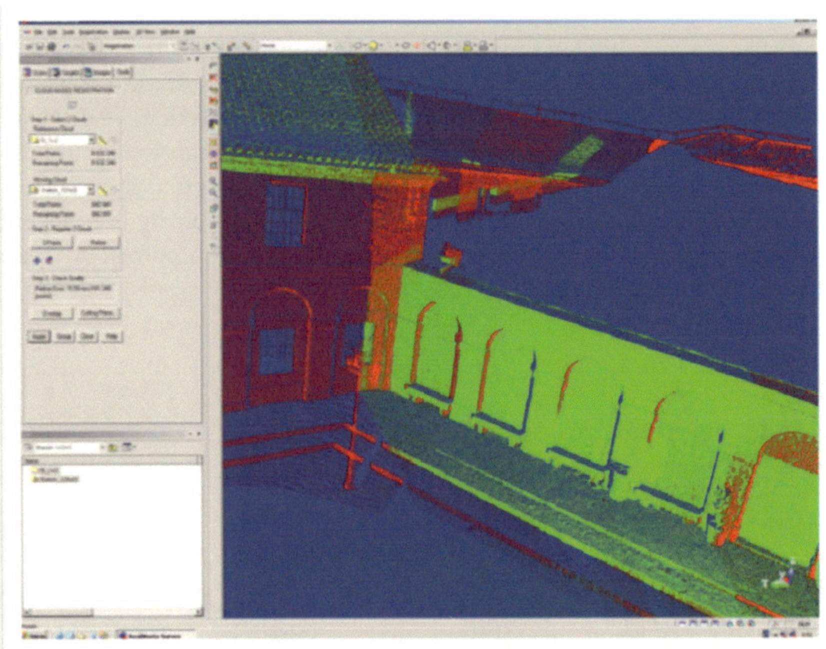

Fig. 5. Joining point clouds.

It would be interesting to have more information on this process, including maps of the differences obtained from the homologous points used to join the sweeps, to obtain information on the way in which different joining methods can deform the geometry of the rest of the model. At the moment, the operator can do little more.

Table 2 shows the standard deviation of the joining together of the sweeps from the different stations.

Stations	Standard Deviation (mm)
1+2	7
1+2+3	12
1+2+3+4	6
1+2+3+4+5	8
1+2+3+4+5+6	6
1+2+3+4+5+6+7	7

Table 2. Standard deviation of the joining together of scans from different stations.

3.4.3 Georeferencing

In order to convert the instrument coordinate system to the official Spanish cartographic system at that time (ED-50), the points had to be referenced in both systems of coordinates. Seventy-five points were measured in the official system by a total station within the working zone. From this total, eleven were selected to calculate the transformation parameters. The dynamics of the georeferencing process was as follows:

- Points were identified in the point cloud that had coordinates in the ED-50 system.
- The coordinates of these points were then manually entered in the ED-50 system.
- When three points had been registered the standard deviation of the adjustment was shown on the screen. As new points were entered, the calculation of this parameter was updated.
- Also available was the option of eliminating any existing points that caused excessive discrepancies in relation to previous adjustments, so that they could be left out of the final transformation.

The geometric distribution of the points to be used with coordinates in both systems must be borne in mind. It is not advisable to have all these points on the same plane and they should be evenly distributed throughout the modelled zone to avoid extrapolations. The location of the points we used is shown below (Figure 6). The standard deviation of the georeferencing adjustment was 19mm.

Fig. 6. Distribution of the points used in the georeferencing transformation.

3.4.4 Data filtering

With even the most powerful computers it is usually considered necessary to reduce the size of the file. In our project there came a time when we found ourselves with a file of around five GBs, so that it took between five and ten minutes simply to save the changes. Fearing the possibility that the computer would not be able to deal with all this information, we decided to drastically reduce the number of points stored from over 15m to 5m by sampling the point cloud with the Spatial Sampling filter and entering the distance required between the different points. For the sampling, we separated the planes in which a sampling of 100mm could be made, e.g. the outside of buildings. In detail zones such as shields, statues and reliefs the sampling was carried out at a distance of 5mm (Figure 7).

Fig. 7. General scan of the façade before and after applying a 100mm spatial sampling.

In spatial sampling points are lost by considering only the single variable Distance to the Nearest Point, so that the metric detail of the model loses clarity. A massive and indiscriminate amount of information is thus lost which may well have an effect on the objective of the project. So then, what is actually being represented? Why were so many points registered? Does the need for spatial sampling indicate that parameters have been incorrectly entered prior to data acquisition? Can the application be used simply to reduce the size of files?

Fig. 8. Complete façade after spatial sampling.

When we acquired the data with the laser scanner we captured a real surface on a mesh at a pre-defined interval. This surface, which could be considered as a first order approximation to the object, has been reduced to only about a third of the original. What part of its identity has disappeared? How much detail has been lost? If we aim for an Xth order approximation to reality, should we not lay down certain criteria during the design phase?

Should we not define experimental control methods for data processing? We should devote a lot of time to deciding what we really wish to represent. For the present, we show the image obtained of the façade in our project after processing the data (Figure 8).

As we have said, after filtering, the initial 15,066,854 points became 4,185,284. This volume made it possible to continue with data processing in the computers available, which were of the latest generation at that time. What had we left out? How did the model obtained relate to the real object? What had remained hidden? What would happen nowadays if we processed the same data with the programs presently available? Would there be any differences in the final models?

3.4.5 Construction and cleaning of the model

This phase consists of constructing a triangular mesh to define the geometry of the object. After deconstructing what has been constructed (i.e. breaking the monument up into a multitude of points with the laser scanner) we have to think about its spatial reconstruction. This is a long process, but to what end? Is it justified for a possible restoration? Should we apply any additional objectives?

We used the *Mesh Creation Tool* to create the solid body, after which we looked for faults in the surface, most of which we could fix by eliminating triangles. Others had to be corrected by reconstructing the solid body in stages after cleaning up the point cloud (Figure 9). The model was divided into three zones: ground, church and museum and each of these was split into objects. It would have been impossible to construct the solid body in a single stage due to the large volume of information.

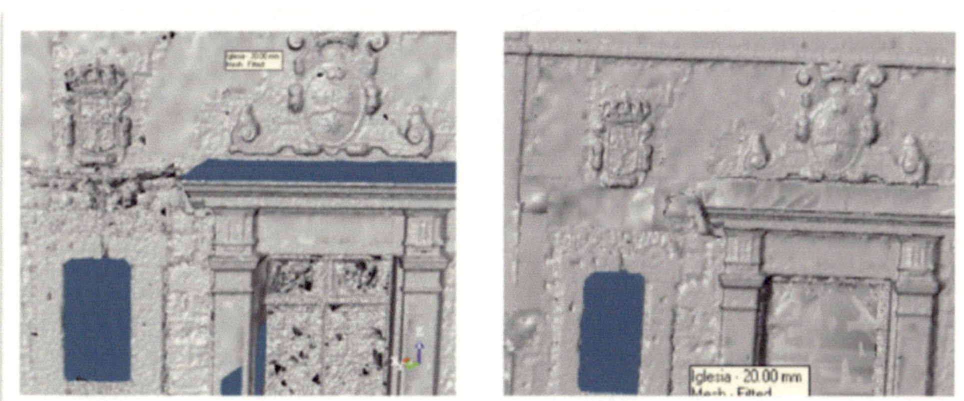

Fig. 9. Untreated and treated solid models, eliminating triangles and superfluous points.

When we created the solid body we found certain problems with hidden zones. As we mentioned before, some zones simply could not be scanned due to being out of the line of sight. Some of these cases were solved by generating a solid body from the existing points (even if they did not comply with the regularity of the spatial resolution) so that the texturing process could be carried out (if there are gaps the model cannot be textured). This criterion was adopted for purely aesthetic reasons. Today, we could question these reasons and, like the points that brought us closer to modelling the real object (in our case, the façade), they are better hidden in a regular, smooth and uniform model.

Fig. 10. Solid model of the door of the Church of the Convent of Santa Teresa.

After generating the solid bodies we had surfaces that represented the surface of the Church of Santa Teresa and its surroundings (Figure 10). We had created a scenario. Would it be of any interest to lay down an experimental solid model control system to suit the sampling conditions? How could the difference maps contribute at this time to validating the representation of the real object? Reality has been covered with a curtain and we still have to discover to what extent it can stand up to metric analysis. We may say there is precision in the point cloud, also in the joining together of the models and in the georeferencing. But, where is the real object? Where did it get lost? We have a representation of the tree, but can we discover the size of its leaves from the digital model we have created? And can we do the same for the details of the façade?

3.4.6 Texturing the model

To give the model a more realistic appearance textures were applied from digital images. As the camera incorporated in the laser scanner was only of 1 megapixel, we used a Konica Minolta DiMAGE E50 de 5.2 megapixels and maximum resolution of 2560 x 1920 pixels. The resulting images were digitally treated with Adobe Photoshop CS2 as follows:

- Elimination of objects between the façade and the camera, such as lamp-posts, benches, waste paper baskets, etc.
- Treatment of brightness and contrast to give all the images similar lighting.

The next step was to apply these images to the model, which involved finding corresponding points in the model and the photograph (Figure 10). The greater the number of points selected, the better the image fitted into the model. No criteria have yet been established for selecting homologous points, the minimum number to achieve acceptable quality or their distribution. It should not be forgotten that the images are not calibrated and are really types of optical effects. The process is usually automatic and we should be highly sceptical of these graphical methods.

Fig. 11. Application of image to the model identifying homologous points.

This was one of the most complicated stages in the project. The photos to be used in texturing had to be selected and given the appropriate treatment to make tonality variations invisible. The image treatment results are shown in Figure 12.

Fig. 12. Final image of the façade obtained from three photographs.

3.5 Results and conclusions

After all the data had been processed a model with the required precision of the Santa Teresa Church Square was obtained, in compliance with the general project objectives (Figure 13).

Fig. 13. Model of the Santa Teresa Church Square with and without texturing.

Let us leave the discussion of these points for now and continue with our attempt to define a future laser scanner project with a design phase that asks the questions *Why?* and *What for?* Also to be established are the objectives by which the project is guided, a decision as to when each laser scanner product is required, for what purpose and with what characteristics, plus the consideration of the definition of images, resolutions, pixels, model precision required, etc.

In the following section we will look back in history to the way in which modelling and survey projects were carried out in the past. For these, a process did exist, but the important thing was that people were given the training to enable them to take decisions in the knowledge that every decision involved a certain loss, even though it may have opened up new possibilities. You could not have everything, but at least you were aware of the data that had been acquired and you were thus equipped to take the appropriate actions. Nothing was lost in the intermediate processes and the iconography was human.

4. Dialogue with traditional surveys: The laser scanner method listens to the traditional method in its search for answers.

I was invited to take part in the work on the *Casa de Taracena* at the Clunia archaeological site in Burgos, Spain (Juez and Martín, 2006). My contribution would be to generate a series of documents to geometrically describe this ancient Roman settlement. *La Casa de Taracena* (Figure 14) is a large rectangular Roman villa that took up a large part of the settlement and had few open patios, due to the cold winters experienced in the region. In this project I used traditional topographical methods assisted by GPS receivers.

Fig. 14. Aerial view of the Casa de Taracena.

In the past, a survey of the site would have been a fundamental step, as would the design of a basic network. The existing official 1/25000 scale maps published by the National Spanish Geographic Institute were used to study the site topography. The definitive signal locations were later selected in the field, the vertices were identified by pre-fabricated markers and note was taken of their locations.

The system of GPS observations was then designed, including reference observations from the National Geodesic Network. When the observations of the basic network were complete, the radiation was carried out to gather information on the details of the survey zone. These details were so diverse that a Code Table had to be compiled. The codes were assigned to points as radiation progressed and provided a relationship between the model and the actual site in order to identify its different features. Point radiation was carried out entirely by GPS in real time with no problems due to screening. With the laser scanner system it is also possible to lose signals and acquire false data, for example due to rain or when representing corners. How can these situations be overcome?

All observations were then processed and coordinates were obtained for all the points making up the site. DXF file extension points were generated that enabled us to later contour and shape all the details in the cartography. The design of the site map was performed with digital cartography methods on AutoCad 2004. The site cartography consisted of a series of personalized sheets with symbols for points, lines and surfaces. Relief lines were shown by 0.5m equidistance curves for 1/500 scale maps and 0.2m equidistance for the 1/200, all of which were obtained by MDT triangulation. In order to obtain reliable curvature, break and inclusion lines, as well as all detail points, were entered. The general process is described below.

4.1 Data capture

As we have seen, data capture was started by designing a basic network. What are the conditions for the design of laser scanner stations? Perhaps specific requisites should be laid down or a set of conditions that seek to achieve something more than a general coverage of the object. Sweep overlaps could be defined (from geometric or other criteria) that make it possible to carry out partial internal control tests. For the survey of the *Casa de Taracena* we located stations in the places with the best visibility and no obstacles to GPS observations. These stations were placed on a map of the site on a scale of 1/25,000 with 5m equidistance.

We then made a survey of the site to ensure that the chosen stations were the best available for placing the receivers. How is a survey performed with a scanner? How is coverage analysed? Can it be quantified? If we follow the traditional method, this can be done according to geometries, intersections, visibility and homogeneity, among others.

For the observations we used GPS300 receivers with dual satellite reception frequency and SR9500 sensors. We bore in mind the following criteria when selecting network points:

- The point network had to completely cover the work zone.
- Points had to be evenly distributed throughout the site, with the highest concentration in the survey zone.
- The points had to be visible from other points, or at least from the neighbouring points, for orientation in case of the possible future use of a total station.
- Good inter-point geometry should be achieved.

A basic network was designed in accordance with the above criteria that covered almost the complete site and took full advantage of its topography. It was composed of 7 vertices with an additional one in the geometrical centre. Laser scanners make it possible to design special controls that ensure a certain degree of redundancy by capturing additional point clouds that can be used for model control. The network inter-vertex distance was between 100 and 200m. Have you made sure that a similar distance range is used for the entire laser scanner survey? Have you analysed any discrepancies in model homogeneity when performing scans, limited only by the device's data capture capacity?

To calculate the network we used three receivers, one of which was fixed and the others mobile. Only one scanning device is used in these projects. The quality of the results therefore depends on the status of this device and no experimental control or verification is carried out. Calibration is normally performed once a year. A process control system is defined for the observation measurements of all topographical equipment after the survey has been completed. Do we have to define a laser scanner verification procedure? For example, you could take direct field observations by another method prior to carrying out the calculations to validate field data. Is there a control program for the device's systems? Topographical observation methods are designed to be used with a working verification system (direct circle/inverse circle, Bessel's Rule, Moinot's Method, double calculation of baselines from different antennas, etc.).

The basic network of the *Casa Taracena* was constructed from a static, simultaneous, long-duration observation (approximately one hour) in order to define the vertex coordinates with high precision. Are the coordinates of laser scanner points guaranteed to be accurate? Are all captures the product of a non-verified radiation?

The equipment set-up phase began with a project configuration that specified the way in which measurements would be carried out. The following parameters were defined:

- Type of operation: Relative Static
- Satellite following parameter: 15° elevation mask. This is the minimum elevation to consider the position of the satellite when calculating GDOB (the parameter that estimates the goodness of the observations according to satellite-receiver geometry).

The laser scanner method also defines initial parameters, such as resolution, scan distance, measurement repetitions, etc., but there are no criteria expressly linked to modelling. It all depends on the operator's skill and experience. Should we establish an elevation mask for laser scanners to avoid angles of incidence greater or less than 20^g?

4.2 Data storage features

The three receivers were set up in the relative static method. A copy was made with all the parameters for inclusion in the Clunia Red Mission, in which we participate.

The points that defined the 2 and 3D cartography were then obtained by the GPS radiation method in real time from WGS84 coordinates. The baseline was calculated to an accuracy of 10mm ±1ppm. The points representing the terrain morphology were then converted to the ED50 reference system and the 3D model was created.

The coordinates are obtained directly by laser scanners, but we are really aware of what is going on inside the devices? Which mathematical formulae are processed? What is the order of redundancy? Is there any residue treatment or error detection system? Are any visual images eliminated during the least squares adjustments? Laser scanners are equipped with associated software in which we blindly trust. If we should try to process data with different software we may find that the information is not complete and so the process cannot be carried out. For example, having information on x, y, z, I, we have sometimes found that for texturing we need the normal information from the triangulated model, which we had failed to import. It would be a good idea to identify the data necessary for each of the processes and import these from the corresponding software. In other words, we should identify the information transmitted with each file extension and each data format.

4.3 Cartographic design

When considering present-day cartography, we should be aware that they consist of 3D documents that metrically reproduce a site or an artefact. In traditional mapping the classification of symbols was used according to the nature of the objects (points, lines and surfaces), roughly corresponding to those defined in the previous sketches made in the field during the survey.

From the data capture and processing stages for the geometric definition of the *Casa Taracena* we produced a point cloud containing:

- A topographical map to a scale of 1/500 of Area 3 in the site, covering an area of around 15 hectares.
- A topographical 1/200 scale map of the centre of the settlement known as the *Casa Taracena.*

The result of the process was a point cloud distributed in a three-dimensional space, which was used as the base material for making maps and generating the digital model of the terrain (Figure 15).

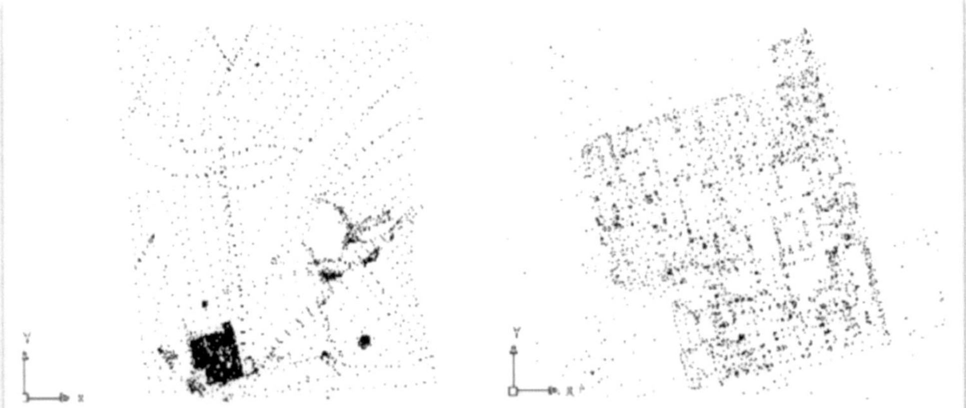

Fig. 15. Point cloud of Area 3 with details of the zone known as Casa Taracena.

Maps were produced from the point cloud files. Other indispensable elements were the field sketches (Figure 16), which not only represented the survey zone but also indicated the correspondence between specific elements in the terrain and the buildings with the points captured in the field.

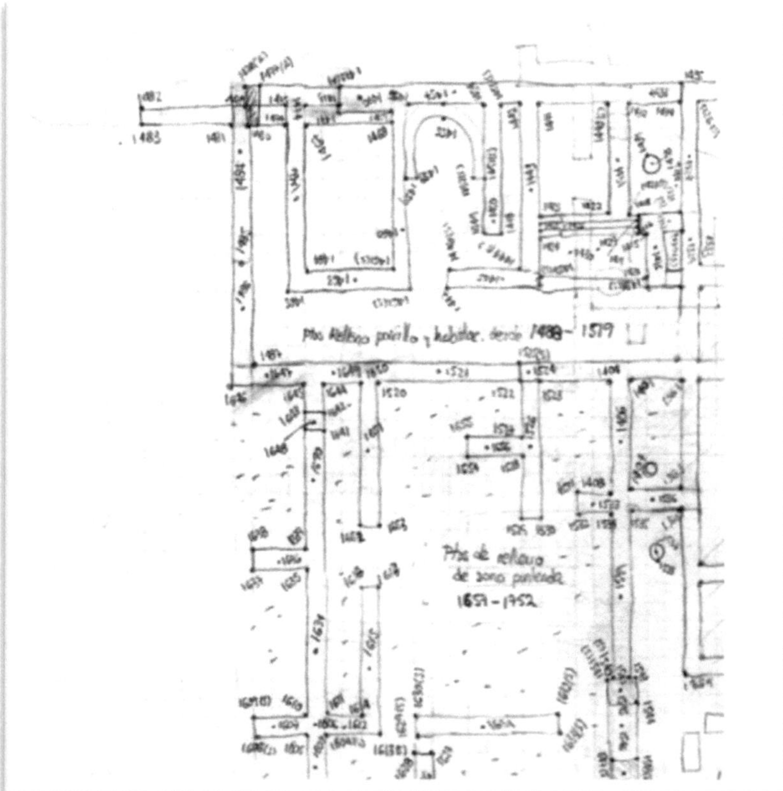

Fig. 16. A sketch of part of the Casa Taracena.

Laser scanner data processing starts with the point clouds. This does not consist of a single visual image but a set of images, which could be considered as an analogue beam containing all the points of which it is formed. Must each of the beams be characterized in some way? What about maximum and minimum angles and distances, the surface covering the target? With GPS, the parameter that characterizes the geometry of the observation is known as *Geometric Dilution of Precision* (GDOP). Would it be any advantage to have something similar in laser scanners? Not all the beams that are emitted by a device are identical with respect to the data captured in the field. Some will always be more important for creating the model (normal-convergent, near-far, angle of incidence on the surface, signal repeatability, the size of the laser beam when it reaches the target, etc.). If, for a given device these questions (after analysis) ever come to be trivial, we will then have to train ourselves to join point clouds from other devices (e.g. beams acquired from time-of-flight equipment at a distance of 1 km or phase-measuring equipment at a distance of 40m).

In the traditional method, each point has a number (in the present registering devices they also have a code) and the structural lines of the model are edited. In the laser scanners, the point clouds are cold. The only value is illuminance or RGB values. Could we consider some type of radiometric treatment or selection that applies image treatment technologies or algorithms, pattern recognition, or others pertaining to photogrammetry or remote sensing?

The software used in modeling the Casa Taracena was *AutoCad* 2004 for cartographic editing (Figure 17), *MDT* for creating the 3D model and *Protopo* for altimetric contouring.

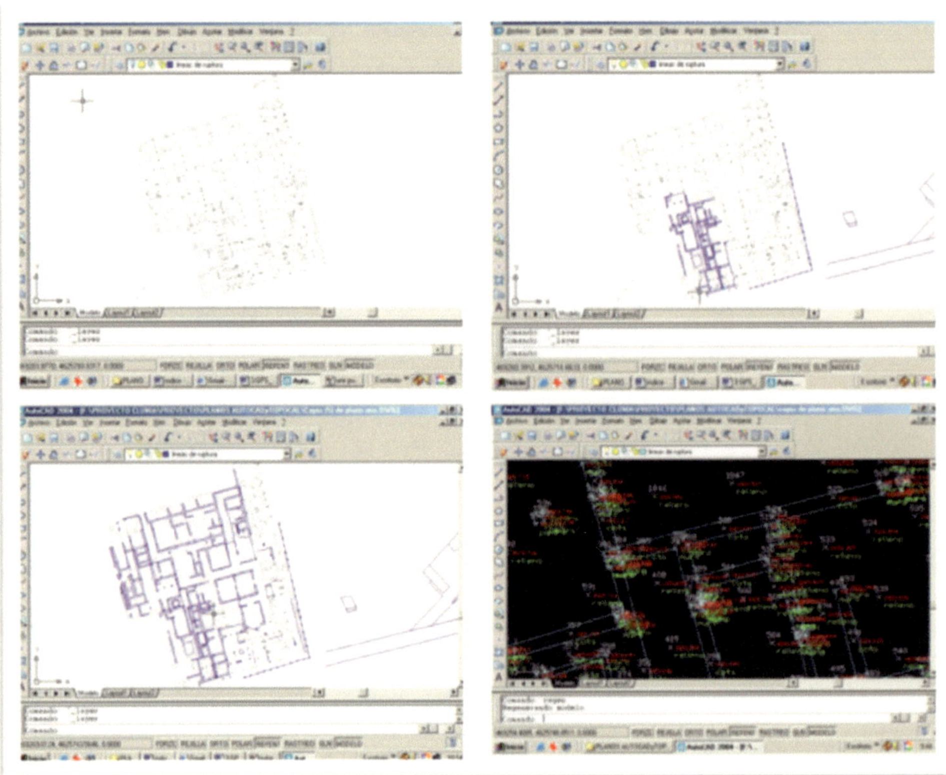

Fig. 17. Different stages of the wall editing process in the Casa Taracena.

With the help of identification points and field sketches, in the traditional method it was possible to reconstruct reality in accordance with a certain iconography represented by the symbols belonging to those generally used for the corresponding scale. What happens with laser scanners? Can illuminance or RGB values be used as internal codes for each point? Could an active parallel system be included containing a new parameter?

In this phase, cartographic editing began with assigning symbols, which consisted of obtaining a simplified and/or exaggerated representation on the appropriate scale of the elements that were not represented at the scale used but were important enough to be included. These symbols aided the reading of the plans by being as clear and intuitive as possible. How could this concept be incorporated into 3D laser scanner models? Do we require reality to be the result of being captured at a certain resolution and point density?

The symbols used were of three types (Figure 18) and referred to points, lines and surfaces. Point symbols were used to indicate elements considered to have position but no extension and included complementary (or filling-in) points and different cells (e.g. trees, waste baskets and reference network points. Linear symbols were shown as one-dimensional and were used for main contour and depression contour lines and also for the handrail. Surface symbols were used for small elements that could be represented by lines on the surface of the map and were applied in zones with lines or points with a common property. The following elements were considered to be surface elements: walls or structures defined by contour lines and textures, raised structures and mosaics, streets, roads, benches and steps, which were given two-dimensional representation on this scale.

Fig. 18. Table of symbols used in the project.

We considered different cartographic variables of the 1/200 and 1/500 scale urban maps when designing the symbols, such as colour and dimensions, in order to make them as conventional as possible.

After structural details had been entered in the model we dealt with contouring the topography, taking into account the results of the previous phases and applied modifications as required. Laser scanners carry out this process with cold information. How can errors be detected in the capture or in the information itself? Why are certain points and triangles eliminated? Are they got rid of to improve surface uniformity? Are there no parallel field notes that refer to specific areas of the model? What should our limits be? Why are we using laser scanner modelling?

4.4 Digital terrain model

We created a digital terrain model after editing the maps. This file allows contour lines to be drawn and generates a 3D model to which shading can be added to facilitate terrain interpretation and hypsometric tints to indicate height.

We used the three coordinates of all the points obtained by GPS as basic project data and could choose the points that best represented the terrain: break lines, dividing lines and depressions. The quality of these lines was fundamental to obtain an acceptable mathematical terrain model with an even distribution of points and increasing point density as necessary. The software used was *MDT*. A triangular network was generated from the original irregular point cloud (Figure 19).

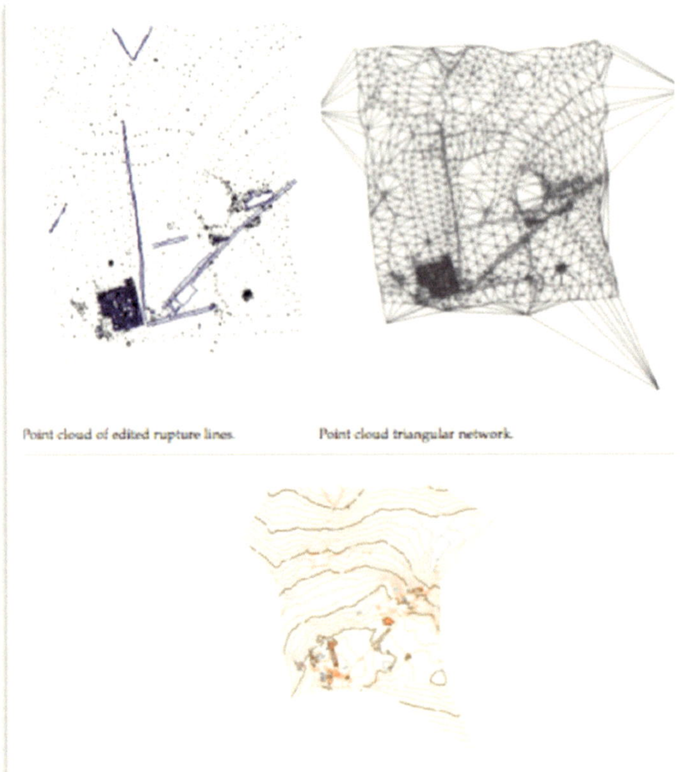

Fig. 19. Process of obtaining contour lines.

At this point there only remained to include the labelling, which was done in accordance with our field notes. We used Arial font for names and contour lines. We paid special attention to the contour lines in the area of the *Casa Taracena* in order to provide the greatest possible amount of height information in this zone.

The maps were then divided up into various sheets for easier handling. Laser scanners are often limited by the available capacity of the computers used by the research team. If file size is reduced, this is done by a method in the form of a pyramid (details are lost on a certain scale and thus also the capacity to analyse them) or by compression, which reduces the quality of the information. We could extend the traditional idea of map sheets to construct a more versatile type of digital file for the final products of modelling processes.

The design of the grid, framework and legend was dictated by the size of the sheet. In this case we opted for the A1 sheet size, with two for the 1/500 scale (Figure 20) and one for the 1/200 scale. Before approving the final product the maps were given another editing in search for any remaining errors.

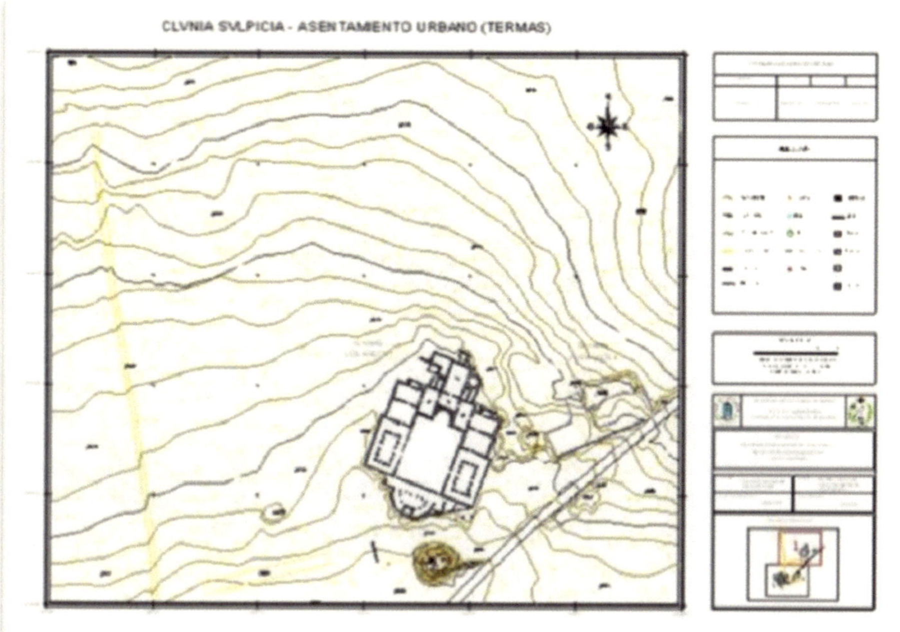

Fig. 20. 2D cartography.

When working with laser scanners, can we incorporate metric control protocols? Studies are at present being carried out on the repeatability of surface representation and modelling according to the perspective of the scans in relation to the object. In our work, we try to maintain control of the model precision and verify results by obtaining points by other topographic methods before we are satisfied with the final product.

5. Conclusions

The question that gave rise to this work sought to establish the real objectives of 3D modelling as to What? *Why? How?* and *How much?* and how these criteria should never be lost sight of in projects involving laser scanner data acquisition and processing. The enormous quantity of data acquired and the excessive amount of time required for its processing has led to a situation in which a large part of our research efforts are now involved with finding new algorithms and improving automatic processes while ignoring the risk of the arbitrary loss of field data and the fact that the final representation in terms of quality and quantity may not be all that it should be.

Faced with the challenge of putting human intervention back in control of the representation of monuments, artefacts and spaces and daring to question the power of machines, we wish to propose the *Theory of the Hidden Reality*. We are technicians and engineers and we handle points (hundreds with the traditional surveying methods and millions with the laser scanner). Following the scientific method of the traditional topographic data treatment, our aim is to open a line of analysis that leaves to one side the automated protocols of laser scanner systems and gives way to a new approach: point clouds that give a partial view of the reality we want to represent, or at least that makes us aware of how it is being represented and its limitations.

Fig. 21. Process products and stages of a laser scanner modelling project.

For too long we have considered laser scanner systems as a new technique that can convert traditional topography into an automatic process, closely related to photogrammetry, analogue representations of surfaces that make up the object we wish to reproduce. At the present time there are two issues we want to make you think about; firstly, we must question the *reality* of the object to be represented and question also the capacity of laser scanners to faithfully capture this reality, an underlying *hidden reality*, rather like a geoid of the surface of the Earth. Laser scanner technology tells us the role we have to play: collecting data, joining sweeps, cleaning up point clouds, triangulation, creating a rigid body, obtaining a realistic image (Figure 21). Specialists in this technology can take part in all these stages, considered as windows open to other products or models of representation. There is no need to follow the entire process, it is enough to analyse to what extent each of them can help us to obtain valid results.

We propose a search for new analytic methods and procedures with the data obtained, new documents for scientific interpretation that create new knowledge from the point clouds acquired in the field.

In traditional surveys, points were captured, sketches were drawn, coordinates were calculated and files of drawings were then edited. Later, digital models allowed triangulated meshes and more or less hyper-realistic 3D or flat 2D textured models to be obtained (Figure 22). However, there is no reason why we should forget one of these techniques when we are using the other.

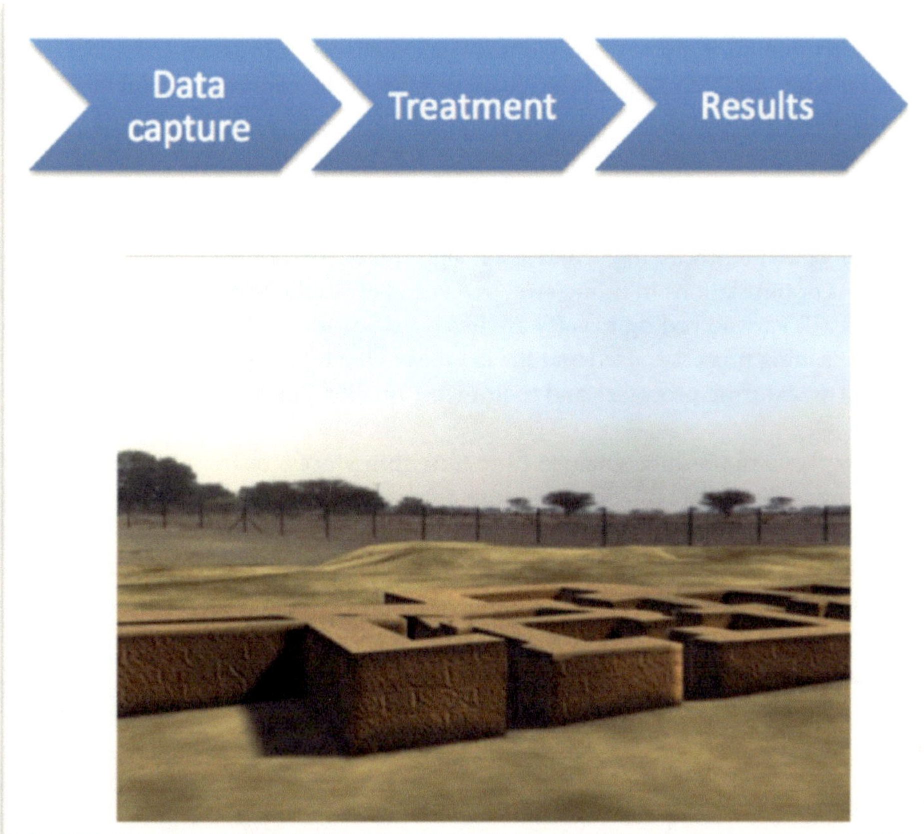

Fig. 22. Virtual representation of the Mleiha archaeological site in the Sharjah Emirate on a scale of 1/500 obtained from GPS receivers.

As specific proposals in laser scanner methodology we would like to propose the following lines of research and critical points to be taken into consideration:

- Equipment check: introduce a systems control to determine to what extent they affect all aspects of the point cloud.
- Control of variables that affect results: e.g. topographic phase measuring equipment incorporates a system of corrections for temperature and pressure that are not included in laser scanners of this type.
- Identification and control of variables that could affect the measurement process: dust in suspension, relative humidity, etc.
- Control of the modelling target zone: surface materials of the object (up to now this problem has not been quantified), identification of structure lines, points or surfaces of

interest that will not be captured in sufficient detail for work requirements (e.g. corners or edges of windows).

- Combinations with other technologies that complement the capture in accordance with the project objectives.
- Analysis of the influence of the angle of incidence on laser scanner data capture.
- Establishing limits in point cloud captures in accordance with geometries.
- Parameterised identification of the characteristics of the capture or point cloud: angles, apertures, distances, etc.
- Analysis of overlap indices between sweeps from different stations and a study of the possibility of performing additional complementary sweeps to improve fits and coverage (as in convergent sweeps for a photogrammetric model).
- Control of data processing: control of algorithms, redundancy analysis, control of rotations or translations in fitting sweeps, control of results in cleaning point clouds, etc. Results will vary according to software used and "distortion" of the model will depend on the cleaning processes used and the degree to which points are removed.
- Control of texturing processes and their metric quality. Analysis of point distribution in the model.
- Study of the final product obtained from the entire process or from intermediate stages, according to the documentation project requirements.
- Identification of the parameters that characterise a laser scanner project: in the survey (number of stations, resolution, distance, etc.), processing (programs used and parameters introduced in the different options, final products (accuracy, image resolution, etc.).

Our aim in the two projects discussed in this chapter was to obtain a complete detailed geometry of the object. Both are examples of the application of metric documentation. We would like to leave to the reader the question of what products or examples are best suited to his/her work. Not only should the model requirements be considered but also one's personal situation, including familiarity with different technologies, the budget available for the project and the training received in the processing of georeferenced digital data. New documentation systems are appearing that make digital reality possible, but what we do not know is just how far they take us away from what we really need or how well they can represent this reality.

Anna Maria Marras, in her doctoral thesis entitled *Topographia e topothèsia: dialogo diacronico su uno rurale nella campagna dell´alto Tell tunisino tra geografia e archeologia* for the Università degli Studi di Siena, introduces the term topothesia in contrast to *topography*. The latter is involved with a detailed description of a place, the former is a description of something fictional or imaginary. We may believe we can achieve absolute reality, but perhaps the laser scanner technology is only giving us a topothesia of reality.

6. Acknowledgements

This project formed part of the R+D project HAR2010-21976, "Segeda y Celtiberia: Investigación Interdisciplinar de un Territorio" financed by the Spanish Ministry of Education and Science with FEDER funds.

We are also grateful for the assistance received from PADCAM (Archaeological and Documentary Heritage of the Autonomous Community of Madrid: Systematization, Management, Evaluation and Diffusion of Local Areas within the European Framework) financed by the Education Department of the Autonomous Community of Madrid.

7. References

Alonso, M.; López Mazo, A.; Farjas, M. & Ayora, F. (2002). Levantamiento de la cúpula de la Basílica del Monasterio de San Lorenzo de El Escorial. Aplicación experimental de la estación total de lectura directa. *Topografía y Cartografía, XIX,* (May-June 2002), pp 19-33.

Bracci S., Falletti F., Matteini M., y Scopigno R. (2004). *Explorando David: diagnóstico y estado de la conservación.* Giunti Press. Italia.

Expósito García, M. (2009). *Levantamiento mediante laser escaner 3D del monasterio de San Miguel de Escalada.* Unpublished degree dissertation. Universidad Politécnica de Madrid (UPM), Spain.

Farjas, M. & Sardiña, C. (2003). Novedades Técnicas: Presentación del equipo Cyrax 2500 de Leica Geosystem". *Topografía y Cartografía, XX, Vol. 116, pp 70-71.*

Farjas, M. (2003). Las ciencias cartográficas en la arqueología: la búsqueda de la métrica en los modelos de divulgación científica. *DATUM XXI.* Year II- No 3, March 2003, pp 4-12.

Farjas, M. & BRAVO, A. (2007). Tecnologías de representación 3D en los procesos de documentación del patrimonio pétreo, In *Ciencia, Tecnología y Sociedad para una Conservación Sostenible del Patrimonio Pétreo,* Restauradores Sin Fronteras (Ed.), pp. 47-57. Madrid.

Farjas, M. (2007). *El registro en los objetos arqueológicos: Métrica y Divulgación.* Ed. Reyferr, Madrid.

Farjas, M. & García-Lázaro, F. J. (Eds.) (2008). *Modelización Tridimensional y Sistemas Láser Escáner.* La Ergástula , Madrid, Spain.

Farjas, M. et al. (2010 a). Virtual Modelling of Prehistoric Sites and Artefacts by Automatic Point-Cloud Surveys, In *Virtual Technologies for Business and Industrial Applications. Innovative and Synergistic Approaches,* Raghavendra Rao, N. (ed.). pp. 201-217. Published by Business Science Referente (an imprint of IGI Global), Hershey, USA.

Farjas, M..; García-Lázaro, F.J.; Zancajo, J.; Mostaza, T. (2010 b). Cartografía en Patrimonio: la métrica en la documentación. ¿Una realidad pendiente? In *La Documentación gráfica del Patrimonio - Presente y Futuro,* Instituto de Patrimonio Cultural de España, pp. 80-89. Ministerio de Cultura. Editado por la Secretaria General Técnica. Subdirección General de Publicaciones, información y Documentación. Retrieved from http://www.calameo.com/read/0000753358b142b1c934c

Juez Alvarez, E. & Martin Martinez, A. I. (2006). *Levantamiento del asentamiento urbano (área 3) del yacimiento arqueológico de Clunia, provincia e Burgos.* Unpublished degree dissertation. Universidad Politécnica de Madrid (UPM), Spain.

Marras, A. M. (2010). *Topographia e topothèsia: dialogo diacronico su uno rurale nella campagna dell´alto Tell tunisino tra geografia e archeologia.* Ph Thesis. Università degli Studi di Siena, Italia. Unpublished.

Prieto Cañal, R. (2008). *Levantamiento mediante Láser Escáner 3D de la fachada de la Iglesia de Santa Teresa, Ávila.* Unpublished degree dissertation. Universidad Politécnica de Madrid (UPM), Spain.

Lerma García, J.L. & Biosca Tarongers, J.M. (2008).*Teoría y práctica del Escaneado Láser Terrestre.* Retrieved from http://www.heritagedocumentation.org/3Driskmapping/Tutorials/Leonardo_Tutorial_Final_vers5_SPANISH.pdf

Zazo, A.; Jimenez, D.; Farjas, M. (2011). *Guía Visual Real Works.* La Ergastula. Madrid. Spain.

9

The Workflows of 3D Digitizing Heritage Monuments

Hung-Ming Cheng
China University of Technology
Taiwan

1. Introduction

Digitizing heritage monuments is a process of spatial data acquisition, geometry modelling, digital archiving and web-based representation. Moreover, there are several survey and digitizing techniques working and developing such as traditional manual methods, topographic methods, photogrammetric methods and scanning methods. The most popular on application and developing is 3D laser scanning technology right now. Especially for outdoor space and historical building, the laser scanner is an appropriate tool for the whole process of 3d digitizing. Therefore in this study, the 3D digitizing heritage monuments take laser scanner as for developing an integrated procedure to digital recording and archiving of cultural heritages.

The whole 3D scanning process involves the three-dimensional digitization, digital data processing, archiving and management, representation and reproduction (Varady et al., 1997). The scanning action of acquiring, survey and recording spatial data for the determination of actual position and existing form, shape and size of a monument in 3D space at a particular given moment in time.

3D digitizing technologies apply in several fields, including manufacturing industry, medical sciences, entertainment industry and, cultural heritage (Addison & Alonzo, 2006; Berndt & Carlos, 2000; Levoy, 1999). In application of manufacturing industry, inspection probes and survey are substituted by non-contact laser scanning equipment that are often used for aero industry and automotive parts design and testing (Willis et al., 2007). And more computer aided design and reverse engineering are employed in a wide range of applications in the field of science and industry, together with animation techniques and WEB application (Fontana et al., 2002; Li et al., 2007; Pieraccini, 2001).

Cultural heritage transmits conventional environment and craftsman's experience in the civilization and cultural progress which needs to consider in specific way. (Arayici, 2007; Shih et al., 2007; Yilmaz et al., 2008) Owing to the longer process of confirming and the limitation of budget and timing, the preservation of historical architecture is not efficiently executed preservation. Moreover, natural disaster, such as earthquake, fire and accident collapse, caused historical architecture disappeared in one moment. Therefore, integrating digitizing techniques could help building preserved works at such limited opportunity.

Traditional methods take time-consuming and present a number of evident limitations (Yilmaz et al., 2008). The way of survey and measuring of historical architecture are using photographing and manual tape method, then manually transferring these discrete numbers into engineering drawings by AutoCAD (Fischer & Manor, 1999; Yilmaz et al., 2008). Although 2-dimensional drawings and graphics provide traditional support documents to rebuild historical architecture. They are not complete solutions to detailing spatial information without truly 3-dimensional model. Cultural heritage needs more advanced techniques to support preservation and conversation in cultural heritage (Arayici, 2007).

Digitizing historical cultural architecture is a trend on international preservation. Traditional tools using survey and probing in manuals couldn't correspond to preservation procedure in effective and efficient (Fontana et al., 2002; Li et al., 2007). It is necessary to consider an integrating mechanism with tools and procedures. Laser scanner is using the reflection and projection of laser beam and probing the difference of time. As the objects scanned by laser beam, the scanner machine calculates the distant of machine and objects. And by calibrating the ejection angle of laser beam, the laser scanner records the spatial data as data of point clouds. This study is employed 3D laser scanner with the high quality digital camera that can reach more than 1000 meters. And the inaccuracy of survey result is less than 5mm. For the thumb of digitizing, acquiring surface data of building are the most important of heritage preservation (Monga & Benayoun, 1995).

For the main purpose of digitizing, the 3D spatial data of cultural heritage needs extracted for 3d document application. Hence, digitizing techniques are considerably procedure in systematic and reasonable than those isolating in machine body or limited to people training. The whole workflow starts an investigation of survey technology and reverse engineering which is more mature field to utilize laser scanning technology. After comparing the techniques of laser scanner and photogrammetry application, laser scanner was chosen for more accurate, efficient, faster and reliable than other documentation techniques.

2. Survey process

Cultural heritage protection is an important issue in the world. The sustainable concept in public awareness and these kinds of monuments constitute an important part of our past. Digital 3d documentation presents a process of restore and protection which integrate the survey technique and building computing between reality and virtual world.

Heritage preservation is a continuous process with many data needs to integrate, acquire and analysis, which means a lot of construction data to be recorded. There are a serial of workflow and operation that include engineering surveying, drafting and design, monitoring for post-construction analysis (Berndt & Carlos, 2000; Shih & Wang, 2004). Recorded data are made by discrete manual 2D drawings have limitations in describing the allocation of geometries in 3D space. 3D data are intuitive, visualizing and continuous representation to simulate space characteristics which will be the best way to manipulate with architectural and historical senses. As such, we choose the 3D laser scanner of survey manner to reach our objective directly.

2.1 Laser scanner and operation system

For reconstructing a 3D environment and architecture, Riegl LMS-Z420i with LiDAR system are chosen, which offers the most powerful in distant effective to capture the spatial data within point clouds format. The machine scans the surface of object to receive spatial data about 11,000 points per second with high precision up to 4 mm. The field of view is 80 x 360 and the range is up to 1000m.

For each scanning mission, scanner machine needs portable computer with bundled operation software package "RiSCAN PRO" (Fig. 1.) to operate the scanner for acquiring high-quality 3D data in the field (Goldberg, 2001) [10]. By mechanically scanning the surface of object with the laser set, different sections of the object are sequentially acquired and cloud data are therefore generated. For the specific parts and detailed objects, scanner machine combines a concentrated shot function on these specific details with condense scanning of laser points.

The laser scanner can take advantage in the territorial field with reference points to connect difference station data. However, Target site have some obstacle views to disable cross scanning. The scanning plan therefore adopts higher view point to solve these limitations.

Fig. 1. Riegl LMS-Z420i & software RiSCAN PRO

2.2 Survey planning and scan works

For acquiring the 3D data of real environments, an integrate scanning work has to planning for the whole archaeological objects from several different scanner stations which can combine a set of order of multiple view (scanning worlds). The major steps in combining multiple views is survey planning and accurate registration that related to information management in multiple views to be combined successfully. Such set of information is organized as token ring or star topology, and captured by multiple views in serial survey with registration targets. (Fig. 2.)

Fig. 2. Survey Planning for Scanner Station

For development of 3D scanner supports 360 degrees field-of-view, the scanner has to move around archaeological objects to complete exterior scan views with registration targets. The laser scanner is difficult to work at those buildings placed closely without appropriate station-locations on scanning and registration entire surroundings. Therefore, several higher locations are selected nearby those buildings. Moreover,, higher scanning locations are used to avoid obstructions problems and reduce numbers of scanning station. (Fig. 3.). These obstructions accessing stations are manipulated by special devices which can raise 3D scanner machine for survey higher and wider. Mobile vehicle carries lift-able device and work with more function controlled.

Fig. 3. The device raising 3D scanner and working on vehicle

For those objects to be scanned in a well-controlled environment (for example: indoor space and no obstruction problem), objects can be oriented scan completely without extra registration targets. However, ranged scanning is often incomplete for obstructions or lack of controlled-high-view scans at outdoor site. There is no way to solve such as problems until a lift-able car and movable support are made up.

The device provides a carrier of 3D laser scanner with lift-able and stable character. It is not only fixing on the ground but also avoid to vibration. Furthermore, it can be controlled in height by expert experience for the best survey. The device with platform on the top can fixed scanner on the lift-pole. The each of multi-supported legs implements dynamic device for lift-up the truck that is for stable survey implementation without shaking by rubber tires. The lift-able device includes scissors-like instrument and platform, which includes a space for 3d laser scanner instrument and multi-action kinematics extending the range of survey and reach proper height for advantage process. Therefore, the creative work includes lift-able and stable character that is not only stability of scanner and also convenient adjusting height for the best survey efficiently. (Fig. 4.)

Fig. 4. lift-able car and movable support

2.3 3D data process

The modelling operation starts to capture the feature of object. These features (for example, points, lines or components) will be extracted from images, enhancing the spatial characteristics and representing the coherence of geometry with mesh or solid of computer graphics algorithm (Datta, 2001; Li et al., 2007). The proposed feature extractors are adapted to solution software that includes CATIA and Rapidform XOR2 (Fig. 5). These softwares are based on the similarity of directional feature vector. For the 3D modelling (shape, surface and volume), there are some other options so called middleware of Reverse Engineering in order to develop Rapid Prototyping in design application (Fontana et al., 2002; Marschallinger, 1998; Monga, & Benayoun, 1995) .

3. The procedure of digitalizing archiving

In digital processing of heritage monuments, inspection probes and survey probably met the difficult operation for physical environments. For application of digitizing archiving, 3D spatial information is transfer to different data formats as fundamental preservation.

Fig. 5. Software Rapidform XOR2 processes the point cloud as mesh modeling

3.1 Workflow of 3D scanning

For the work flows of 3D data acquisition of cultural heritage, the quantity and density of point cloud control the detail of 3d modelling by reverse engineering. In the registration of laser scanning, both software and hardware influence the representation and accurate of point cloud. Therefore, a procedure of laser scanning for different size and distant of heritage's objects is developed. In 3d data acquisition of laser scanning, not only specific machine on manual to acquire 3D data is process but also different machine and software to registration point cloud are integrated. For the digital archiving of cultural heritage, these precious 3d data can provide an original digital format for reconstructing physical heritage in future.

On the post operation phase, 3D modelling depends on the software of reverse engineering which is a from the survey technology for production. The software of reverse engineering (Rapidform XOR2) are manipulated for modelling the digital building which is a key process of industry utilization in physical form design. The whole processes are through surface analysis, holes filling, and rebuild the mesh models. The purpose of 3D modelling is building up full scale physical heritage in digital form that can apply to visualization and virtual reality on the Internet.

For the web representation, proprietary VR authoring tool is introduced to demonstrate the reconstructed results. The comprehensive virtual scene is then laid out in VR tool that is the web-base browser for navigating in 3D virtual world. For more application of VR, the data format of 3D point file is transformed to 3D boundary curve, parametric surface or replaced by adjusting points in original representation. In digital manipulation, many 3D forms and shapes are merged into virtual reality for representing new visual experience and immersive digital environment. As a result, rich illustrations (multimedia) are carried out through interactive manipulation in the Web. (Fig. 6).

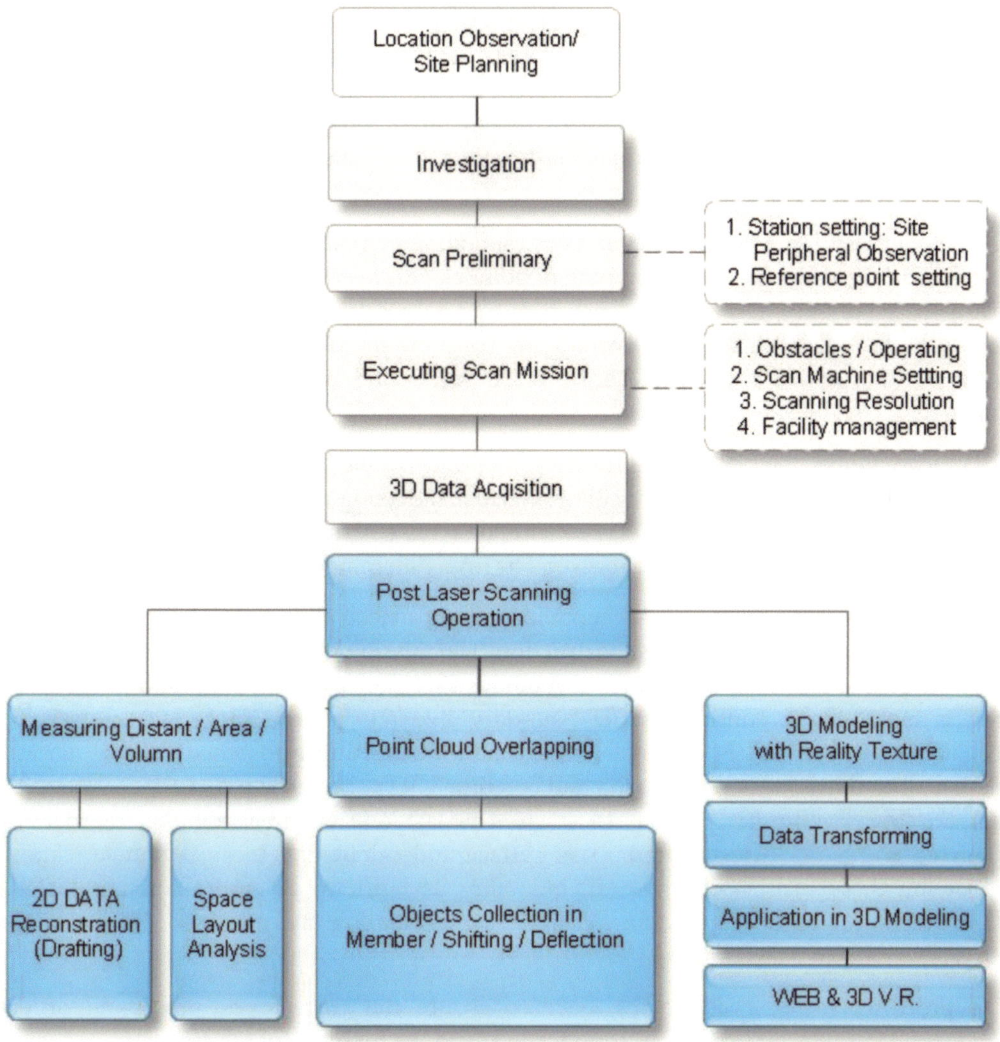

Fig. 6. Workflow of Digitizing

3.2 Determining 3D data

Determining 3D data is challenges work for digitizing cultural heritage whose purpose is to generate accurate model for more applications. 3D modelling of laser scanning is using the techniques of reverse engineering for processing points cloud overlapping (raw data), creating parametric surface and image texture mapping on objects. The main works are addressed on data transformation and 3D geometry rebuilt which are through automatic registration of points cloud, shape and form fitting and digital modelling.

The point clouds are generated in local station centred coordinate systems whose limited field of views are captured from different viewpoints in order to obtain a full coverage of object surface. For reconstructing 3D model of object surface, these data of points cloud and images need to be registered in a single global coordinate system. For registration mission, point clouds from several different views have to align with target marks (at least 3 marks between 2 points data set merge) in iterative registration. (Fig.2.) Registration of image also determines the camera space parameters bringing one data set into alignment with the other.

On the other hand, the normalizing and corresponding on points cloud is another challenge. For registration assignment, overlapping of point cloud are critical processes for application of the laser scanning techniques which could be solved by increasing efficient ability of computer. However, point cloud overlapping consume more time on the normalizing and corresponding points cloud by filtering noise points and reducing points of some part manually.

For practice works, the operation software packages such like "RiSCAN PRO" are chosen for automatic registration of overlapping 3D point clouds. Dealing with post operational software, overlapping point cloud registration is easier to determine correspondences between different data sets representing the same free form objects from different scanning world.

Moreover, the other issues of 3D modelling have object surface modelling, 3D object recognition and texture mapping construction. The mission of surface reconstruction from scanned data and images are proposed methods as 4 stages (Fischer & Manor, 1999). The first is extracting a continuous 3D boundary line (curve) by image post-processing techniques. The second is parameterizing a non-intersecting grid based on a numerical method. The third one is recognizing and creating a 3D parametric base surface whose 3D boundary was computed by the 1st stage. The last one is projecting the internal 3D digitized points on the parametric base surface and calculating mesh parameterization. Following these processes, the mesh modelling is built up by approximating the 3D digitized points as a B-Spline surface while interpolating the extracted 3D boundary curve.

The processes of mesh modelling have to ensure the correct post-processing of the reconstructed mesh surface in CAD applications and the topology of the reconstructed mesh surface has to be correct. The mesh modelling method fall into three categories are current approaches in reversed engineering of point cloud reconstructed model. The methods for reconstructing triangle mesh surface from normalizing point cloud. Starting with a seed triangle, the 3D modelling builds a partially reconstructed triangle mesh by selecting a new point based on an intrinsic position of the point cloud.

The reconstructed mesh is essentially base for 3D modelling application, two-dimensional drafting and 3D web observation. For the practice assignment, package software "Rapidform XOR2" is adopted as tool to operate the 3D modelling missions. The process of 3D modelling is dealing with filtering noisy, sampling and smoothing, 3d modelling analyze through mesh build up, surface analysis with region group and auto segment, mesh sketch with CAD manipulating, 2D boundary drafting, 3D editing tools and file export. (Table 1.)

Stages	Items	Illustrations	Description
File import	representation of points cloud	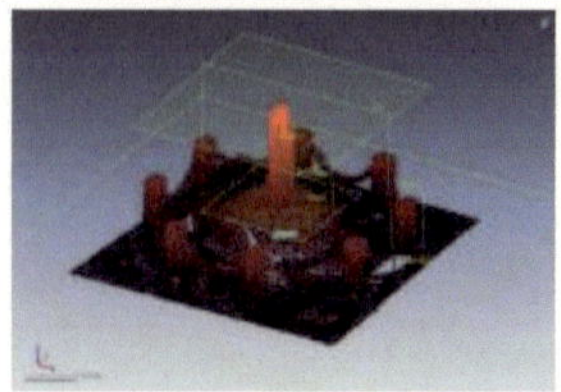	1) Import .dxf file . 2) Units choosing, model scaling. 3) zoom, viewpoint, observation type
Editting	Filtering noise points		Tools for points cloud: Filter noisy, sampling, smooth, filling hole...
Surface	Surface forming and analysis	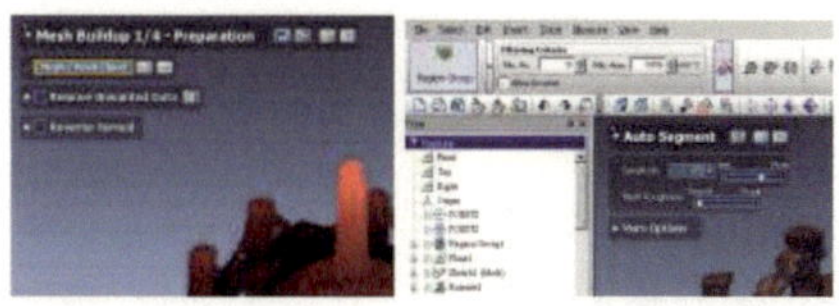	1) Transforming surface thru. Mesh buildup. 2) Analysis tools: region group/ auto segment...
3D modelling	Surface manipulating/ 2D boundary drafting		1) Buildup 3D boundary thru. Mesh sketch. 2) Based on the points cloud, sketch 2D boundary.
	3D editing tools	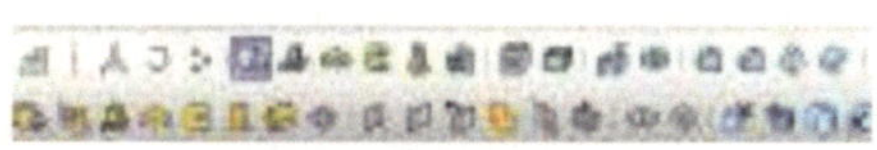	1) generating geometry by automatic recognizing 2) 3D tools: extrude, loft…
File output	Export 3D data file		Thru. File/Export export file format "IGES".

Table 1. 3D modelling for "Rapidform XOR2 "

3.3 Phase works of 3D scanning

In this project, 3D laser scanner techniques are used to rebuild the objects of heritage, which includes an upright stone tablet, pavilion, bridge and historical building. (Table 2.) According to past experiments, this integrated approach is efficient and accurate. LiDAR (Laser Radar) technology could precisely digitize fine details and sculpted surfaces which are essential for modelling monuments and historical buildings. In the procedures of texture mapping, the raw data are processing from real picture and some statistics numbers of stations into images of point clouds, images of 3d modelling. Furthermore, 3d data representations can present several types of heritages which are interior, garden, monuments, building and bridge.

Interior of Fort San Domingo, Danshui			
Real picture	Image of point clouds	Image of 3d modeling	Stations and point clouds (statistics)
			Stations: 8 40 Million Points
Lin Family Gardens			
Real picture	Image of point clouds	Image of 3d modeling	Stations and point clouds (statistics)
			Stations: 9 17 Million Points
Tomb of Dr. Mackay			
Real picture	Image of point clouds	Image of 3d modeling	Stations and point clouds (statistics)
			Stations: 4 2 Million Points
Shanchia Station			
Real picture	Image of point clouds	Image of 3d modeling	Stations and point clouds (statistics)
			Stations: 18 168 Million Points
Shanhsia Arch Bridge			
Real picture	Image of point clouds	Image of 3d modeling	Stations and point clouds (statistics)
		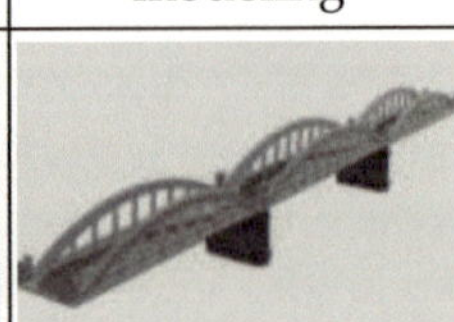	Stations: 7 12 Million Points

Table 2. Five Types of digitizing cultural heritages

4. Application in scanning technology

Digital archiving has been applied on several purposes in heritage preservation recently. There are two issues for processing of digitizing project. The first is monitoring of cultural which is following currently reconstruction work. The second is trying to find out the data characters for decision supported making.

4.1 Documentation and monitoring of cultural heritage

Virtual preservation takes advantage of the expansion and long-term saving. Cultural heritage needs these data for sustainable monitoring. In particular, the heritages in real environment are irreversibly damaged by environmental disaster or atmospheric damages. Those damages sometime were discovered too late. High accuracy 3D scanning, at regular times, could detect deformations and cracks. These data of monitoring are the fundamental knowledge for reconstructing heritage.

The 3D documentation archives the spatial data of cultural heritage. Documentation should be considered as an integral part of a greater action in general documentation of the cultural heritage. Those data includes text, picture, music, and more other format media. These multi-format media with historical documentation, architectural documentation store a whole picture of cultural heritage.

On the other hand, cultural heritage protection is a key issue around the world today. Those issues evoke public awareness over recent years which address some important monuments of our past. The documentation and display of ancient artefacts and antiquities is an essential task in the field of cultural heritage preservation (Yilmaz et al., 2008). Digital archive of high quality three-dimensional models would give improvement in the restoration science field. Digital archives thus can be used as reference for monitoring and restoration of cultural heritages (Pieraccini et al., 2001). (Fig. 7.)

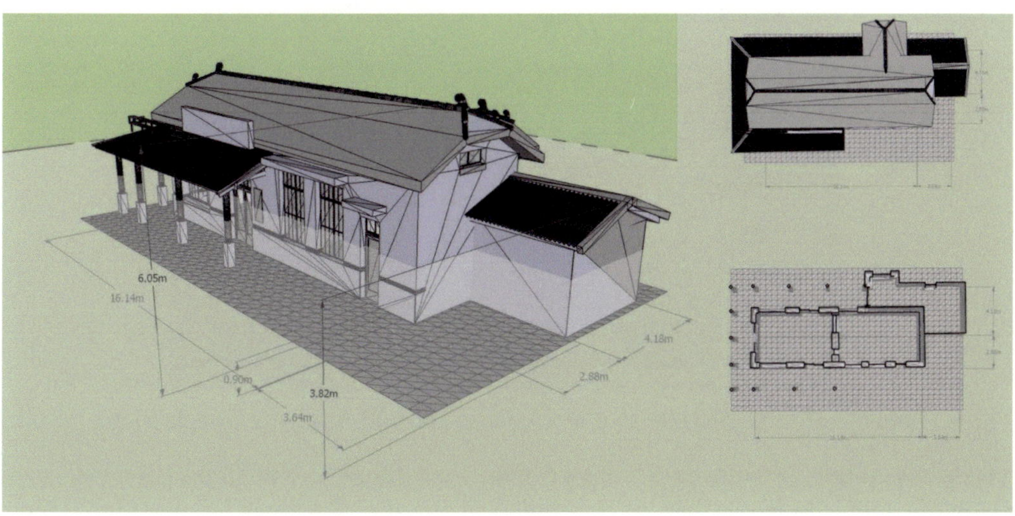

Fig. 7. 3D Documentation (Jing-Tong Train Station)

4.2 Data management and multimedia

Multimedia (such as 3D scenes navigation, animation) has developed for two purposes. The first offers in showing the sequence of possibilities image sequences, where each image represents an option based on a set of parameters. As architects are interested in specific image, that shows the closest match between model and the virtual measurement. The 3D navigation also reveals an insight of spatial design thinking that cannot possibly be revealed through flat and analogue representation.

Fig. 8. Photo Picture and Texts Present in Webpage and GIS Applying Google Map

This project produces heritages in Taipei County establish archive of all the survey material combined with the work, that are integrated in the digital format for next application. Heritage materials are combined with historical document, physical situation and contemporary observation. Digitally managing materials have been essential programs that need to organize comprehensive representation in terms of the web environment to display accurate information (Fig. 8.). On the other hand, text, image and drawings are importance in design studio for representation issues. Consulted these data through the digital archives, the design can continuously work. Multimedia has been used to assemble these data for specific purposes including as-build environment representing or VR tour. These 3D digital data help design teamwork for next design decision making.

5. Discussion

Preserving and representing cultural heritage motivates the new technology for producing complex and heterogeneous data. For managing these data, digitalizing process is an essential task for the use and the diffusion of the information gathered on the field. These issues are discussed in terms of the concepts of digital archiving, 3D documentation and WEB application for the emergence of digitalization in recent years.

5.1 Digital archiving in cultural heritage

The jobs of digital archiving include several phases of the preserving, organizing and retrieving on the cultural heritage. 3D laser scanning technology evolves in the most diverse fields, an increasing number of cultural institutions take into consideration the need to capture 3D datasets of heritage assets. However, the digitizing procedures with 3d laser scanning are very heterogeneous and complex, including not only the economic management and the logistic activities which take place in the computer labs, but also on-stage artistic production and craft-made activities in workshops.

The whole process involves the three-dimensional digitization, digital data processing and storage, archival and management, representation and reproduction. The preservation methods for three-dimensional digitization are briefly reviewed that are applicable to cultural heritage recording.

5.2 Survey technologies and other applications

With recent developments in computer and information technologies, this well-known traditional method has been replaced with digital survey and Internet technology.

These new methods offer us new opportunities such as automatic orientation and measurement procedures, generation of 3D point-clouds data, digital surface modelling and WEB representation in cultural heritage. These methods and equipments commonly used for digitalizing buildings are: traditional manual surveying methods, photogrammetric methods, scanning methods and WEB database application. These methods determine the main data representation, the instruments limitation and the key points of each cultural heritage.

On the other hand, 3D laser scanning is an essential technology for territory survey that has become increasingly popular for 3D documentation. These techniques provide very dense 3D points on an object surface with high accuracy. In addition, the 3D model within digital image can be easily generated using generated 3D point cloud and recorded as vector measured drawing data. These large data must be processed by proper procedures for solving the work flow in order to connect human-machine interface and operation software.

Current digital archiving has been carried out for the digital preservation and treatment of Cultural Heritage information. The development of computerized data management systems to store and make use of archaeological datasets is a significant task. For such application in the Internet, 3D WEB representation is a broadcasting platform for highlighting 3D spatial information browsing which should be processed on the phases of captured, structured, and retrieved in order to transform multimedia performances in cultural heritage for other application. The whole digital system need to compliant to every kind of Cultural Heritage site and allows management of heterogeneous data.

6. Conclusion

The laser scanning workflow develops three phases to initialize digitizing works. The first phase is 3D data acquisition which using the 3d laser scanner to rebuild the surfaces of environment which are several stations of view to registration. The second phase is 3D modelling which is using the reverse engineering software to process those raw data (points of cloud). This phase mainly is data transformation and 3D geometry rebuild. (3) Web representation and others application: We procedure digitizing process to integrate 3D data and others media format (i.e. text, picture…) for navigating in World Wide Web. These 3D data are integrating into 2D graphic drawing and specific derails to present rich culture heritage.

Concluding current works, roadmap of digital archiving is proposed as 3 possible directions: The first direction considers Web visualization is an essential communication platform to represent 3D data in the world. These data are easy access and retrieve by Internet technology which also are unlimited time and place. The second direction is 3D laser scanners become an essential application from industry engineering to cultural heritage. The tools and theory are contributed to the heritage preservation and conservation. Furthermore, strategic applications on heritage preservation are developing for cultural creation industry. For the last direction is digital archiving, these data are formed a kind of "3D documentation of digital heritage". In the wide definition of conservation and preservation, 3D laser scanners grab the most formal appearance of heritage which includes more than primitives of geometry, shape, colour and texture.

In the future works, the scanning support tools of lift-able car and movable already integrate to the workflow of 3D laser scanning on heritage and historical architecture which is the best way to overcome visual (scanning accessing) obstacles. These support mechanisms in scanning workflow help digitizing physical environment successfully. In the process of digital archiving, managing and applicable technologies extend 3D data information system for digital museum in heritage preservation. Therefore, the spatial digital data with 3D character of culture heritage become virtual heritage.

7. Acknowledgment

This study is sponsored by the National Science Council for 2007-2008 National Digital Archive Program (NSC-96-2422-H-163-001, NSC-97-2631-H-163-001) and 2010 (NSC-99-2632-H-163-001), Taiwan.

8. References

Addison, A.C. and Alonzo, C. (2006). The Vanishing Virtual: Safeguarding Heritage's Endangered Digital Record, in T. Kvan and Y. Kalay, (eds), *Proceedings of New Heritage: Beyond Verisimilitude,* pp.36–48, University of Hong Kong,

Arayici, Y. (2007). An approach for real world data modelling with the 3D terrestrial laser scanner for built environment, *Automation in Construction,* Vol.16, No.6, pp.816-829,

Berndt, E. and Carlos, J. (2000). Cultural heritage in the mature era of computer graphics, *IEEE Computer Graphics and Applications,* Vol.20, No.1, pp.36–37,

Bhatti, A. Nahavandi, S. and Frayman, Y. (2007). 3D depth estimation for visual inspection using wavelet transform modulus maxima, *Computers and Electrical Engineering,* Vol.33, No.1, pp.48-57,

Bosche, F. and Haas, C. (2008). Automated retrieval of 3D CAD model objects in construction range images, *Automation in Construction,* Vol.17, No.4, pp.499-512,

Datta, S. (2001). Digital reconstructions and the geometry of temple fragments, *The Proceedings of the 2007 international conference on digital applications in cultural heritage,* pp.443-452, National center for research and preservation of cultural properties, Tainan, Taiwan,

Dorai, C., Weng, J., Jain, A. K. and Mercer, C. (1998). Registration and integration of multiple object views for 3D model construction, *IEEE Transactions on Pattern analysis and Machine Intelligence,* Vol.20, No.1, pp.83-89,

Fischer, A. and Manor, A. (1999). Utilizing image Processing Techniques for 3D Reconstruction of Laser-Scanned Data, *CIRP Annals - Manufacturing Technology,* Vol.48, No.1, pp.99-102,

Fontana, R. Greco, M. Materazzi, M. Pampaloni, E. and Pezzati, L. et. al. (2002). Three-dimensional modelling of statues: the Minerva of Arezzo, *Journal of Cultural Heritage,* Vol.3, No.4, pp.325-331,

Goldberg, H.E. (2001). Scan Your Would with 3D Lasers, *CADALYST Magazine,* pp.20-28

Levoy, M.A. (1999). The Digital Michelangelo Project, *Computer Graphics Forum,* Vol.18, No.3, xiii-xvi(4)

Li, J. Guo, Y. Zhu, J. Lin, X. Xin, Y. Duan, K. and Tang, Q. (2007). Large depth-of-view portable three-dimensional laser scanner and its segmental calibration for robot vision, *Optics and Lasers in Engineering,* Vol.45, No.11, pp.1077-1087,

Marschallinger, R. (1998). A method for three-dimensional reconstruction of macroscopic features in geological materials, *Computers & Geosciences,* Vol.24, No.9, pp.875-883,

Monga, O. and Benayoun, S. (1995). Using Partial Derivatives of 3D Images to Extract Typical Surface Features, *Computer Vision and Image Understanding,* Vol.61, No.2, pp.171-189,

Pieraccini, M. Guidi, G. and Atzeni, C. (2001). 3D digitizing of cultural heritage, *Journal of Cultural Heritage,* Vol.2, No.1, pp.63–70,

Shih, N.J, Wang H.J., Lin, C.Y. & Liau, C.Y. (2007). 3D scan for the digital preservation of a historical temple in Taiwan, *Advances in engineering software,* Vol.38, No.7, pp.501-512,

Shih, N.J., and Wang, P.H. (2004). Point-cloud-based comparison between construction schedule and as-built progress - a long-range 3D laser scanner's approach, *Journal of Architectural Engineering,* Vol.10, No.3, pp.98-102

Varady T., Martin R. & Cox J. (1997) Reverse Engineering of Geometric Models -- an Introduction, *Computer-Aided Design,* Vol. 29, No. 4, pp.255-268,

Willis, A. Speicher J. & Cooper D.B. (2007). Rapid prototyping 3D objects from scanned measurement data, *Image and Vision Computing,* Vol.25, No.7, pp.1174-1184

Yilmaz, H.M., Yakar M. & Yildiz, F. (2008). Documentation of historical caravansaries by digital close range photogrammetry, *Automation in Construction,* Vol.17, No.4, pp.489-498

10

Multiple Hypothesis Tracking Implementation

Angelos Amditis[1], George Thomaidis[1], Pantelis Maroudis[2],
Panagiotis Lytrivis[1] and Giannis Karaseitanidis[1]
[1]*Institute of Communications and Computer Systems,*
[2]*Institut Supérieur de l' Aéronautique et de l' Espace,*
[1]*Greece*
[2]*France*

1. Multiple hypothesis tracking

Laserscanner sensors deliver distance measurements from the reflections of objects in the environment. In most automotive applications, the output of the sensor should consist of a list of detected objects. A common architecture for object detection in a laserscanner processing system is the following:

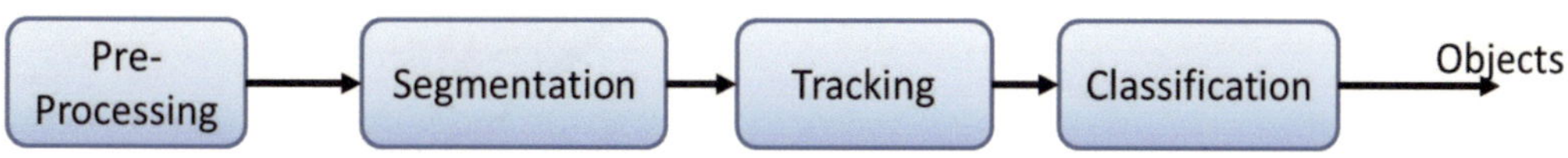

Fig. 1. Laserscanner measurements processing architecture

Tracking algorithm takes as input a list of clustered laserscanner measurements (objects), which for simplicity will be called measurements. The tracking algorithm forms a track whenever there is "enough" evidence that a sequence of measurements represents a real target. Additionally, by using the appropriate filtering techniques the tracking algorithm estimates the kinematic state of the formed track.

A basic step of a tracking algorithm is the measurement to track association. This is a very important procedure since wrong association could mean updating a track with a wrong measurement, initialization of a false track or deletion of a real track if it has erroneously not been associated with a measurement for one or more scans.

When a new set of measurements is outputted by the sensor, each measurement can i) be assigned to existing track, or ii) initiate a new track or iii) be considered as a false alarm. The simplest and most widely spread approach for measurement to track and track to track association, is the global nearest neighbour algorithm (GNN) (Blackman 1999). This approach formulates the most likely track to measurement and new track hypotheses. In the Joint Probabilistic Data Association (JPDA) algorithm (Fortmann T. 1983), multiple track to measurement hypotheses are generated. Hypotheses probabilities are calculated and then assignment hypotheses for each track are merged. With this method, track state is updated using all the measurements that are within the track gate by using a weighted sum of each hypothesis track estimate. These two methods form one track estimate for each track hypothesis. So, GNN and JPDA methods make a "hard" decision on the current scan in cases of conflicting measurement to track association hypotheses.

On the other hand, Multiple Hypotheses Tracking (MHT) methods form alternative association hypotheses in case of observation to track conflict situations. The set of hypotheses is propagated in the next scans with the anticipation that future observations will resolve assignment ambiguities. This method is divided into two approaches, the hypothesis oriented MHT (HOMHT) and the track oriented MHT (TOMHT).

In order to demonstrate the difference between these tracking techniques a practical example will be shown. Assuming there are 2 tracks on scan N and 3 measurements are received in the same scan. The gate formation is illustrated in Fig. 2.

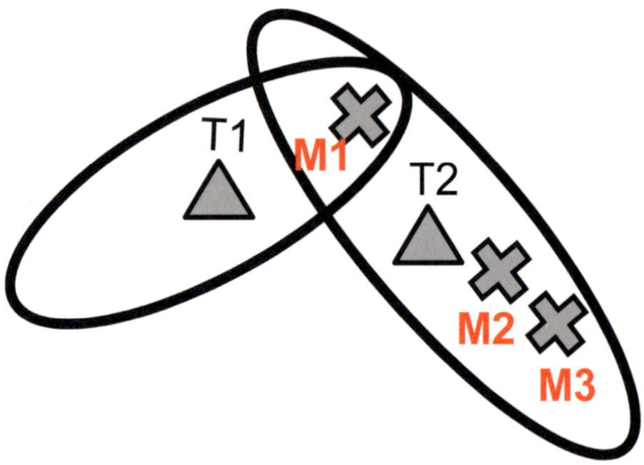

Fig. 2. Gating procedure, tracks are drawn with triangles and measurements are drawn with crosses.

If d_{ij} is the statistical distance between track i and measurement j, and this distance is less than a threshold value (gate size), this pair is candidate for association. Gating is necessary for elimination of unlikely observation to track pairs. Usually a cost function is used for defining the cost of assigning track i to measurement j. The GNN algorithm would use an optimization algorithm for solving this problem, so by using this method T2 would be updated using M2, and T1 would be updated using M1. M3 would initiate a new track. On the other hand, by using JPDA, T2 would be updated by using all measurements and T1 would be updated using M1. The MHT will form different hypotheses by taking into account all the possible sources of a measurement: i) new track, ii) false alarm and iii) existing track.

This chapter aims into giving the reader an overview of the Multiple Hypothesis Tracking approach. This technique is a complicated approach that has a strong mathematical formulation and many variations regarding proposed implementations. This chapter aims to give the reader an overview of the Multiple Hypotheses Tracking philosophy along with some practical examples regarding techniques used in MHT. Reader is encouraged to use this chapter as a starting point of getting familiar with this technique and investigate further with other more focused publications on this topic. Since this chapter aims in making the reader familiar with MHT, complicated mathematical expressions were avoided

2. Hypothesis oriented MHT

Hypothesis oriented MHT presents an exhaustive method of enumerating all possible assignment track to measurement combinations. When a new measurement set is received, observations that fall within a track's validation region set a possible measurement to track assignment.

The idea of propagating multiple assignment hypotheses was first presented in (Singer et. al.,1974), however Reid is considered as the first to present a systematic approach using a hypothesis oriented approach for implementing a multiple hypothesis tracker (Reid 1979). In this approach, when a new set of measurements is received, an existing hypothesis is expanded to a set of new hypotheses by considering all the possible assignments of the tracks contained in the original hypothesis. Each hypothesis contains a set of compatible observation to track assignments, leading to an exhaustive approach of enumerating all the possible assignment combinations

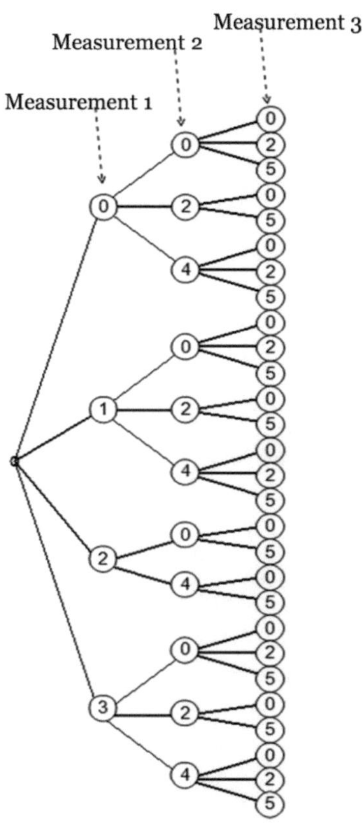

Fig. 3. Tree representation of the formed hypotheses

Taking as example the assignment problem in Fig. 2, the formed hypotheses can be represented in tree (Fig.3) or table representation (Fig. 4). The root of the tree is the original hypothesis - in the current example the original hypothesis is that there are two tracks. The first expansion of the hypothesis tree is done by using all the possible assignments of the first measurement. The numbers in the nodes indicate to which track the measurement is assigned. Regarding the origin of measurement 1, 4 hypotheses can be made: i) false alarm (assignment to the zero track), ii) new track (assignment to track number 4) , iii) and iv) assignment to existing tracks 1 and 2. In the same way the rest of the tree is created. It can be seen that the tree depth is equal to the number of measurements in the current scan. From an implementation perspective, it is more practical to represent the set of hypotheses in a matrix form. After taking into account the first measurement, the hypotheses matrix is a 4x1 matrix. Some lines of the hypothesis matrix are marked in red. These are hypotheses that don't contain compatible assignments (a track is assigned to more than one measurement), thus these lines (branches) of the hypotheses matrix (tree) have to be deleted.

0	0	0
1	0	0
2	0	0
3	0	0
0	2	0
1	2	0
2	2	0
3	2	0
0	4	0
1	4	0
2	4	0
3	4	0
0	0	2
1	0	2
2	0	2
3	0	2
0	2	2
1	2	2
2	2	2
3	2	2
0	4	2
1	4	2
2	4	2
3	4	2

0	0	5
1	0	5
2	0	5
3	0	5
0	2	5
1	2	5
2	2	5
3	2	5
0	4	5
1	4	5
2	4	5
3	4	5

Fig. 4. Matrix representation of the formed hypotheses

It has to be mentioned that according to the association that has been considered as correct, the state estimation of the a track is not the same in different hypothesis. For example, track 2 estimates in hypotheses [2 4 5], [1 2 5] and [1 0 5] are different since in each case the track is updated with a different measurement.

2.1 Implementation example

In Fig. 5 a basic architecture of the MOMHT is given. This section will give an overview of the basic processing steps of the method.

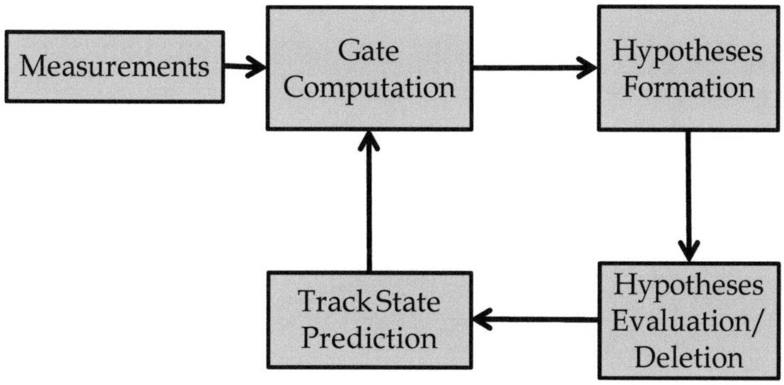

Fig. 5. MOMHT basic architecture

Assuming that the laserscanner processing in scan k provides M possible objects each having a measurement vector z_m, $m = 1...M$. Each hypothesis contains a set of tracks $x_i(k)$, $i = 1...N$, with respective covariance $P_i(k)$. The criterion that a measurement z_m is inside the track gate of size G is based on the innovation vector $z_m - H \cdot x_i$, where H is the measurement matrix:

$$(Z_m - Hx_i)^T B^{-1}(Z_m - Hx_i) \leq G \tag{1}$$

where $B = HP_iH + R$ and R is the measurement noise covariance matrix.

Then the hypothesis formation step is taking place. The gating procedure indicated the compatible track-measurement pairs of the current hypothesis. From an implementation point of view each line of the hypothesis matrix of scan k-1, which represents a hypothesis, is expanded into a new sub-matrix. This recursive procedure is repeated for each line of the hypothesis matrix resulting in the hypothesis matrix of scan k (Fig. 6). The new sub-matrix is expanded by inserting a new column until each measurement is added. When a new measurement is taken into account all the possible origins of this measurement will form the corresponding hypotheses. In any case, each measurement will form two hypotheses, one for assuming the measurement as false alarm and one considering the measurement as new target. Then, according to the gating results, additional hypotheses will be formed in case the measurement is inside one or more track gates. In the first scan, where there are no prior tracks, measurements are considered only as false alarms and new tracks. Attention has to be made that the associations within the hypothesis should be compatible. As a result a track cannot be assigned to two measurements in the same hypothesis.

Gating can be a time consuming process since a measurement set has to be check with the track set of each hypothesis. If the statistical distance calculation involves a matrix inversion, the gate calculation can cost a great part of the total MHT calculation time. During the implementation, the least complex distance measure should be chosen (for example Euclidean), that delivers the required results. In addition, distance measure calculation is affected by the sizes of the used matrixes. For example a four state track vector takes less time for gate computation than a six state vector.

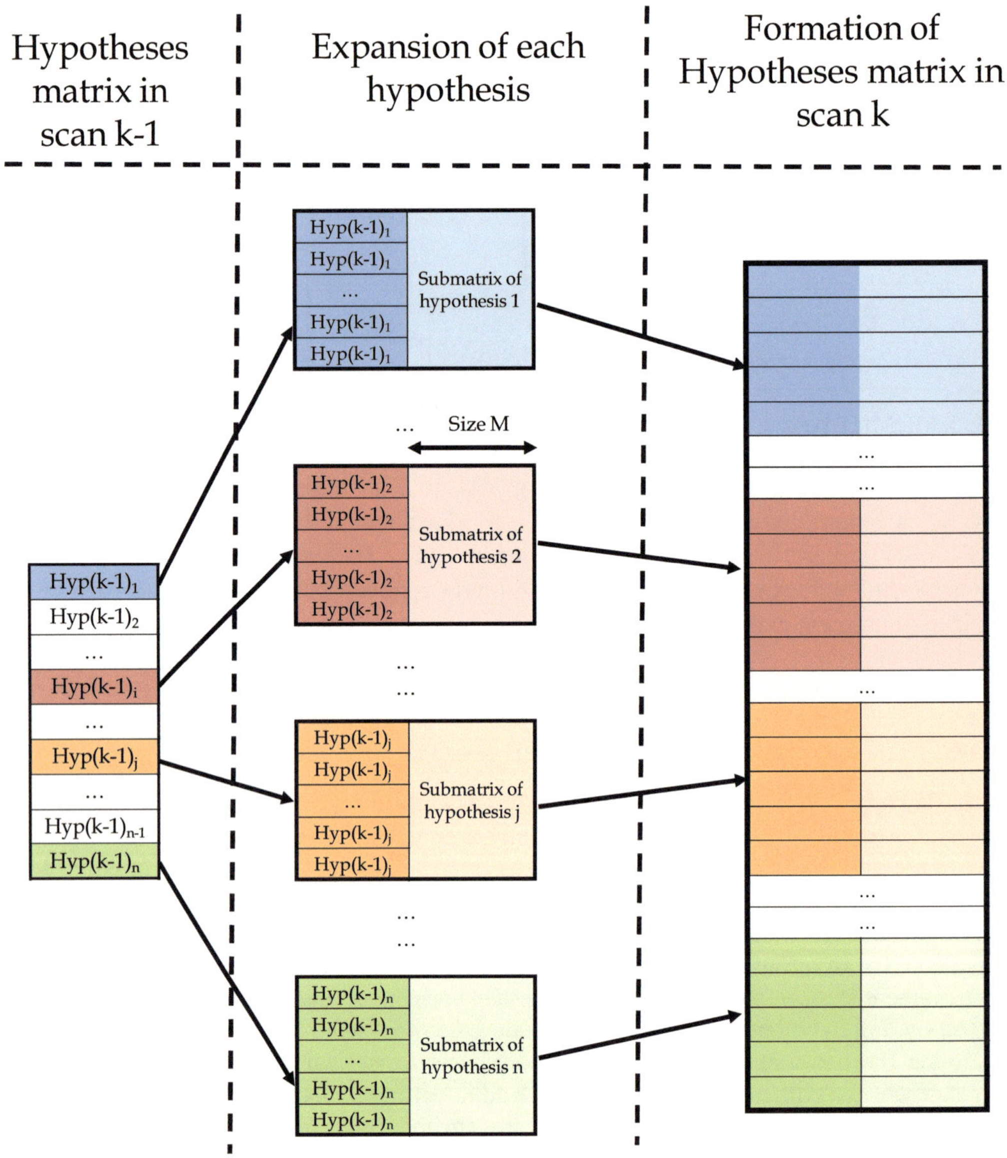

Fig. 6. Expansion of the hypotheses matrix

In parallel to the hypothesis formation, the track states are also updated. If a track has been associated with a measurement in the current scan, the state update is performed using the Kalman filter:

$$\hat{x}(k) = \bar{x}(k) + K[z(k) - H\bar{x}(k)] \tag{2}$$

$$\hat{P}(k) = \bar{P} - \bar{P}H^T\left(H\bar{P}H^T + R\right)^{-1} H\bar{P} \tag{3}$$

where K is the Kalman gain. As it has already been mentioned, a track with the same ID will have different state vectors in each hypothesis since it has been associated with a different measurement. In addition to the track state update, the hypothesis probability is also calculated and is described in section 4.2.

The next step of the algorithm is the reduction of the number of hypotheses. Since this method uses all the possible assignment possibilities, many of the hypotheses will have a very low probability and will be deleted after they have been created. There is a variety of hypotheses reduction techniques which will be discussed in Ch. 5.

An important aspect of the algorithm is to output a file of confirmed tracks. As already mentioned the essence of MHT is to resolve assignment ambiguities in the future. Usually the number of scans that hypotheses are propagated is fixed, called also as window size. This means that we go a number of scans back in the hypothesis tree and make a decision about the origin of the measurement. Assuming we have a window size equal to 3, on scan N we go back to the columns of the hypothesis matrix corresponding to scan N-2. Each column represents the possible origins of the measurement. If the entries in a particular column are the same, this means that the measurement has a unique origin, so the track that is associated with this measurement has a probability equal to one and can be considered as confirmed. If the measurement does not have a unique origin the authors use the following procedure. In the case mentioned there are two options: a) no track is confirmed from this particular measurement and we wait for future scans in order to confirm a target or b) if a track is more likely to be associated with the measurement then this track can be considered as confirmed. An illustrative example can be seen in Fig. 7. If measurement 2 has a unique origin, track 1 is confirmed. On the other hand if there are two assignment options with probabilities P_1 and P_2, depending on the implementation neither track 1 or track 2 are confirmed or if $P_2 > P_1$ and $P_2 > P_{threshold}$ track 2 is confirmed. Usually, it is useful to keep a list which indicates the hypotheses that contain each track. In this way, the track probability and state vector can be retrieved from each hypothesis.

After track confirmation the question is which is the track state and covariance that will be given to the output. The simplest method is to take the most probable hypothesis and use the state vector from this particular hypothesis. Another option is to combine the estimates, by using for example JPDA or a simple weighted sum using track probabilities, in order to combine the estimates from all the hypotheses that contain the track. In addition to the track confirmation, the implementation should incorporate a track deletion scheme. As mentioned a track's probability is the sum of the probabilities that contain this track. So if a track has a low probability it should be deleted from all the hypotheses. An additional heuristic measure is to delete a track if it has no associations for a continuous number of scans.

As it can be seen the track management scheme is based on the calculation of a probability of each hypothesis. The correct tuning of the parameters in the probabilistic expressions and the right choice in deletion and confirmation thresholds is crucial. Otherwise, wrong parameters can result in premature or delayed track confirmation and deletion. In addition, the propagation of a large number of hypotheses and tracks as a result of bad parameterization can compromise the real time efficiency of the algorithm.

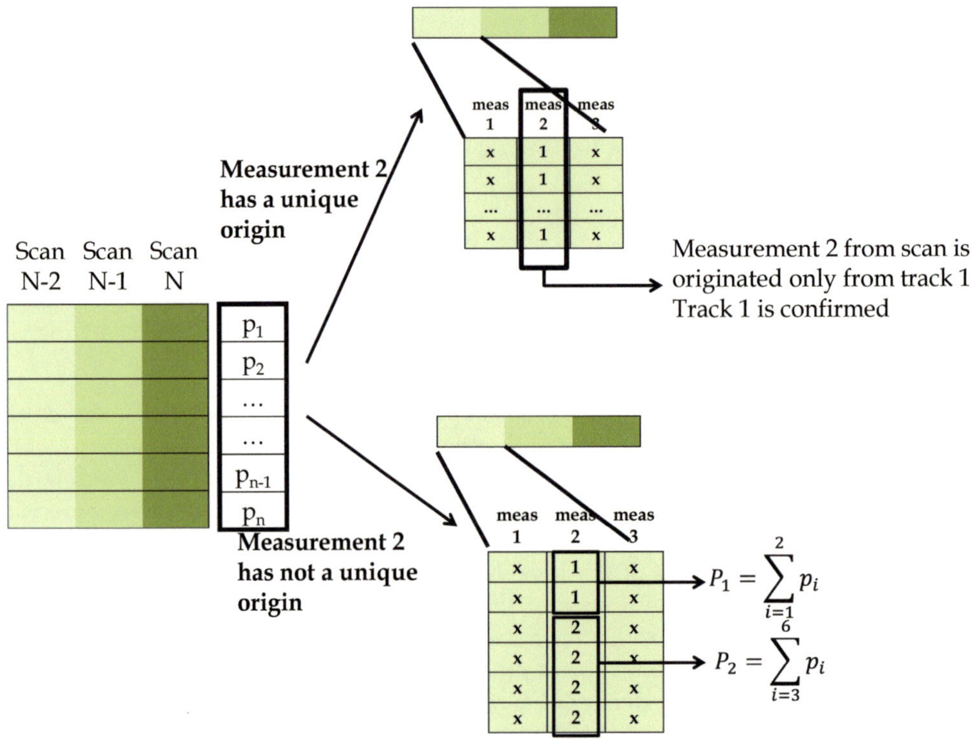

Fig. 7. Selection of the best measurement- track association

3. Track oriented MHT

Track oriented MHT approach was introduced in (Y.-B. Shalom 1990) and in contrary to the hypothesis oriented approach, it does not maintain hypothesis from scan to scan. When a new measurement set is received, the existing track set is formed into an expanded new track set using all the possible assignments. This method does not maintain a hypothesis set between scans, but it maintains a set of tracks which are not compatible. This set of incompatible tracks is used to form the hypotheses.

In order to point out the differences with HOMHT, the hypothesis formation method will be explained with the same example given in section 2. The original track set contains two tracks. These tracks are represented by the two leftmost trees in Fig. 8. The nodes are numbered according to the measurement that the track has been associated in the current scan. Let's assume that Track 1 is associated with measurement 2 and Track 2 is associated with measurement 1 at the previous scan. Recalling the gating results (Fig. 2), there are 4 possible assignments for Track 1. Assignment to the dummy measurement is indicated with 0 and declares that the track is not assigned to any measurement in the current scan. A matrix representation of the formed tracks is depicted in Fig. 9. The difference with the hypothesis oriented approach is that each track is represented by a tree, and the alternative association hypotheses are represented by the different nodes in the target tree. Each node within the tree is incompatible with all the other tracks of the tree, so the set of tracks in the same tree can represent one target at the most.

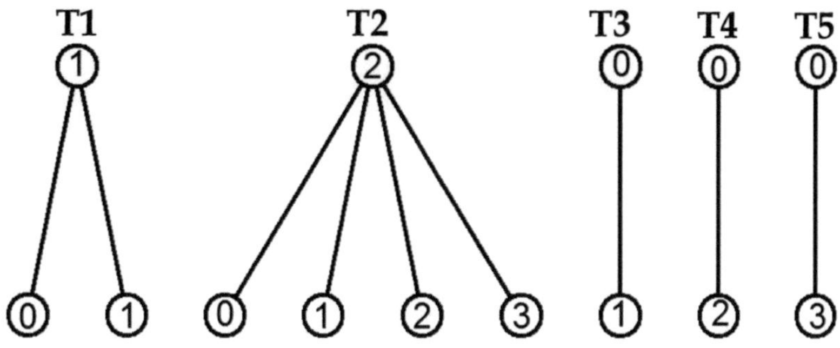

Fig. 8. Tree representation of he formed tracks

0	1
0	2
0	3
1	0
1	1
2	0
2	1
2	2
2	3

Fig. 9. Table representation of the formed tracks

3.1 Implementation example

In this section a general description of the basic TOMHT steps will be described. An overview of sample architecture is depicted in Fig. 10. This is a sample logic and the sequence can be changed or more blocks may be added according to the implementation.

As in section 2.1, the first step of the tracking algorithm is the gate computation. The gates are calculated using for example equation(1). All the possible assignments result in the formation of new tracks or updates of existing ones. The target tree expansion begins with the assumption that tracks are updated with the dummy measurement, meaning that the track is not updated with any measurement (indicated as node "0" in the target tree). Additionally all measurements spawn a new track. Then all the possible measurements that can be associated with a track are used to update the track state. This is indicated as different branches that are formed in the target tree. Taking as example track T2 in Fig. 8, the track may be assigned to the dummy measurement and also to measurements 1, 2 and 3. In parallel to the track formation step, track probabilities are also calculated (sect. 4.2).

The next step is the reduction in the number of tracks. Each track is assigned with a probability. Tracks that have a probability lower than a defined threshold will be deleted. Additionally similar tracks (sec. 5.2) can also be merged, thus the overall number of tracks can be further reduced

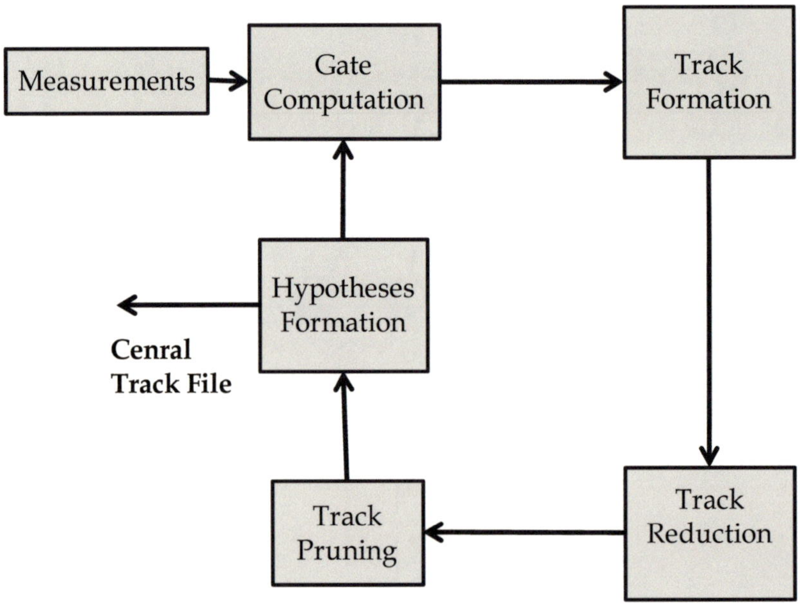

Fig. 10. Architecture of TOMHT algorithm

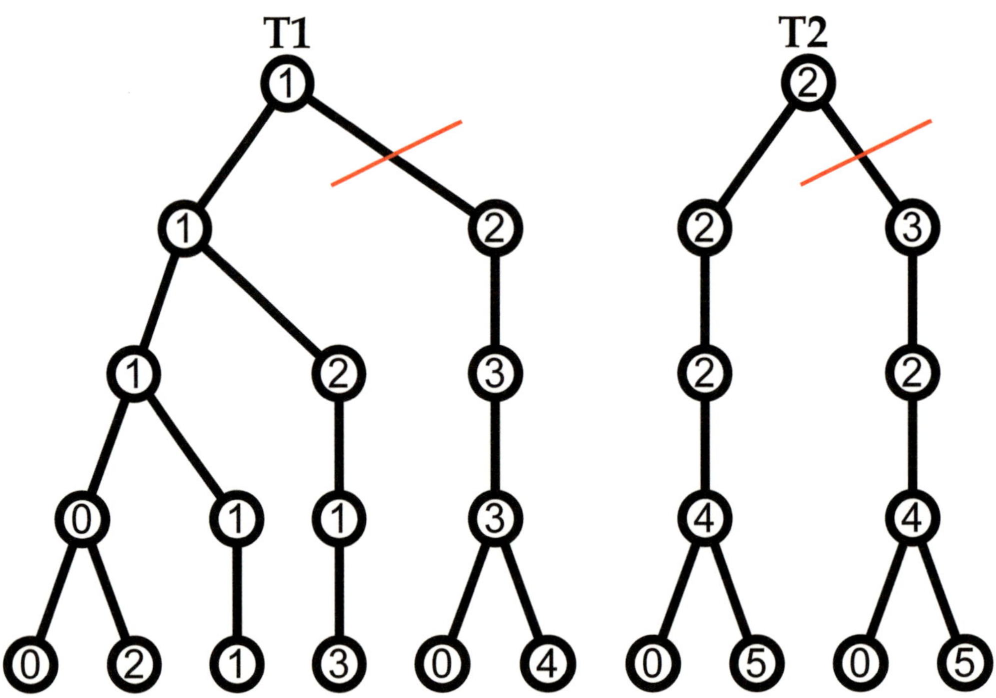

Fig. 11. Target tree pruning after solving the multi-dimensional assignment problem

After the track reduction step, the target trees have to be pruned. Similarly to HOMHT we have to go back a number of scans in the past and decide which track-measurement assignment is correct. For the rest of the section the following example will be used. Let's assume we have a 5 scan window and the track hypotheses trees are formed as illustrated in Fig. 11. In this particular association problem we have to decide if T1 is associated with measurement 1 or 2 in scan N-3 and whether T2 is associated with measurement 2 or 3. Assuming that the correct track pairs are T1-M1 and T2-M3, all tracks that do not stem from these associations will be deleted. In order to perform the tree pruning, the multi-dimensional assignment problem has to be solved. A possible solution that also has been used by the authors is the use of Lagrangian relaxation algorithm (Fisher 1981). The output of the algorithm will be a set of S-tuples that represent a set of compatible tracks. In our example the two 5-tuples will be [1 1 1 1 1] and [2 2 2 4 5]

The final step of the presented TOMHT architecture is the formation of the most probable hypothesis. The number of trees indicates the maximum number of possible tracks. The total number of tracks is equal to the number of nodes in the target tree that correspond to the last scan or equal to the number of rows in the hypotheses matrix. As mentioned before, this set of tracks is incompatible. In order to form the most probable hypothesis that contains a set of compatible tracks, the tracks are arranged into descending probability order. The most probable track is added to the hypothesis. Then, the next most probable track is added. This track should be compatible with the rest of the track in the hypothesis. This procedure is continued until there are no more compatible tracks that can be added to the hypothesis. The steps of composing the most probable hypothesis in our example are depicted in Fig. 12. Apart from the best hypothesis, the m-best hypotheses may also be formed (Popoli et. al.,2001;Fortunato et. al., 2007). Tracks that are not contained in the m-best hypotheses can also be deleted.

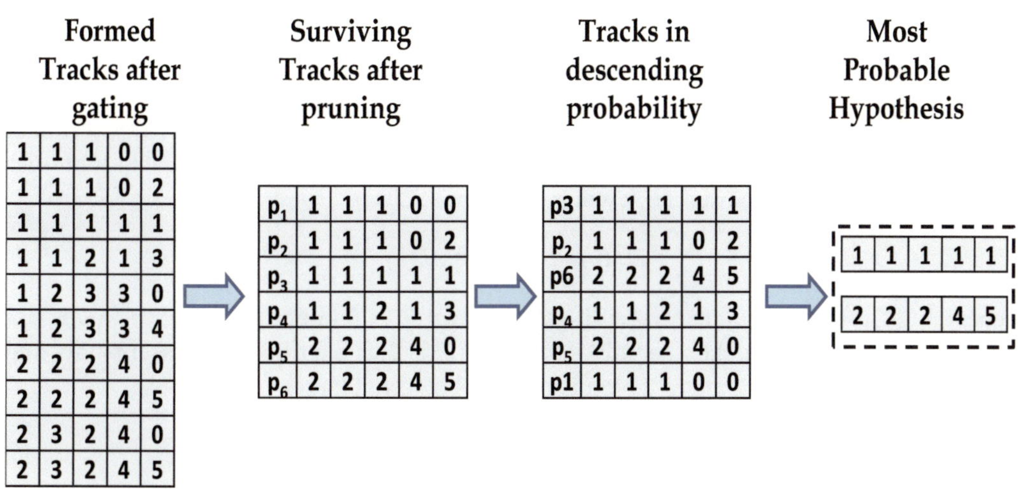

Fig. 12. Hypothesis formation in TOMHT

4.1 Window size

An important parameter of the MHT algorithm is the number of scans that the track history is kept, called also the window size. The larger the window size, the larger the number of hypotheses that are created. If the window size is N and n_i is the number of measurements in scan i, the tree depth in HOMHT will be $\sum_{i=1}^{N} n_i$. TOMHT consists of multiple trees, each one having depth equal to N. Window size is a design parameter and should be chosen so that it is large enough to resolve assignment ambiguities but also not result in an unmanageable number of hypotheses.

4.2 Hypotheses probability

As mentioned, for each hypothesis a probability has to be calculated. In Reid's original work the probability P_i^k of hypothesis Ω_i^k formed from hypothesis Ω_i^{k-1} is calculated as follows:

$$P_i^k = \frac{1}{c} P_D^{N_{DT}} \left(1 - P_D\right)^{(N_{NGT} - N_{DT})} \beta_{FT}^{N_{FT}} \beta_{NT}^{N_{NT}} \times \left[\prod_{m=1}^{N_{DT}} N\left(Z_m - H\bar{x}, B\right)\right] P_g^{k-1} \tag{4}$$

Where:

P_D = Probability of detection

β_{FT} = Density of false targets

β_{NT} = Density of previously unknown targets that have been detected

N_{DT} = Number of measurements associated with prior targets

N_{FT} = Number of measurements associated with false targets

N_{NT} = Number of measurements associated with new targets

$Z_m - H\bar{x}$ and B are the innovation vector and innovation covariance matrix.

Log Likelihood Ratio (LLR) is a more convenient value to be calculated. Many formulas exist for the calculation of this value. The basic advantage is that if $L(k-1)$ is the LLR of track at scan $k-1$ the LRR $L(k)$ at scan k will be (Blackman 2004):

$$L(k) = L(k-1) + \Delta L(k) \tag{5}$$

where $\Delta L(k)$ can be calculated using the original work in (Sittler 1964) or expressions like (Bar-Shalom et. al., 2007). Another advantage of LLR is that is a dimensionless quantity and it can also be easily converted to the probability P_T of a true target

$$P_T = e^{LLR} / \left[1 + e^{LLR}\right] \tag{6}$$

5. Hypothesis and track management

Multiple Hypothesis Tracking can result in an exponential increase of hypotheses, making this approach impossible without the use of hypotheses reduction techniques. In this section a number of techniques regarding hypotheses and track management will be presented, including some practical application examples.

5.1 Hypotheses clustering

Clustering is a well explored topic in computer science (Bouguettaya & Le-Viet, 1998; Steinbach et. al., 2000) whereas it has also been applied to target tracking (Czink et. al., 2006). Reid had indicated the need of a clustering scheme in order to improve the efficiency of his HOMHT algorithm implementation (Reid 1979). As a result, clustering was integrated into the HOMHT algorithm architecture

Clustering is not a direct way of reducing hypotheses number in HOMHT. However it presents a way of dividing a large tracking problem into smaller tracking sub-problems. Each cluster consists of a separate hypothesis tree, so the number of hypotheses that are managed in each cluster is smaller than it would be in the case that the tracking problem was handled in one hypothesis tree. For example, it is not necessary to gate a measurement with all the tracks, but only to those tracks that belong to a particular cluster. In addition, with the use of multi core processors, it is possible to process each cluster separately.

A cluster management algorithm for MHT mainly consists of three sub-routines: cluster initialization, cluster merging and cluster splitting. Cluster management should not allow clusters to grow uncontrollably. In this case, the additional processing overhead of cluster management algorithm will not be covered but the gains from the reduction in MHT processing time.

The first step is to distribute the measurements into existing clusters. For every measurement i the distance d_{ij} from each cluster j is calculated. The calculation of the distance can be the defined as the statistical distance between the measurement and the cluster centroid for example. If p_i and x_i are the probability and the estimate of track i in the cluster, the cluster centroid can be defined as follows:

$$x_c = \frac{\sum_i p_i x_i}{\sum_i p_i} \tag{7}$$

Alternatively, d_{ij} can be considered as the closest distance between the measurement and the cluster's tracks. If d_{ij} is smaller than a threshold value then the measurement is assigned to the particular cluster. If the measurement is assigned to more than one cluster, these clusters are merged, a process that will be managed by the cluster merging routine. Finally, if the measurement is not assigned to any cluster, a new cluster is initialized.

Cluster merging is the second routine of the clustering algorithm. If a measurement is associated with two clusters, these two clusters have to be merged. The merging method should combine the hypotheses that are contained in the two clusters. This involves the combination of the tracks and measurements that are contained in these two separate clusters. Additionally, a combined hypothesis matrix and tree has to be formed while the combined hypotheses probabilities have to be calculated. Assuming we have the case of Fig. 13 where clusters A and B have to be merged. Cluster A has a two row hypothesis matrix whereas cluster B has a three row matrix. Numbers above each column represent the scan of the corresponding measurement. The final matrix combines the rows and columns of the initial matrixes, whereas the probability of the merged hypothesis will be the product of the probabilities of the merged hypotheses.

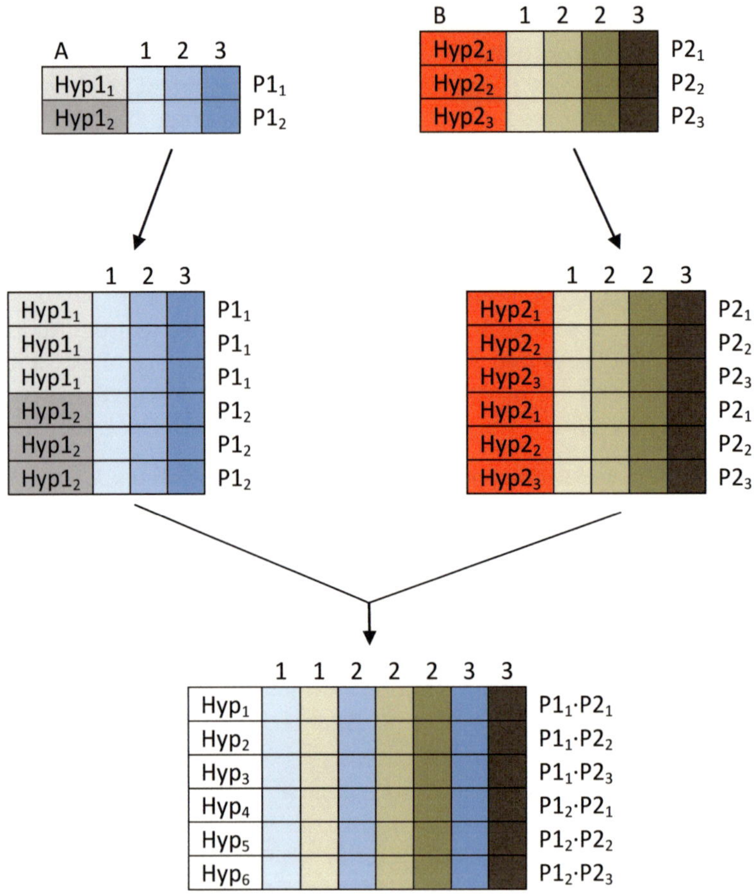

Fig. 13. Merging of two hypotheses matrixes

Cluster splitting is the process where one track of a cluster is removed from this cluster and initiates a new cluster. For this purpose we have to set a similarity criterion for tracks within the cluster. A simple method is to use the Euclidian or Mahalanobis (8) distance in order to calculate the distance d_{ij} between the tracks.

$$d_{ij}^2 = \left(\hat{x}_i - \hat{x}_j\right)^T \left[\left(P_i^T + P_j^T\right)^{\mathrm{T}}\right]^{-1} \left(\hat{x}_i - \hat{x}_j\right) \tag{8}$$

where $\hat{x}_i$, P_i and $\hat{x}_j$, P_j are the state estimate and covariance of tracks i and j respectively.

If this distance is greater than a define value, these tracks are considered to be candidate for splitting. Let's assume we have 3 tracks, we form a 3x3 matrix which contains binary values. A value of "1" at entry (i,j) indicates that these two tracks are "far away" so they can be assigned to different clusters. A value of "0" indicates the opposite. This example is illustrated in Fig. 14. All the columns are examined sequentially, if the sum of the column is greater than zero, then this track is removed, along with the corresponding row and matrix, and a new cluster is initiated. In addition, all the hypotheses that contained the particular track are removed from the hypotheses matrix.

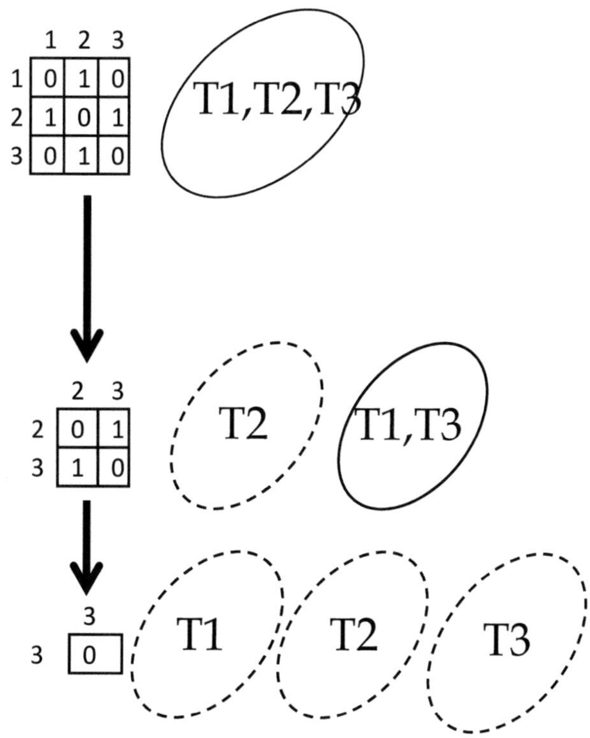

Fig. 14. Cluster splitting procedure

A final routine that a clustering algorithm should include is the cluster deletion. Clusters that do not contain any tracks should be deleted. Similar clustering strategy applied to tracks and not hypotheses can be used also in TOMHT.

5.2 Hypotheses reduction

In this section some methods for hypotheses reduction will be presented. As it has been described hypotheses are reduced when a decision for an assignment in the past is made.

In addition, hypotheses deletion is an efficient measure for keeping algorithm calculation time under control. Hypotheses reduction techniques for HOMHT and TOMHT will be described separately.

5.2.1 Hypotheses reduction in HOMHT

The simplest way of reducing the number of hypotheses is to delete those hypotheses that have a probability below a certain threshold. Since HOHMT is an exhaustive method that enumerates all the possible assignments, a large part of those assignments will be highly unlikely, so they can be deleted. However, this presents the danger of deleting some useful hypotheses also. For example when a track has just been created, the hypotheses that contain it will have a low probability. Some of these hypotheses have to survive deletion and wait for the next scans in order to calculate whether the probability will be increased or not. A useful practice according to authors is that each column in the hypotheses matrix that corresponds to the last scan should contain not only "0" but at least one track number. As a result, we will allow alternative hypotheses for this measurement to be propagated and when we decide about the origin of the measurement the correct assignment will be made. Of course this can result in an unacceptable high number of hypotheses. Authors suggest that a maximum number of hypotheses and tracks should be set in the algorithm. If the hypotheses or track number exceeds these thresholds, additional hypotheses deletion should take place. This will ensure that the system will always be able to run real time, even if tracking performance is sacrificed. Apart from hypotheses, the number of tracks that are contained in the hypotheses can be reduced. A track probability is the sum of hypotheses probabilities that contain it. Low probability tracks are deleted from the hypotheses that contain them.

Apart from hypotheses deletion, similar hypotheses can be merged in order to reduce the total number of hypotheses. Two hypotheses are merged if they satisfy one or more similarity criteria. In general, two hypotheses can be merged if they contain at least the same number of tracks and these tracks are "close enough". Assuming we have 3 hypotheses as indicated in Fig. 15. Hyp_1 and Hyp_2 contain 2 tracks whereas Hyp_3 contains 2 tracks.

	1	2		3
Hyp_1	1	1	2	3
Hyp_2	1	4	2	2
Hyp_3	1	0	2	0

Fig. 15. Hypotheses merging example

The first two hypotheses are candidate for merging. The distance measure $d_{ij}, i = j = 1...3$, between the tracks of the two hypotheses is calculated. If all the tracks of the first hypotheses have a distance from a track of the second hypotheses below a defined threshold, these two hypotheses can be merged. The probability of the merged hypotheses will be the sum of the individual hypotheses probabilities. The estimates of the merged tracks can be calculated using JPDA which results in a weighted sum of the individual estimates.

5.2.2 Track reduction in TOMHT

As in MOMHT, a first step is to delete low probability tracks. In TOMHT, low probability tracks can also be deleted before and after the hypothesis formation step. Deletion thresholds should be adapted carefully since new tracks have a low probability; it may happen that hypotheses which contain new tracks may be deleted. A useful practice is to have each measurement of the current scan associated with at least one track. In this way, if the measurement represents a real target, the track will have an increased probability after the next scan, otherwise it will be deleted. On the other hand, tracks with a long history have relatively high probabilities, so a large number of "old" tracks may describe the same target

Similar tracks can also be merged. Again, a similarity criterion has to be set. For example, two tracks can be merged if they share N measurements in the past and are closer than a defined threshold are merged. If one track is associated with a measurement and another track is associated with the dummy measurement, these two tracks are considered as compatible in the particular scan. In Fig. 16, for example tracks 1 and 3 are considered as compatible, so they can be merged. The same applies to tracks 5 and 6. On the other hand, tracks 2 and 3 are not compatible since they are associated with a different measurement in the last scan.

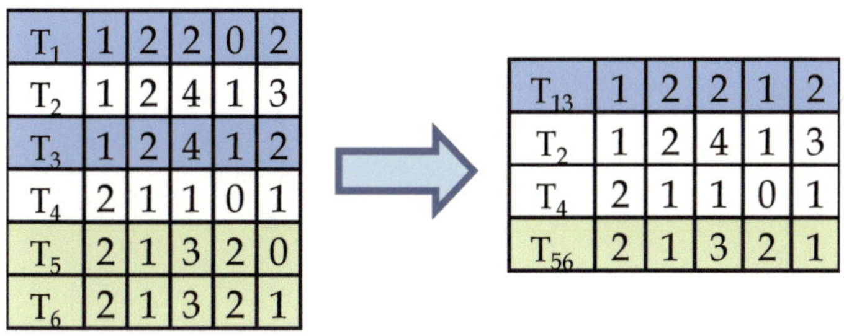

Fig. 16. Track merging example

A possible danger in track merging is the situation depicted in Fig. 17. If the similarity criteria are satisfied, the five tracks will be merged into two tracks. These two tracks are not compatible, so in the track extraction process only one track is outputted. However the correct hypothesis may be that there are two real targets represented by measurements 3 and 5 of the last scan. This example is used in order to show that the similarity criteria have to be set carefully or that the necessary measures have to be taken so that enough tracks are maintained.

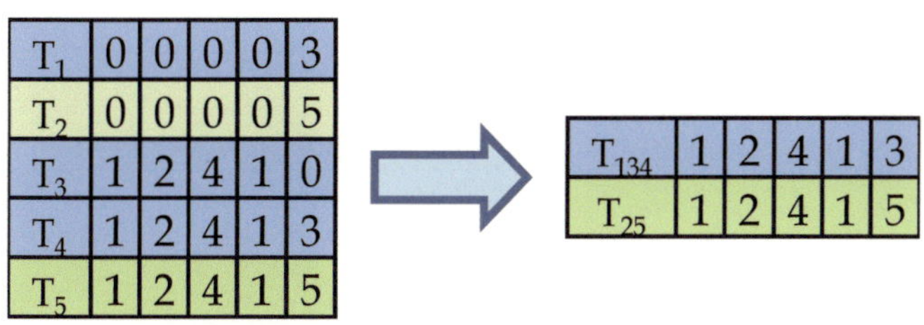

Fig. 17. Track merging resulting in one track

6. Results

In this section some indicative results of MHT implementation will be presented in order to demonstrate the performance gains of the MHT algorithm.

In the first set of tests, five scenarios recorded from an IBEO LUX laserscanner system in highway scenarios were used. The scenarios differed in the amount of traffic that was encountered. Theses series of tests were conducted in order to measure the savings in execution time after applying a clustering algorithm in HOMHT. The sensor setup is illustrated in Fig. 18. Laserscannner network topology. The laserascanner raw data are processed by a fusion module (Fig. 19) whose output consists of a list of detected objects. This list is then given as input to the HOMHT algorithm.

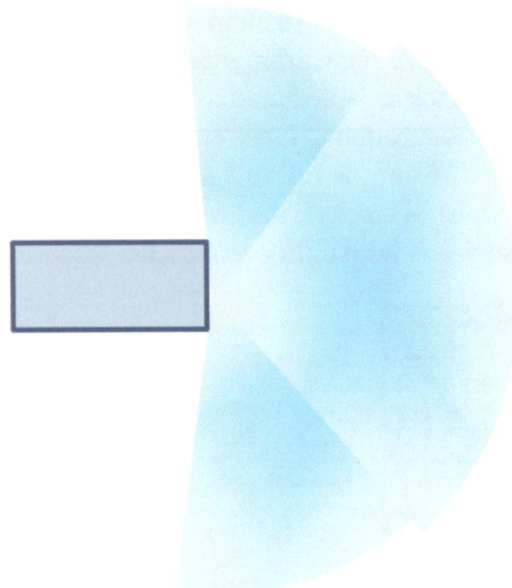

Fig. 18. Laserscannner network topology

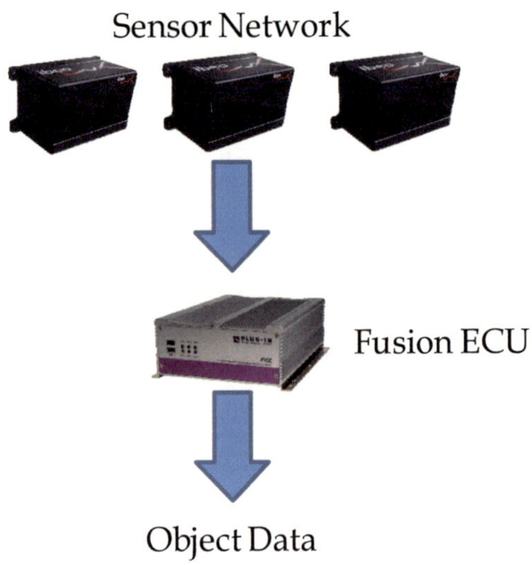

Fig. 19. Laserscannner processing architecture

Two algorithms were implemented in a C++ environment, the first one without clustering and the second one with the clustering scheme presented in section 5.1. The use of a profiling software made possible to extract the calculation times indicated in Table 1

The results indicated that the average execution times can be improved by almost 92%. Of course, this improvement heavily depends on the sensor used and also by the scenario. However, since a multilayer laserscaner network has a great range, it is not unusual to receive up to 40 targets in a highway scenario, or even double this number in urban environments. As it can be seen, the algorithm without clustering is impossible to run real time, since the calculation time exceeds the refresh rate of a usual sensor. It has also to be mentioned, that the tracking performance of the two algorithms does not differ. Clustering does not deteriorate at all the performance gains of the MHT technique.

	No Clustering			Clustering			Difference		
	Total	Mean	Max	Total	Mean	Max	Total	Mean	Max
Scen 1	5437	3	199	1573	1	9	**71.08%**	**71.03%**	**95.46%**
Scen 2	20311	20	447	1618	2	7	**92.04%**	**91.98%**	**98.42%**
Scen 3	51100	51	21127	17447	18	80	**65.86%**	**65.86%**	**99.62%**
Scen 4	81740	47	37345	25633	14	150	**68.64%**	**69.62%**	**99.60%**

Table 1. Execution times in milliseconds for the MHT algorithm with and without clustering

Another interesting test was to measure the difference in calculation times by changing the width of the observation area. For a given scenario, only the measurements that had a lateral distance below a certain limit were given as input to the MHT. In this way, it is possible to measure, roughly, how the calculation times increase as the number of measurements increase. The comparison chart is depicted in Fig. 20. This chart shows that execution time of a Multiple Hypotheses Tracker that incorporates clustering processing steps is increased by far less compared to an MHT without clustering. However, efficiency of the clustering mainly depends on the density of the measurements.

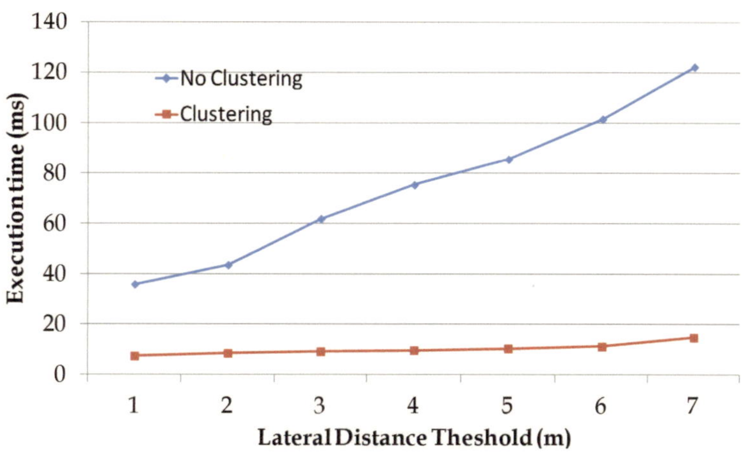

Fig. 20. Execution time vs observation area chart

In order to test the performance of the HOMHT algorithm and identify the benefits over the conventional GNN, a series of simulations of a typical driving scenario were performed. The scenario to be examined is: two parallel moving targets. In simulated measurements (122 in total), a variable random Gaussian noise was added, with standard deviation varying from 0.1 to 2 meters. In addition, random clutter (40% of the total measurements) was added, simulating measurements from non-vehicle objects. The measures that were calculated are i) ID losses, ii) Correct Associations ratio, iii) False Alarms ratio and iv) Estimation error. Tracking results for both methods are shown in Fig. 21 and Fig. 22. These figures were extracted from the scenario with measurement noise of 1.2m. Red circles correspond to true observations while black squares represent measurements originated from random clutter. The estimated position for each target is shown with colourful crosses. Consecutive crosses of the same colour correspond to the same track ID. The plots clearly show that MHT maintains the track throughout the scenario while GNN fails to provide a good tracking result. A detailed analysis of these scenarios and results can be found in (Spinoulas 2010) and (Thomaidis et. al.,2010).

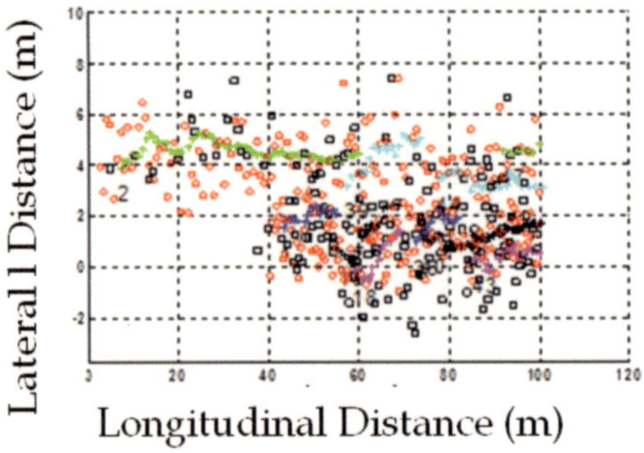

Fig. 21. Tracking output of GNN

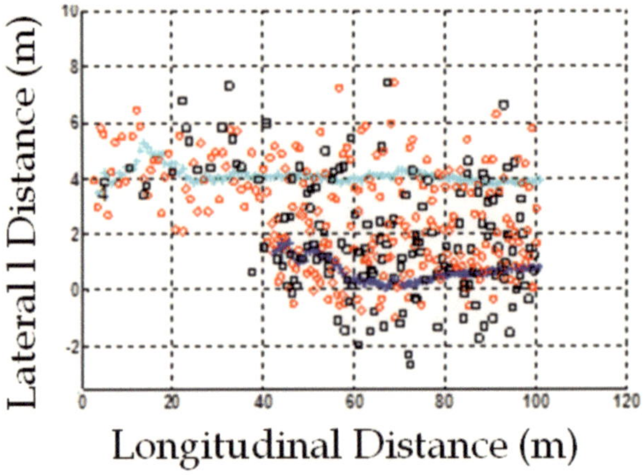

Fig. 22. Tracking output of MHT

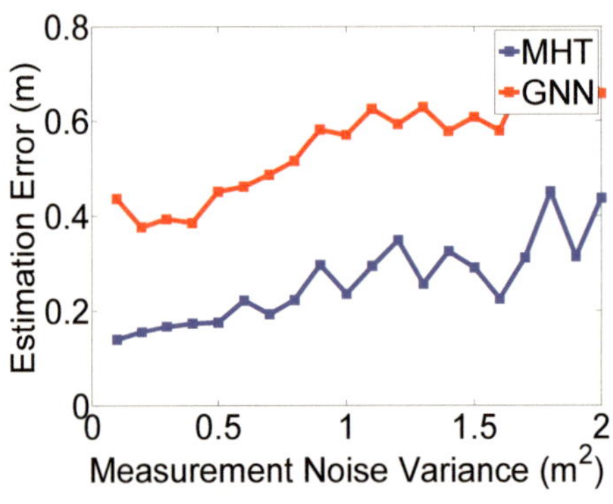

Fig. 23. Estimation error of MHT and GNN

In Fig. 23, the estimation error of both algorithms is plotted. As it can be seen MHT provides more accurate estimation than GNN. This is explained by the fact that MHT has better association results; it associates more often a track with the correct measurement. If a track is updated with the wrong measurement, then the track estimation will degrade or the track may be lost at all.

Fig. 24 presents the percentage of correct associations and the ID changes of the two algorithms. Correct associations indicate the percentage of scans that a track was associated with the correct measurement. ID changes present the number of times that the ID of the track was changed; indicating the number of times that a track was lost and re-initialized. As it can been seen, MHT has a higher percentage of correct associations over GNN. In addition the two algorithms have a degrading performance as noise levels increase. The second plot of Fig. 24. indicates that MHT has less track losses than GNN method.

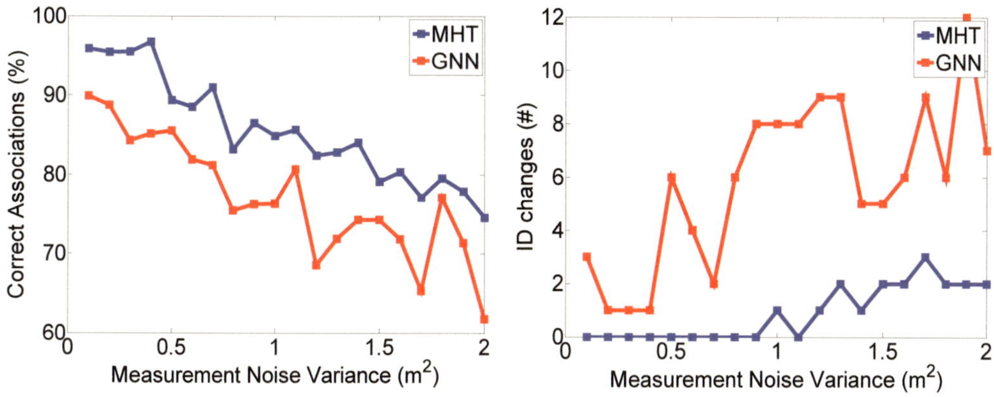

Fig. 24. Comparative results of MHT and GNN

7. Conclusions

This chapter presented practical aspects in the implementation of a Multiple Hypothesis Tracker. The two basic algorithmic approaches where presented, but of course alternative algorithms such as the Bayesian MHT (Koch 1996) have been proposed.

The MHT tracking algorithm yields better results than other methods which propagate only one association hypotheses. Undoubtedly this performance advantage comes at a cost. This method is much more complicate in the implementation than other tracking methods since it involves many processing steps. In addition each step can be implemented in a variety of ways. The propagation of many hypotheses as it has been mentioned also implies increased needs in terms of computational time. However, with the advent of modern fast CPUs and the use of optimization methods like the ones presented in this chapter, the implementation of a real time MHT is feasible.

8. Acknowledgment

The authors would like to thank Leonidas Spinoulas for his contribution in the work presented in this chapter.

9. References

Blackman, S., Multiple Hypothesis Tracking for multiple target tracking, *IEEE Trans. Aerospace and Electronic Systems,* vol. 19, no.1, pp. 5-18, 2003

Bouguettaya, A. Le-Viet, Q. Data Clustering Analysis in a Multidimensional Space, *International Journal on Information Sciences,* Vol 112, Iss. 1-4, pages 267-295. 1998.

Bar-Shalom, Y. (Ed.)(1990), *Multitarget-Multisensor Tracking: Advanced Applications,* ArtechHouse, 096483122, Nonvood, MA.

Bar-Shalom, Y. Blackman, S., Fitzgerald, R. J. Dimensionless Score Function for Multiple Hypotheses Tracking, *IEEE Trans. Aerospace and Electronic Systems* vol. 49, no.1, pp. 392-400, 2007

Blackman, S. Popoli, R.(1999), *Design and analysis of modern tracking systems,* Arthech House,1580530060, Nonvood, MA.

Czink, N. Del Galdo, G. Mecklenbräuker, C. A novel automatic cluster tracking algorithm, *Proceedings of IEEE International Symposium on Personal, Indoor and Mobile Radio Communications (PIMRC),* Helsinki, Finland. 11.09.2006 - 14.09.2006.

Fisher, M. L. The Lagrangian relaxation method for solving integer programming problems. *Management Science,* vol. 27, no.1 pp. 1-18, 1981

Fortmann, T. Bar-Shalom. Y. Scheffe. M., Sonar tracking of multiple targets using jointprobabilistic data association, *IEEE Journal of Oceanic Engineering,* Vol. 8, Iss. 3,pp.173 - 184, July 1983

Fortunato, E. Kreamer, W. Mori, S. Chee-Yee, Chong. Castanon, G. Generalized Murty's Algorithm With Application to Multiple Hypothesis Tracking, *Proceedings of 10th International Conference on Information Fusion, 2007,* 978-0-662-45804-3, Quebec, 9-12 July 2007

Kock, W. Retrodiction for Bayesian multiple-hypothesis/multiple-target tracking in densely cluttered environment, *O.E. Drummond (Ed.) Proc of SPIE Signal and Data Processing of Small Targets,* pp.429-440, vol. 2759, May 1996

Popoli, R. Pattipati, K. Bar-Shalom, Y. m-Best S-D Assignment Algorithm with Application to Multitarget Tracking, *IEEE Trans. Aerospace and Electronic Systems.* vol. 37, no.1, pp. 22-39, 2001

Reid D. , An algorithm for tracking multiple targets, *IEEE Trans. on Automatic Control,* vol. 24, issue 6, pp. 423-432, Dec.1979.

Singer, R. Sea, R. Housewright K., Derivation and evaluation of improved tracking filter for use in dense multitarget environments , *IEEE Trans. Information Theory,* vol. 20, pp. 423–432, Jul. 1974.

Sittler, Robert W. An Optimal Data Association Problem in Surveillance Theory, *IEEE Trans. on Military Electronics,* vol. 8 iss.2, pp. 125 – 139, April 1964

Spinoulas, L. Object tracking using Multiple Hypothesis Tracking for road environment perception. *Diploma Thesis,* National Technical University of Athens, Feb. 2010.

Steinbach, M. Kapyris, G. Kumar, V. A comparison of Document Clustering Techniques. *Proc. of KDD-2000 Workshop TextMining.* Aug. 2000

Thomaidis, G. Spinoulas, L. Lytrivis, P. Ahrholdt, M. Grubb, G. Amditis, A. Multiple hypothesis tracking for automated vehicle perception, Proc. IEEE Intelligent Vehicles Symposium (IV), 2010, 978-1-4244-7866-8, San Diego, 21-24 June 2010

11

The Illuminating Role of Laser Scanning Digital Elevation Models in Precision Agriculture Experimental Designs – An Agro-Ecology Perspective

Jeffrey Willers, Darrin Roberts, Charles O'Hara, George Milliken,
Kenneth Hood, John Walters and Edmund Schuster
Genetics and Precision Agriculture Research Unit,
USDA-ARS, Mississippi State, Mississippi,
Department of Plant and Soil Sciences, Mississippi State University, Mississippi,
Geosystems Research Institute, Mississippi State University,
Spatial Information Solutions, Starkville, Mississippi,
Department of Statistics, Kansas State University,
Milliken Associates, Manhattan, Kansas,
Perthshire Farms, Gunnison, Mississippi,
InTime, Inc., Cleveland, Mississippi Aggeos, Inc., Fulton, Mississippi,
Laboratory for Manufacturing and Productivity,
Massachusetts Institute of Technology, Cambridge, Massachusetts,
USA

1. Introduction

In his essay, The New Organon, Bacon (1561-1626) wrote *"So must we likewise from experience of every kind first endeavor to discover true causes and axioms; and seek for experiments of Light, not for experiments of Fruit. For axioms rightly discovered and established supply practice with its instruments, not one by one, but in clusters, and draw after them trains and troops of works."* (Donner et al., 1968). While Bacon's use of English is a bit opaque by today's writing styles, his statements are still very relevant and hold true for any significant area of inquiry when a key discovery or application is uncovered. Therefore, this chapter endeavors to indicate how laser scanning data streams, a 'light' based technology, enable the art, practice, and implementation of diverse investigations of agricultural systems, gaining insight into the various ecological processes involved. Our goal is to provide insight for others to similarly develop their 'trains and troops of works' according to their interests, which will, in turn, enrich all investigators of agricultural systems through the spread of shared knowledge and techniques.

Laser scanning data streams, when linked with multi-spectral, hyperspectral, apparent soil electro-conductivity (EC_a), or other kinds of geo-referenced data streams, aid in the creation of maps that allow useful applications in agricultural systems. These combinations of georeferenced information provide an opportunity to include several types of statistical analyses, permitting the best interpretation of the information conveyed by such maps, and provide the capability of building new, detailed, and more informative maps. Such maps have enabled a past, present, and, probably, future explosion *'of works'* leading to remarkably innovative methods for solving the problems of agricultural systems.

Several illustrations are presented to demonstrate a few of the numerous kinds of applications enabled by laser scanning data streams in agriculture. These illustrations focus on a Mississippi cotton field and a Nebraska corn field. Topics considered include (1) describing an approach to statistically evaluate impacts upon production by site-specific management practices (such as seeding rate, nitrogen and potassium applications, irrigation, and other farming operations), including the assessment of interactions among these practices and the topographical characteristics of crop fields, (2) assessing the accuracy of laser scanning data products, (3) evaluating the spatial distribution of the abundance, dispersion and other characteristics of agricultural variables as abstractions of agro-ecological populations of interest, and (4) a partial topographical analysis of yield involving two topographical attributes: laser scanning elevation data and the shallow apparent soil electrical conductivity (EC_a) measured by the Veris® cart (Veris Technologies, Salina, KS, USA), which is a proximal sensor system.

Global Positioning System (GPS) equipped hand-held loggers are another technology useful for obtaining geo-referenced scouting information such as crop phenology, soil fertility, and pests. These 'on the ground' measurements have extremely sparse sample sizes in comparison to the very dense pixel counts and small ground spatial distances (GSD) provided by laser scanning, proximal, or remote sensing sensors (Willers & Riggins, 2010). At the end of the production season, harvest yield monitors geo-spatially measure crop yield. Collectively, all of these kinds of information can be superimposed by a geographic information system (GIS) on a digital elevation surface built from a laser scanning mission of the agricultural field. Once assembled into a geo-database by additional geographic information system and remote sensing processing (de Smith et al., 2007; Jensen, 2000; Lillesand et al., 2008; Pouncey et al., 1999; Richards & Jia, 1999; Theobald, 2003; Willers et al., 2004), these data sets can be analyzed by advanced statistical methods (such as count model regression (Long, 1997; Willers et al., 2009b), general linear mixed analysis of covariance models (Gotway et al., 1997; Gotway & Hartford, 1996; Gotway & Stroup, 1997; Littell et al., 2006; Milliken & Johnson, 2002; Milliken et al., 2010; Willers et al., 2008b) or other geostatistical approaches (Oliver, 2010; Piepho et al., 2011; Schabenberger & Pierce, 2002). Such efforts by geographically supported experiments bring *'light'* — illuminating novel solutions to agricultural challenges and tasks. Laser scanning information is the key advance in such spatial experiments involving agricultural production systems.

1.1 Historical context

More than 15 years of on-farm research by the authors' on site-specific insect pest management from a precision agriculture (PA) perspective are beginning to lead toward ways of geographically evaluating whole-field and site-specific management practice combinations in commercial cotton fields (Burris et al., 2010; Milliken et al., 2010; Willers et al., 2004, 2008b). These different forms of geographically-based experimental designs are extensions of numerous concepts found in traditional experimental designs (e.g., completely random design (CRD), randomized complete block (RCB), Split-Plot, Lattices, etc.), yet they differ from traditional designs in several ways since they (1) can utilize the entire field or only certain parts of the field on demand, (2) are not restricted to the use of symmetrical, similarly sized small plots (or strips of plots), (3) excellently partition sources of variability due to both the planned treatment structure and the unplanned treatment structure (which includes one or more sources of field topography) and (4) exploit the geographical content of the site-specific experiment, especially the characteristics and travel paths of the farm equipment which apply site-specific applications. The inclusion of laser scanning data streams is one fundamental technology enabling this advance in statistical evaluation of PA practices.

Laser scanning maps field elevation at sufficient spectral, spatial and temporal resolutions which are typically smaller than the areal extent of a single swath (or harvest) element logged by the yield monitor. Since this scale of spatial resolution is possible, laser scanning information can resolve issues related to the modifiable areal unit problem (MAUP). This problem is comprised of two aspects (Gotway & Young, 2002). In the first instance, many statisticians have learned that different inferences are obtained when the same set of data is grouped into increasingly larger areal units. In the second instance, they have also found that variability in analysis results arise simply due to alternative specifications of areal units which create differences in their shapes at the same or similar scales.

Whenever the ground spatial distance of the pixels describing field topography are smaller than the swath element, then geographically-based methods of statistical analysis exploit the following characteristics of the variable-rate equipped machinery and the differential Global Positioning System equipped harvest yield monitors: (1) The travel path of the variable-rate sprayer (or largest implement) occurs in long strips (polygons) whose paths are polylines following either topographical contours or property boundaries, (2) Precision agricultural prescriptions are formulated to be applied to the polygon or polygons of interest coincident with the travel path of the applying machinery and can be spatially varied along that path, and (3) Demographic characteristics of these polygons of land are available at the level of a "plot", defined as areas which are geographically describable in interspersion, size, shape, and continuity. It follows that yields and other responses can be measured along harvest paths parallel to application paths. The precision agricultural practices are evaluated using analysis of covariance models to obtain regression effects describing the site-specific plot and control plot demographics with respect to a dependent variable such as yield. The process takes advantage of the fact that commercial fields are heterogeneous with respect to soil types, elevation, drainage patterns, and other characteristics. Digital topography maps describing these uncontrollable sources at sufficient spatial and temporal resolution are covariates to improve the statistical assessment of planned treatment effects on yield, or other crop responses. An illustration of this kind of analysis follows.

1.2 Illustration of laser scanning contribution to site-specific analyses

It is conceptually possible to establish a system of plots where standard management practices are applied and to insert within them smaller plots where an alternative management practice is applied (Figure 1). These plots assigned non-standard management treatments are referred to as "floating plots" and are also imbedded within the variable-rate application equipment paths and centered on the mid-line of the harvest equipment paths. The defining information of these floating plots can be collected from the prescription files created and used by the variable-rate controllers that treated them. The minimal size is defined by the variable-rate controller's reaction time to change discharge rates per georeferenced instructions, as well as other mechanical behaviors of the application equipment and additional preset experimental actions. This geographical overlapping process provides the necessary control data to perform a statistical evaluation of the efficacy of site-specific management decisions. All of the information is co-registered to a common geographical coordinate location which is the centroid of each harvest swath element (Willers et al., 2004).

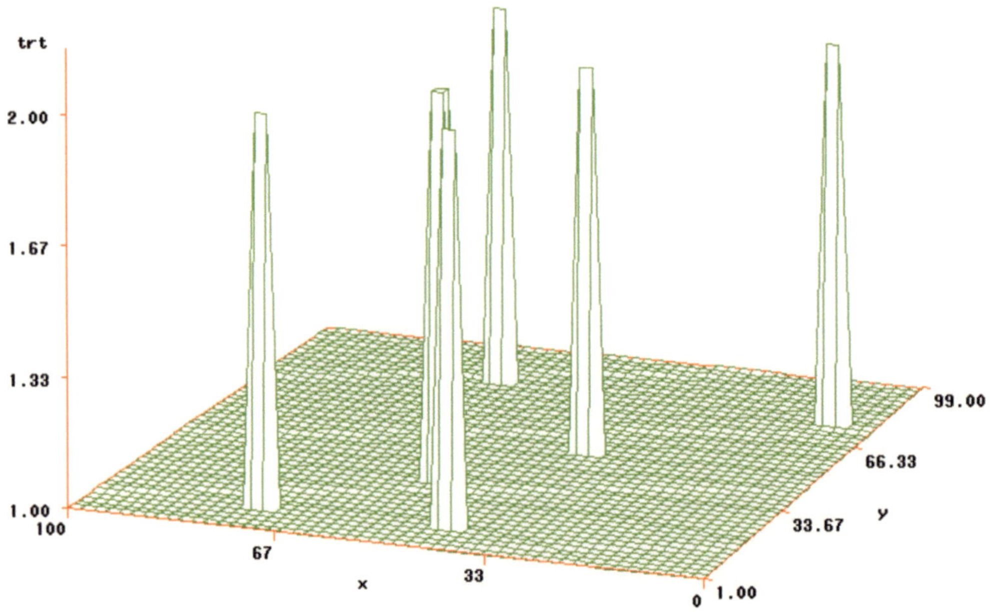

Fig. 1. Simulated treatment mean responses for a new management practice that is 2 times larger than the standard management practice (e.g., 1 unit) at several spatially selected locations in a simulated field.

While the actual 'plot layout' of any particular site-specific experiment can be quite diverse, a simplistic layout is proposed as illustrated in Figure 1, where the simulated 'field' is apportioned into a 100x100 grid of sub-plots in the '*x*' (i.e., Longitude or Easting) and '*y*' (i.e., Latitude or Northing) directions. Six floating plots were embedded to represent where a new treatment (or management practice) of two units will be spatially applied, while the rest of the field receives the standard (or traditional) management practice of one unit. Simulated response surfaces of two field topography characteristics that affect the yield (such as fertility levels (*e1*) and laser scanning elevation (*e2*)) are shown in Figs. 2A and 2B at the same spatial scale as the 'field' plots.

In this simulation, each sub-plot in the field grid was also modeled to contain a 6x6 lattice of points representing yield from harvest swath elements as measured by a yield monitor (Milliken et al., 2010; Willers et al., 2004, 2008b). Each yield point included a random error effect and the simulated yield for the experiment is shown in Figure 2C. The effects of both the conventional and new management practices were further modeled to interact with the topography characteristics (*e1* and *e2*), that were also simulated to have effects on the yield response.

A regression model was applied to these simulated data to estimate the yield response surface (Figure 2D) as a function of the levels (i.e., amounts, rates, elevation values, brightness values, etc.) of (1) the two environmental factors, (2) the conventional management practice applied to the standard plots, and (3) the new management practice applied to the floating plots. The simulated analysis shows that for some combinations of these two topography covariates, the new management tactic was not very effective at some locations (as indicated by troughs or absence of peaks in the yield response), while it was effective at other locations (as shown by the small peaks rising above the yield response).

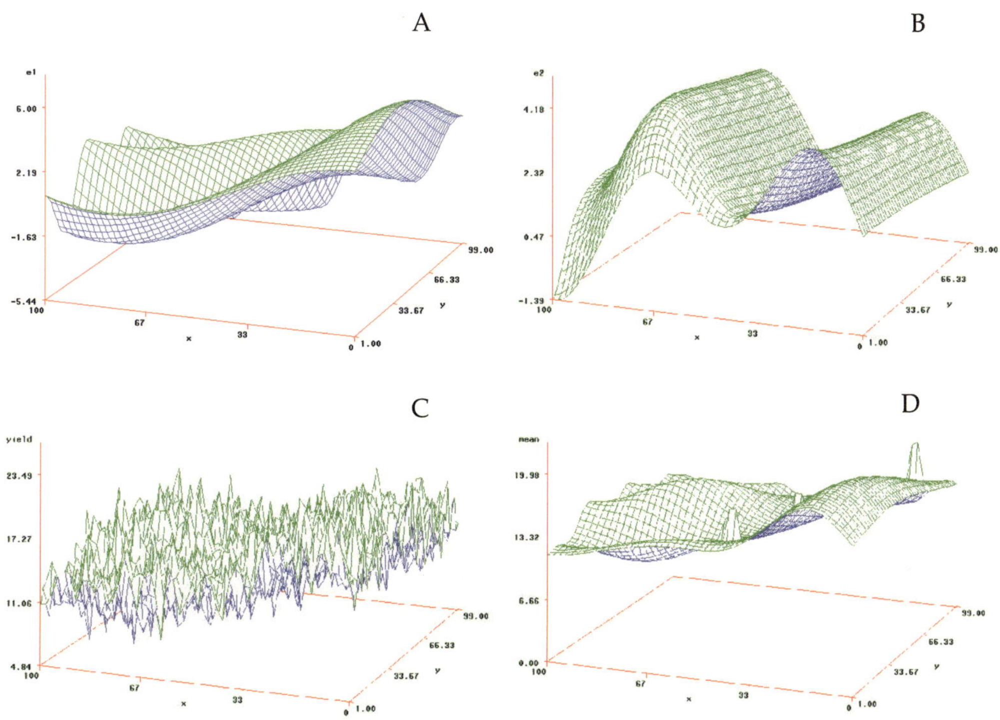

Fig. 2. Simulated response surfaces of A) first environmental factor (*e1*), B) second environmental factor (*e2*), C) mean yield estimates from a yield monitor obtained for each grid cell in both the *x* and *y* directions, and D) modeled yield response surface as a combination of the two environmental factors, the old management practice, and the new management practice which was spatially applied at different locations in the *x* and *y* directions.

There are several general forms of regression models useable in analyzing site-specific experiments. Investigating the best choice of statistical model for a particular geo-spatial combination of conventional and site-specific treatments and choice of applicator and harvester equipment configurations and which choice of topography covariate to use, is a large frontier for research. A key lesson learned in our research to date is that there are several constraints. The difficulty of defining optimal units of replication (Mead, 1988) includes (1) *a priori* definition of an adequate group of floating plots to serve as controls and where to place them, (2) the effects of uncontrollable spatial-temporal variability that is known but remains unmapped, and (3) effects of management practices that may differ for adjacent fields owned by different producers, especially with respect to insect and weed control (Anonymous, 2000; Dupont et. al., 2000). Another constraint is effectively projecting all of the data into the same coordinate system, such as the Universal Transverse Mercator grid (Bugayevskiy & Snyder, 1995). Different variable-rate or harvest equipment and sensor systems log their coordinate information in different formats as unprojected or projected values. A considerable amount of time is involved with co-registration of all data to a common coordinate system. As the number of farm fields increases, the process of resolving numerous data layers into a standard coordinate format becomes too excessive (Willers et al., 2009a).

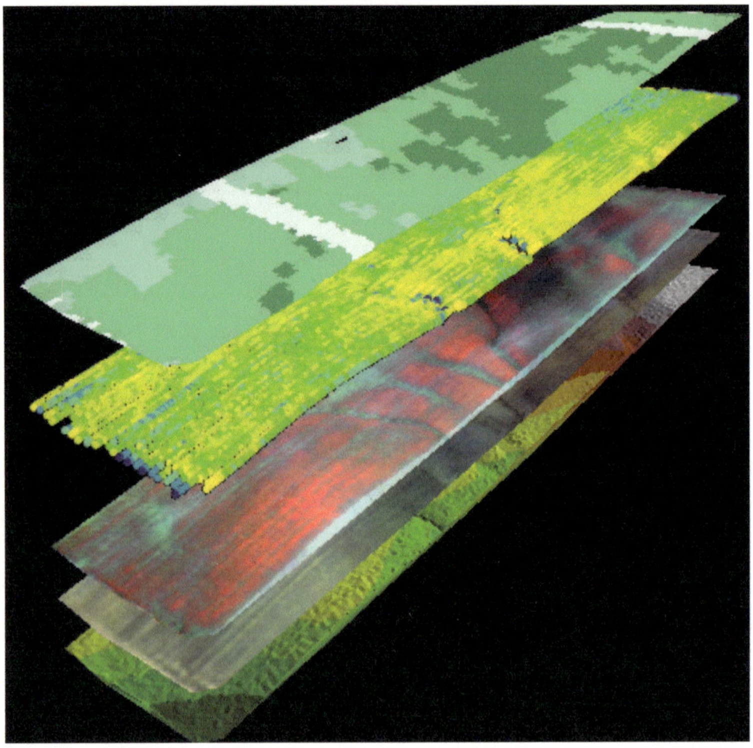

Fig. 3. Real-world data layers are shown to provide contrast to the hypothetical layers presented in Figures 1 and 2 (see Willers et al., 2004, 2008b).

In actual practice, as shown in Figure 3, the variable-rate controllers, yield monitors, and other types of sensors mounted on farm equipment or airborne platforms generate spatial information about agrichemical application rates, yields or other crop or soil attributes useful in analysis (Birrell et al., 1996; de Smith et al., 2007; Jensen, 2000; Kennedy, 1996; Pouncey et al., 1999; Richards & Jia, 1999; Sadler et al., 1998). The lowest layer of this figure is the laser scanned, digital elevation model. The layer above it is the multispectral bare-ground image (January 2002) and the next layer is the image of the crop development (June 2003), followed by the yield map. The topmost layer is the variable rate prescription map of three rates of plant growth regulator (PGR) (applied July 2003). This top layer also shows the embedded control strips (white bands) where no PGR was applied. As indicated in Figure 3, the digital elevation model from a laser scanning system is a foundational layer for the statistical analyses of precision agriculture management methods (Milliken et al., 2010; Willers et al., 2004, 2008b).

2. Solving questions of interest aided by laser scanning information

Questions of interest involved with applications of site-specific experimental designs can be classed into at least three types – does the question relate to (1) the evaluation of a single management tactic at a specific time or (2) a comparison between two or more management tactics at a specific time or (3) the evaluation of differences between two or more management tactics at different specific times of the season and/or different locations in a field. Methods of analyses for the latter two types of questions are not well developed; however, these kinds of questions are likely to be the most important to a commercial farm.

In site-specific experiments, it is likely that unplanned questions (that is, questions not specified *a priori* at the start of the production season) will arise, such as with the occurrence of dramatic, unexpected conditions during the crop production season. Farming operations can also be causes. Some possible operational causes are herbicide drift, mechanical injury to the crop during cultivation operations, or ruts caused by harvest equipment during wet soil conditions (which can cause effects lasting several seasons). If the effects of these unplanned causes can be mapped, then they can be included as effects in a statistical analysis.

Controlling the experiment-wise error rate (Milliken & Johnson, 2009) for planned or unplanned questions is another topic requiring deeper examination. It is likely that these error rate probabilities are going to be affected by the spatial, temporal, and spectral resolutions of the sensor systems involved. The major point with respect to these error rates is that laser scanning digital elevation models excellently support (Figure 3) on-farm experiments as well as other kinds of proximal and remote sensing data products. Utilization of such data streams shine *'light'* into the darkness of reality; otherwise, even if variable-rate controllers and harvest monitors are utilized, the results are only the *'fruit'* of the experimental exercise and, as a consequence, will have a small inferential space.

2.1 Sources of error and standards for laser scanning data streams

To function in the developing world of laser scanning data streams, the agricultural ecologist needs to have a working knowledge of how laser scanning systems acquire data, and how such data are processed and prepared for delivery to clients. There are many references (i.e., Lillesand et al., 2008) to provide such background. Nevertheless, it is necessary to briefly discuss sources of error and standards for laser scanning data to provide a common starting point. This discussion is anchored to a specific agricultural landscape (Figure 4), where more than 22 years of research on site-specific crop and site-specific insect pest management involving laser scanning, proximal and remote sensing systems, and crop yield monitors has been accomplished (Anonymous, 2000; Campenella, 2000; Dupont et al., 2000; Frigden et al., 2002; McKinion et al., 2009, 2010ab; Milliken et al., 2010; Willers et al., 1990, 1992, 1999, 2004, 2000, 2005, 2008ab, 2009ab; Willers & Riggins, 2010).

2.1.1 Lidar (laser scanning) background

Over the past decade, laser scanning (or light detection and ranging (lidar)), has become a primary method for collecting very dense and accurate elevation values. For data collected by a laser scanning system, the reflected pulses create a point cloud of elevation returns from the bare earth, vegetation, buildings, or any other features above the ground. Modern laser scanning systems are capable of recording multiple returns for each emitted lidar pulse reflected from a surface feature. In areas absent of any vegetation, only one pulse return would be recorded. Point clouds are comprised of lidar reflected returns which are processed and classified based upon whether the points represent ground or non-ground reflected returns. Non-ground returns may be further classified into feature categories such as grasses, shrubs, trees or woodlands, urban, withheld, noise, blunders, or any number of categories useful to a particular application.

Laser scanning has significant advantages over other methods of elevation data collection, including higher spatial resolution, vertical accuracies measured in centimeters, and penetration through forested and other vegetated areas. Laser scanning missions are typically acquired from aircraft which collect data in strips or swaths which comprise pulses rapidly collected at a rate that exceeds 150,000 pulses per second across large collection areas. Data acquisition may also be conducted by laser scanners mounted on mobile terrestrial platforms. However, in most applications, laser scanning data are processed to calibrate the data, classify ground and non-ground returns, and ultimately to produce high resolution, high accuracy digital elevation models. For agriculture applications, acquisition from aerial platforms provide data of sufficient pulse spacing and density for adequate terrain characterization providing multiple pulse returns per square meter of ground.

2.1.2 The LAS standard

The American Society of Photogrammetry and Remote Sensing (ASPRS) maintains data standards for remote sensing data through committees of subject matter experts from industry, government agencies, and academia. The lidar standard is copyrighted, maintained, and evolved by the ASPRS committees. The current standard for lidar data sets is the ASPRS Lidar Data Exchange Format Standard (or LAS) (ASPRS, 2004). Each LAS data file is a binary file that includes encoded information subdivided into three parts including the public header block, variable length records, and point data records. The LAS file format was developed to standardize the interchange, use, and implementation of 3-dimensional point cloud data between data producers and among users. The LAS standard was developed primarily for exchange of lidar point cloud data; however, the LAS data type supports the exchange of any 3-D collection of x,y,z data.

The public standard LAS binary file format is an interoperable file format well suited to encoding lidar data and has many advantages over proprietary data types that preclude interoperability, or over generic ASCII files which are characterized by large file sizes, inefficient implementation, non-standard structures, and slow processing. With the recent explosion in lidar technology and use, ASPRS has created a Lidar Division to keep abreast of data standards and implement new versions of the standards needed to support new hardware, sensor, and data technologies. The latest version of the LAS standard is version 1.4 which is pending final approval after public review. The updating of standards and creating new versions of the standard to accommodate the advance of technology is published on the ASPRS web site for the LAS working group at the following link:
http://www.asprs.org/Division-General/LAS-Working-Group.html

2.1.3 Laser scanning accuracy, guidelines, and base specifications

Three publications, NDEP (2004), ASPRS (2004) and FGDC (1998), provide guidance and formulas for determining elevation data accuracy. The Federal Emergency Management Agency (FEMA, 2003) has an early document that describes accuracy and quality assurance guidelines for laser scanning data. More recently, the United States Geological Survey (USGS, 2010) produced a document that has been circulated throughout the industry and is rapidly becoming the standard by which data are being evaluated whether for local, county, state, or national purposes. This document, commonly called the "version 13 specification", embodies an unprecedented emphasis on analyzing and understanding the lidar point cloud, including quantifying the sources of error in lidar data from initial acquisition to final delivery.

Prior to the version 13 specification, standards typically emphasized testing the final digital elevation model for accuracy; whereas, the version 13 specification addresses a sweeping range of aspects of error and uncertainty in the lidar data set. Considerations range from the initial coverage, to flight line overlap, and calculating and minimizing the relative error between adjacent laser scanning strips in their areas of overlap (Aguilar et al., 2010; Maas, 2002; Willers et al., 2008a). It is this relative error discrepancy (or step error) between adjacent strips that precludes or makes problematic the generation of a highly accurate continuous elevation surface for large agricultural landscapes.

Some of the common terms (NOAA, 2008) employed to describe lidar data as well as the errors that are encountered include the following:

- RMSE Z– abbreviation for root mean square error; a measure of the accuracy of the data similar to the measure of standard deviation if there is no bias in the data.
- Accuracy Z, Fundamental Vertical Accuracy (FVA) - a measure of the accuracy of the data in open areas at a high level of confidence (95%), calculated from the RMSE using the formula RMSE Z x 1.96 = FVA.
- Classification - data that have been processed to define the type of object that reflected the pulses; such can be as simple as unclassified (i.e., point not defined) to buildings and high vegetation. The most common is to classify the data set for points that are considered "bare earth" versus those that are not (i.e., unclassified).
- Return Number (First/Last Returns) - many lidar systems are capable of capturing the first, second, third, and ultimately the "last" return from a single laser pulse. The return number can be used to help determine what the reflected pulse is from (e.g., ground, tree, or understory).
- Point Spacing - how close the laser points are to each other, analogous to the pixel size of an aerial image; also called "posting density".
- Pulse Rate - the number of discrete laser "shots" per second that the lidar instrument is firing. Common systems used in 2008 are capable of 100,000 to 150,000 pulses per second. More commonly, the data are captured at approximately 50,000 to 70,000 pulses per second.
- Intensity Data - when the laser return is recorded, the strength of the return is also recorded. The values represent how well the object reflected the wavelength of light (for example, 1,064 nanometers) used by the laser system. These data resemble a black and white photo but cannot be interpreted in exactly the same manner.
- Real Time Kinematic Global Positioning System (RTK GPS) - satellite navigation that uses the carrier phase (a waveform) that transmits (carries) the Global Positioning

System signal instead of the Global Positioning System signal itself. The actual Global Positioning System signal has a frequency of about 1 megahertz, whereas the carrier wave has a frequency of 1500 megahertz, so a difference in signal arrival time is more precise. The carrier phase is more difficult to use (i.e., the equipment is more costly); however, once it has been resolved, it produces a more accurate position reading.

- Digital Elevation Model (DEM) – a surface created from elevation point data to represent the topography. Often a digital elevation model is more easily used in a geographic information system than the raw point data it is constructed from.

Building the DEM from the laser scanned point cloud can employ techniques that are quite diverse and are limited only by the creativity of the developers of any particular application. For example, Wang et al. (2008) process the point cloud to conduct vertical canopy structure analysis and 3D single tree modeling. Vu et al. (2009) developed a multi-scale, mathematical morphology approach to extract building features. Methods to reduce the processing time of these data intense points cloud are also keen areas of research (Han et al., 2009). Whatever the processing method employed for a specific application of the point cloud, the techniques exploit the *xyz* attributes for each return after filtering out blunders and random errors, employ various mathematical models to correct for systematic errors, and then employ various interpolation algorithms to produce the 3D surface of elevation at the appropriate spatial resolution for the intensity. Depending on the purpose of the DEM, that is, a bare earth DEM which describes elevational relief with features such as trees and buildings filtered out, or a digital surface model (DSM) which includes objects that are non-ground, the choices involved require specification of which return to use, be it the first return, the last return, or all returns.

For the agricultural DEMs used in this paper, two were processed by commercial vendors and made available thru either state or federal agencies (i.e., the background layers in Figures 4 and 7 (Mississippi) and Figures 5 and 16 (Nebraska). So, processing details for these 3D surfaces cannot be summarized. But, for the agricultural DEM in Figure 3 and the inset in Figure 4, as well as the DEM used for analyses in Figures 6 – 8 and 10 – 14, the point cloud processing can be summarized. First, the vendor removed systematic errors using proprietary procedures and orthorectified the point cloud returns to the vertical datum, NADV83 and the UTM Transverse Mercator grid for Zone 15 (North). Then a team of investigators (Willers, O'Hara and others) utilized the LAS file provided by the laser scanning vendor to (1) employ Terrascan® software to remove extreme instances of blunders and other random errors and then export the information as a comma delimited test file to upload into ArcMap® software for conversion into a set of point vector shapefiles for each strip (or line), including a tie-line strip acquired by the vendor. Next, these shapefiles were corrected for steps errors using the following algorithm and procedures.

The elevation data points in the overlap area of a tie-line strip were categorized into K groups indexed by k based on their coordinate and strip positions. Each group was characterized by $SubX_k$, $SupX_k$, $SubY_k$ and $SupY_k$ to define the set of points in group k as S_k so that $S_k = \{(i,j) \mid SubX_k \leq x_{ij} < SupX_k \text{ and } SubY_k \leq y_{ij} < SupY_k\}$. The number of points in strip (or line) i in group k was denoted by n_{ik}, so the total number of points in group k was:

$$n_k = \sum_{i=1}^{L} n_{ik} = |S_k| \tag{1}$$

The steps errors among the lidar flightline involved biases initially estimated by eye (also using Terrascan® software) to be about 15-20 cm. Therefore, to remove these step errors by mathematical programming, the variances of the adjusted elevations of points were minimized by determining the best values for a set of decision variables a_i. Let M_k be the mean elevation in group k before adjustment and A_k be the mean elevation in group k after adjustment ($k \in K$). Then, these mean values were found using:

$$M_k = \frac{\sum_{(i,j)\in S_k} e_{ij}}{n_k} \text{ and } A_k = \frac{\sum_{(i,j)\in S_k} (e_{ij} + a_i)}{n_k}. \quad (2)$$

And, then let V_k be the variance in group $k \in K$ after adjustment:

$$V_k = \sum_{(i,j)\in S_k} (e_{ij} + a_i - A_k)^2. \quad (3)$$

In order to minimize the error, the sum of the variances in each group was minimized by determining the values of decision variables a_i, according to the following unconstrained optimization problem:

$$\min_{a_i} \sum_k V_k = \sum_k \sum_{(i,j)\in S_k} (e_{ij} + a_i - A_k)^2. \quad (4)$$

Since (4) has a convex cost function, existing optimization solvers worked well to obtain a unique optimal solution for each strip. Once a line is adjusted, the estimated decision variable (a_i value) for flight line i was treated as a constant in subsequent iterations for the remaining strips. A custom C program supplemented the non-linear optimization routines found in Excel® to allow the estimation of the decision variables with respect to the tie-line. Once the step errors were adjusted among the point clouds of each strip, the non-linear surface tool of ERDAS® Imagine derived the 3D surface grid. See Willers et al. (2008a) for other details.

2.1.4 Sources of error in agricultural laser scanning data

A commercially prepared bare earth digital elevation model from 2009-2010 (feet Mean Sea Level (MSL)) provides the background layer in Figure 4, while a portion of a research derived, step error corrected digital elevation model (Willers et al., 2008a) from 2003 (m Height Above Ellipsoid (HAE)) is the smaller surface inserted near the top left of Figure 4. With some laser scanning data for at least one agricultural landscape now in hand, we further discuss sources of error for laser scanning data streams and data products.

Sources of error in laser scanning data involving agricultural landscapes can be generally grouped into three categories: systematic errors, random errors, and blunders. Systematic errors in laser scanning data are largely caused by biases in the measurements of bore-sighting parameters that relate to the system components and biases in the measurements made by the system that include Global Positioning System information, timing information, inertial measurements as well as potential biases in the scanner angles and ranges. Random errors arise mostly from the accuracy of the systems measurements including the position and orientation measurements from the Global Positioning System /Internal Navigation System (GPS/INS) component, mirror angles, and ranges. Blunders refer to gross errors that may be caused by the sensor system detecting something in the air (a bird) or some other measurement criterion that causes a very large discrepancy between the real-world surface and the lidar data. Blunders are often detected by identifying points which are statistical outliers in which the offsets between the points in consideration exceed the magnitude of normal random or systematic bias.

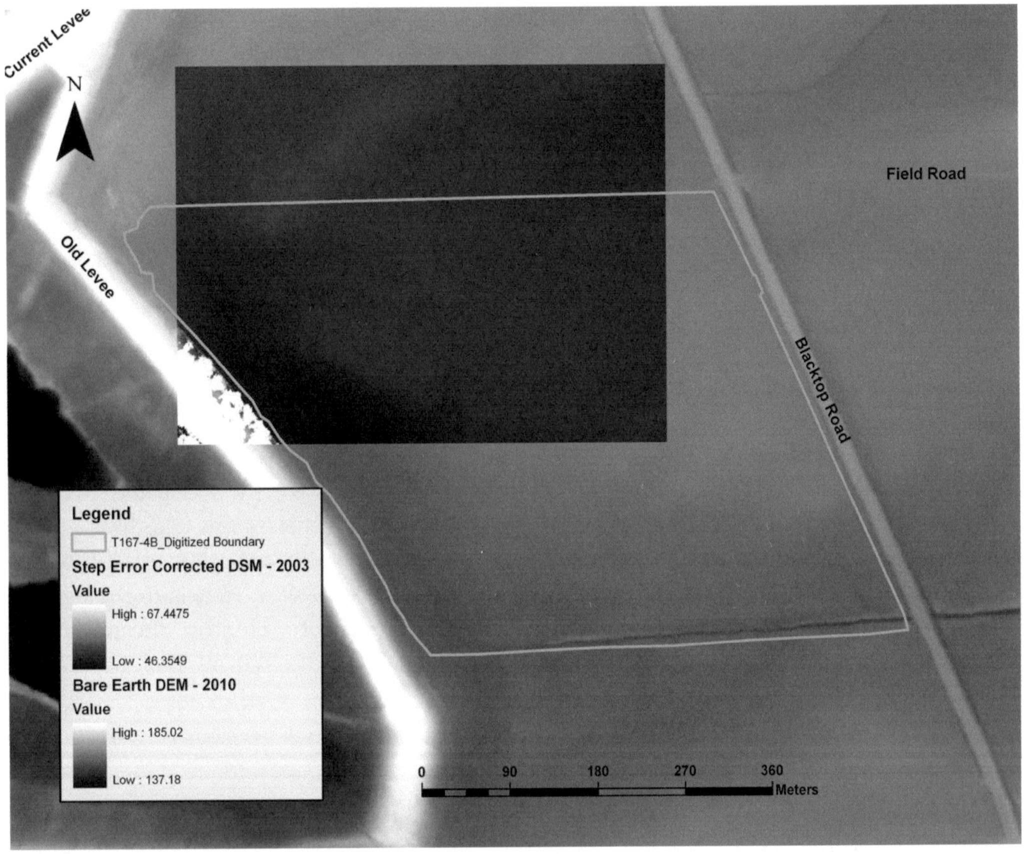

Fig. 4. Geographical detail of the areas of interest (AOI) involving a field location in Bolivar County, Mississippi, USA.

The reader should keep in mind that the literature on sources of error and standards is rapidly changing and quite detailed compared to this simple presentation on these topics (Baltsavias, 1999; Fritsch & Kilian, 1994; Huising & Pereira, 1998; Skaloud & Lichti, 2006; Vosselman, 2002). Nevertheless, our brief examination of sources of error in agricultural laser scanning missions provides a foundation upon which to build support for some '*trains of work*' that comprise other goals of this chapter. A second commercially supplied DEM of another agricultural landscape (Figure 5), located hundreds of kilometers away, is also utilized in this effort.

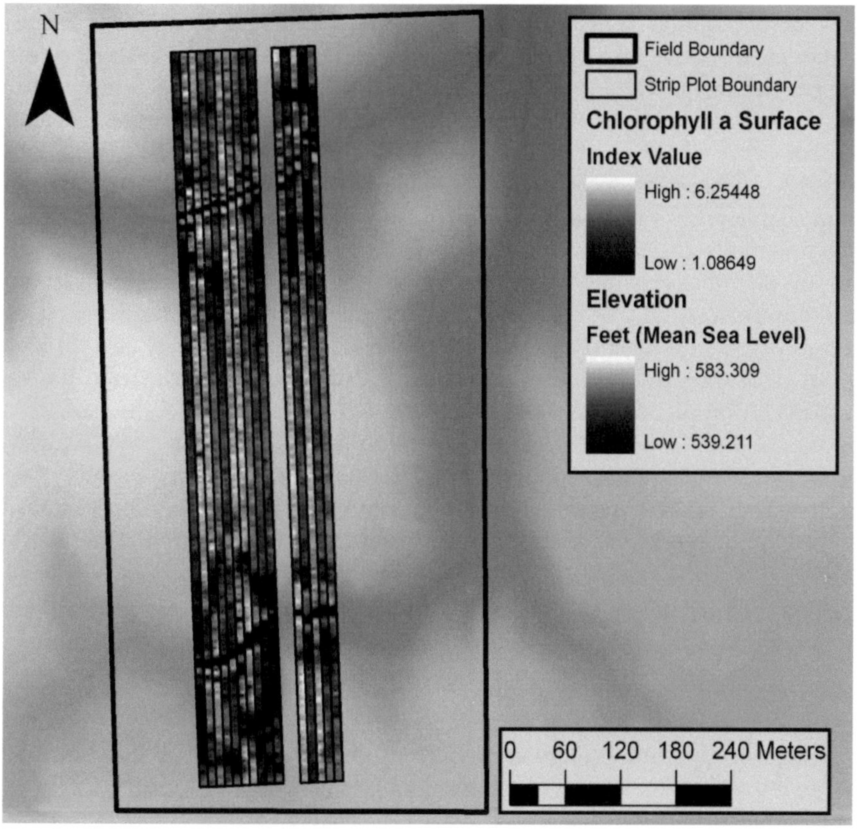

Fig. 5. Illustration of some of the geo-spatial relationships for a site-specific nitrogen experiment in a Nebraska corn field. The background is a laser scanning surface of elevational relief (Feet Mean Sea Level).

Section 2.2 is the first junction for the main journey path of this chapter; the previous discussion points were only collections of *'works'* meant to prepare the reader for some travel across a *'train'* of ideas. This journey covers several concepts involving laser scanning and agriculture and builds toward Section 2.5 as the final junction.

2.2 The population ecology interpretation of pixel attributes

Graphical techniques addressing the resolution of mixed populations of data distributions and other statistical properties of data distributions are discussed in D'Agostino and Stephens (1986) and King (1980). Many of these techniques are valuable in quality control methods and available in various software packages and have great value in image processing, including evaluations of laser scanning digital elevation models.

2.2.1 Population ecology experiments utilizing the attributes of pixels as abstractions of agro-ecosystems

High-resolution, laser scanning and multi-spectral imagery, when resolved with appropriate spectral and temporal resolutions, provide an opportunity to avoid errors in estimation of population statistics (such as mean abundance and interspersion). These digital raster layers permit sample data of an ecological population of interest to be collected from distinct habitats of crop growth (Willers et al., 2005; Willers & Riggins, 2010). When appropriately processed, the raster information can be linked to the ground sample data, allowing for the creation of additional data products describing the population characteristics as a geo-referenced map. The value of being able to build descriptive maps of population characteristics was elegantly discussed by Fleischer et al. (1999). Unlike Fleischer et al. (1999) methods to build their maps, this work, in an elementary fashion, considers the attributes of the image pixels to be surrogates of several important characteristics of biological populations through classification of the imagery attribute values, typically expressed as digital numbers (DN) or brightness values for each pixel of each band in the raster product. These pixel attributes are discrete abstractions of (primarily) the variability in the landscape or plant community structure across the crop. In the case of a digital elevation model, these pixel attributes are a continuous abstraction of the elevation relief of the laser scanned landscape. Therefore, just as is true for traditional data sets of ecological populations obtained by extensive ground survey samples, the collection of raster layer pixels of the agro-ecosystem of interest can have multiple populations of data distributions.

2.2.2 Applications of the probability plot with laser scanning elevation (surface) models

Using laser scanning information for the agricultural landscape contained within the field boundary shown in Figure 4, some issues regarding the step error (Crombaghs et al., 2000; Luethy & Ingensand, 2001; Willers et al., 2008a) are examined by a technique known as probability plotting (D'Agostino & Stephens, 1986).

Inclusion of several local heuristics (e.g., planting date, soil topography, and crop phenology) is useful to best interpret the information provided by the probability plot. For example, since there are up to seven years of time between the two laser scanning missions (Figure 4), a potential question of interest to the producer owning these fields is "What are the estimates of soil erosion rates at different geographical areas in these fields?" However, before answering this question, the prudent analyst should first ask and answer another question "How comparable are the two digital surface elevation models given that different vendors and laser scanning systems produced them?" The probability plot is a useful tool for examination of the second question which leads then to other kinds of decisions involving the first.

It is a small exercise (in a spatial software package) to load, subtract, and then save a new raster layer of the elevational differences between the two laser scanning missions. The difference raster is then exported as a flat file for use in a statistical software package to build the probability plot. Presented in Figure 6 is a probability plot of the difference in laser scanned elevations between 2003 and 2009-2010. The occurrence of several bends and a sharp discontinuity of the attribute values of the output raster created by the subtraction of the two parent rasters clearly show that several unique populations of differences are present, even though the parent rasters of elevation share a common field boundary. Of interest is that the metadata provided by the vendors claims vertical accuracies on the order of 9 cm and 15 cm. The probability plot indicates differences in elevational relief which negatively and positively exceed the maximum tolerance of 15 cm.

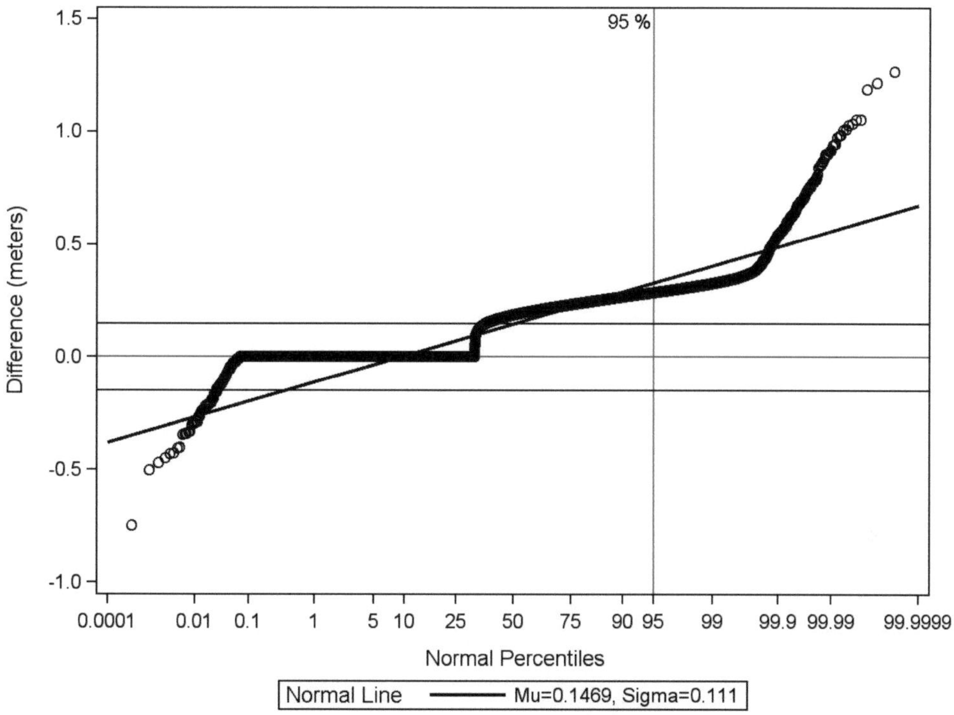

Fig. 6. Probability plot of the elevation difference between a 2009-2019 digital elevation model and a step error corrected (Willers et al., 2008a), 2003 digital elevation model.

In Figure 7 are indications of various geometric patterns in a surface map of change between these two digital elevation models. The east-west linear bands of different widths and intensity are due to different tillage operations between the acquisitions. While each laser scanning data set may have met the standards for each separate mission, the effects of textural change due to tillage and remnant step errors within the 2009-2010 data product, combine to cause combinations of both random and systematic sources of error. The producer's question of interest cannot be effectively answered until (at least) the step error effect in the most recent mission is corrected. The probability plot served a useful purpose in showing sources of different kinds of errors between the two elevation layers.

Traditionally, the description of ecological populations by image analysis is accomplished by applications of one or more classification (Richards & Jia, 1999) procedures to the raster image acquired over the agricultural landscape of interest. However, we have found that it is best to first analyze the raster data content by conversion of the raster image product into a flat file format, which can be statistically processed into a probability plot. Since raster layers can have large numbers of individual pixels, a straightforward way to demonstrate the existence of multiple populations of data distributions is to examine the shape of the probability plot constructed from the flat file of the respective raster layer. If multiple data distributions are present, the plot will not be a straight line (D'Agostino & Stephens, 1986) under the assumption of a single distribution, which is typically the normal distribution (other distributions, such as the exponential, can also be specified). If more than one data distribution is indicated, the next task is to find the meaningful groupings of these populations of data distributions in a manner which relates to the ecological structure of the crop. These meaningful groupings are established through concurrent geographic information system and statistical operations in both the data and geographical spaces of the respective agricultural landscape.

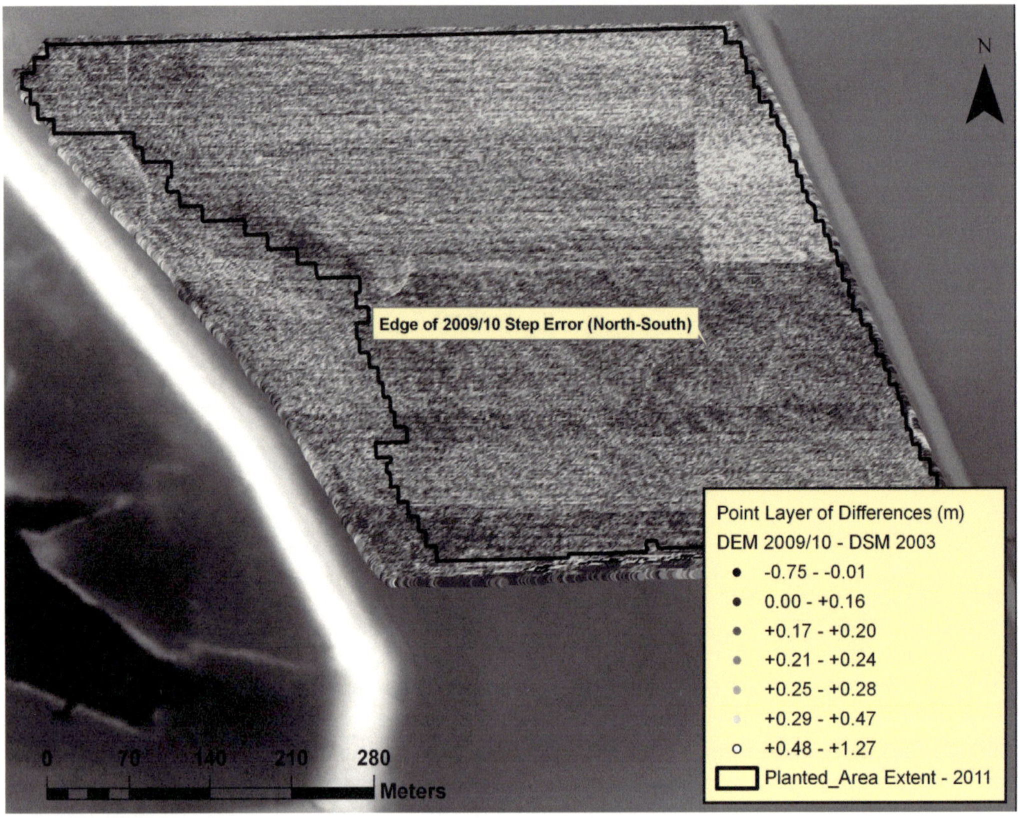

Fig. 7. Potential step error revealed by differences of the 2009/2010 bare earth DEM and the step corrected 2003 DSM.

2.3 Geographical space and data spaces and the Pearson correlation statistic

Previous geographical analysts have elaborated on the concepts of the data space and the geographical space (Berry, 1998; Hargrove & Hoffman, 1999). These concepts merit a brief review at this time and both involve the Cartesian coordinate system (Pignani & Haggard, 1970) as the basic tool for their construct. The elegance and utility (Hogben, 1968 ; Stewart, 2008) of a Cartesian coordinate system can too frequently be undervalued by the agricultural analyst due to too much familiarity. However, with the data density and the spatial resolution obtained by laser scanning digital elevation models, the planar Cartesian coordinate systems referred to as the data space and the geographical space are exceptionally *'illuminating'*.

In Section 2.1, it was discussed that laser scanners create an x, y, z point cloud which can be processed into a surface, or raster layer (de Smith et al., 2007; Lillesand et al., 2008), of elevational relief known as a digital elevation model. Using this surface as an illustration for definitions, the digital elevation model is a map forming a continuous surface, where the system of pixels creates a regular grid of cells over an area (Berry, 1998). The x and y axis position of each pixel cell in the surface grid of elevation represents the information in the geographical space. The z axis of each pixel represents the continuous numeric value of elevation in the data space. If more than one layer of remote sensing information exists for a given area, the data space across the geographical space can be described in more than one dimension (or layer). In such cases, the scale of support (Gotway & Young, 2002) or the congruency of the ground spatial distances of the pixels among the different surfaces (elevation, crop vegetative index, yield, etc.) is an important consideration to guard against source of measurement errors in a geo-spatial analysis (Berry, 1998).

Pearson Correlation Coefficients, N = 428,825 Prob > |r| under H0: Rho=0	b1	ndvi_04	ndvi_11
b1	1.00000	0.30410	0.47413
Corrected Elevation (meters HAE)		<.0001	<.0001
ndvi_04	0.30410	1.00000	0.47168
ATAN NDVI (August 2004)	<.0001		<.0001
ndvi_11	0.47413	0.47168	1.00000
ATAN NDVI (August 2011)	<.0001	<.0001	

Table 1. Tabular representation of the Pearson correlation coefficients describing relationships among three raster layers for the field T167-4B (Figure 4) using information only in the data space without concomitant application of information in the geographical space of these mapped features from the agricultural landscape.

The Pearson correlation statistic is one metric many analysts seem most interested in using with geo-spatial analyses. For many investigators of agricultural systems, Pearson correlation values, such as those presented in Table 1, are typical. In such instances, low values of correlation, while significant, often do not generate an immense level of confidence in using either laser scanning elevation data or imaging data as resources to create a site-specific prescription, or especially build confidence to also go through the expense of preparing one to upload to the controller of a variable-rate equipped farm implement. One reason for reluctance is the sample size (Table 1, top line) involved with raster layers. One traditional dogma is that whenever sample sizes are large enough, significance can be obtained almost anytime. When analyzing raster layers, this traditional view needs careful consideration. Further, such reluctance is particularly acute if the examination of the scatter plots between pixel pairings of two sensor layers is especially non-informative; that is, the scatter plot is without clear representation of either liner or quadratic trends (Figure 8). It is obvious from results found in Table 1 and Figure 8 that without concurrent application of information from the geographical space, the utility of discerning features for site-specific applications is quite limited if information from only the data space is examined.

Conceptual perceptions derived exclusively from the examination and interpretations of only the data space become other extensions of the modifiable areal unit problem (Gotway & Young, 2002). Therefore consequences of an overemphasis upon only the data space of proximal and remote sensor system data streams is unbalanced – it is best to strike a balance among the information content provided from both the geographical and data spaces of these sensor layers (Berry, 1998; Hargrove & Hoffman, 1999). The demonstration of ways to strike such a balance between the data and geographical space domains are the topics of Sections 2.4 and 2.5. In fact, by clever processing in both the data and geographical spaces, the occurrences of experimental evidence of the kind presented in Table 1 or Figures 6 - 8, are actually indicators of opportunities for discovery and progress, particularly if good quality laser scanning DEMS are available.

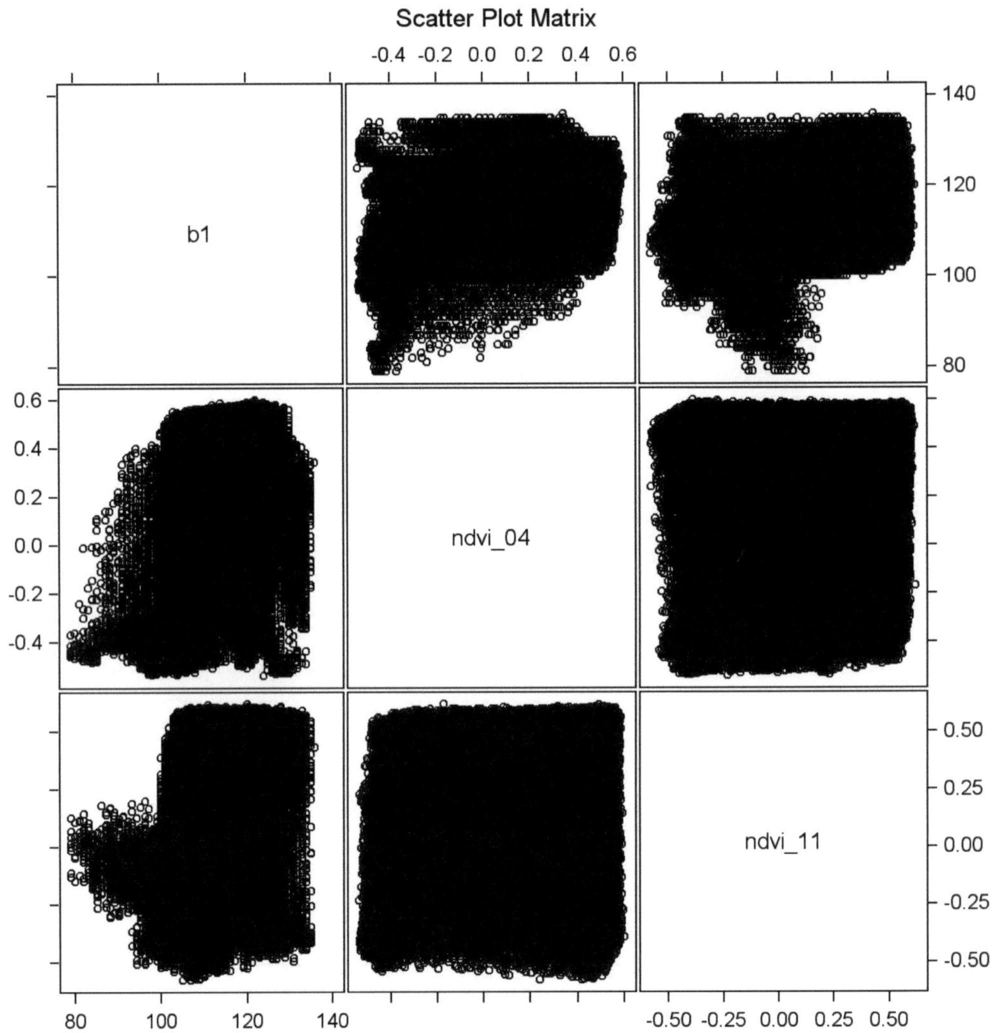

Fig. 8. Scatter plot of the three raster layers for the field T167-4B depicting a graphical representation of the data space of these three raster layers.

2.4 Building crop management zones with laser scanning and remote (or proximal) sensing data streams – development of a categorical, pseudo-likelihood classifier

The task here is to shed '*light*' on how laser scanning digital elevation models contribute an important role in agricultural data analyses of numerous kinds of geo-referenced data streams. Specifically, it is desired to convey an important advance in understanding how to complete statistical analyses of the kind introduced in Section 1.2; especially, whenever both the data attribute and geographical spaces of geo-referenced data streams are concurrently put to work, despite smudged scatter plots or small Pearson correlations. These efforts begin with a terse examination of a technique known as 'maximum likelihood classification' (Strahler, 1980). Additional details and refinements are presented in Willers et al. (2012).

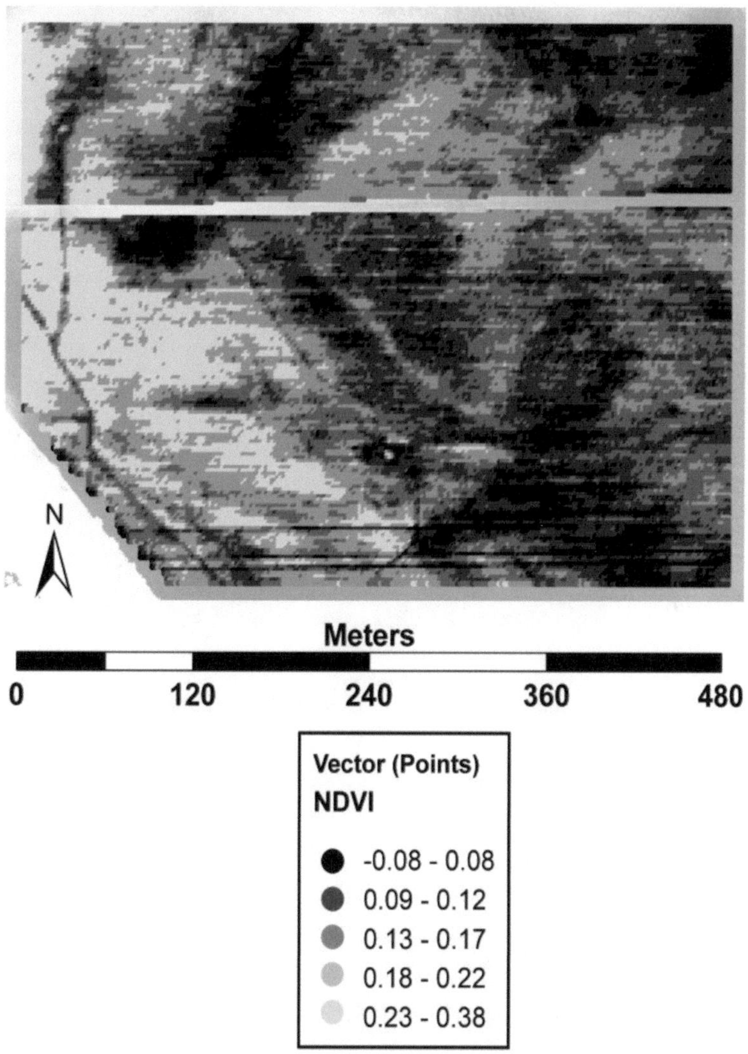

Fig. 9. NDVI ranges as a point vector layer for 2004 subset of two cotton fields.

Figure 9 shows the normalized difference vegetation index (Rouse et al., 1974) representation of the crop conditions for the sub-region previously delineated (Figure 4) during late June of the 2004 production season. In addition to the cotton portion, tall trees are the lightest gray tones beneath the north arrow at the lower left.

Figure 10 presents the laser scanning elevations for the equivalent landscape sub-region shown in Figure 9 (and Figure 4). It was derived from a much larger digital elevation model (Willers et al., 2008a) used to extract this subset for exploratory analyses with the Strahler (1980) algorithm. (Note the trees in the lower left corner, which were excluded in the bare-earth digital elevation model (2009-2010 acquisition) which is the background layer in Figure 4.)

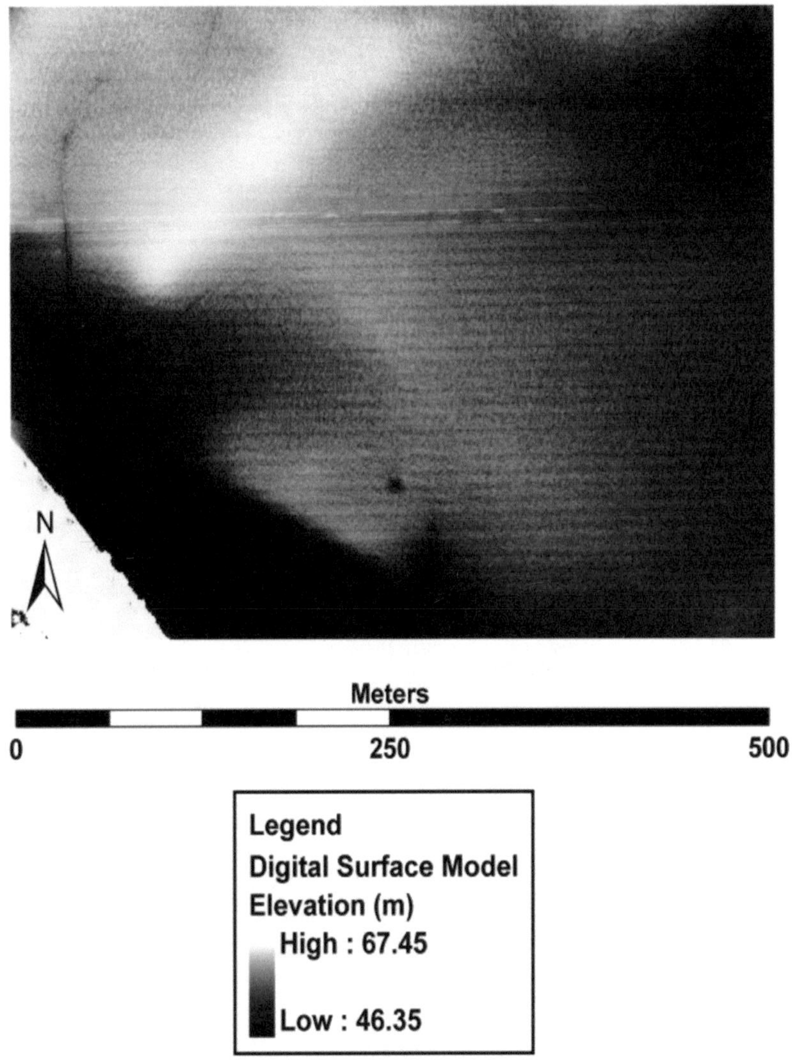

Fig. 10. Step error corrected (Willers et al., 2008a) laser scanning elevations from 2003 for the same region of the two cotton fields.

By making several modifications to the maximum likelihood classification function of Strahler (1980), it is possible to create a new raster layer where the attributes of each pixel are predicted improper probability values, as shown by (5):

$$\hat{p}(\mathbf{x}:\hat{\boldsymbol{\mu}},\hat{\boldsymbol{\Sigma}}) = \left(\left\{ \begin{bmatrix} x_1-\hat{\mu}_1 \\ x_2-\hat{\mu}_2 \end{bmatrix}^T \hat{\boldsymbol{\Sigma}}^{-1} \begin{bmatrix} x_1-\hat{\mu}_1 \\ x_2-\hat{\mu}_2 \end{bmatrix} \right\} \right) \tag{5}$$

where $\hat{\mu}_1$ = the estimated mean normalized difference vegetation index value of that input raster (Figure 9), $\hat{\mu}_2$ = the estimated mean elevation of the laser scanning input raster (Figure 10), and $\sum$-hat provides the estimated covariance parameters between x_1 and x_2 for each pair of input pixels. From an inspection of expression (1), it is obvious that the means or the covariances for normalized difference vegetation index and elevation can be influenced by values from pixels that involve non-crop features. Consequently, it is an important point to remember while processing of the pixels in each input raster layer by (5), that one important geographic information system pre-processing step is to exclude pixels for non-crop features (i.e., trees and field road) that may occur within the field boundary polygon.

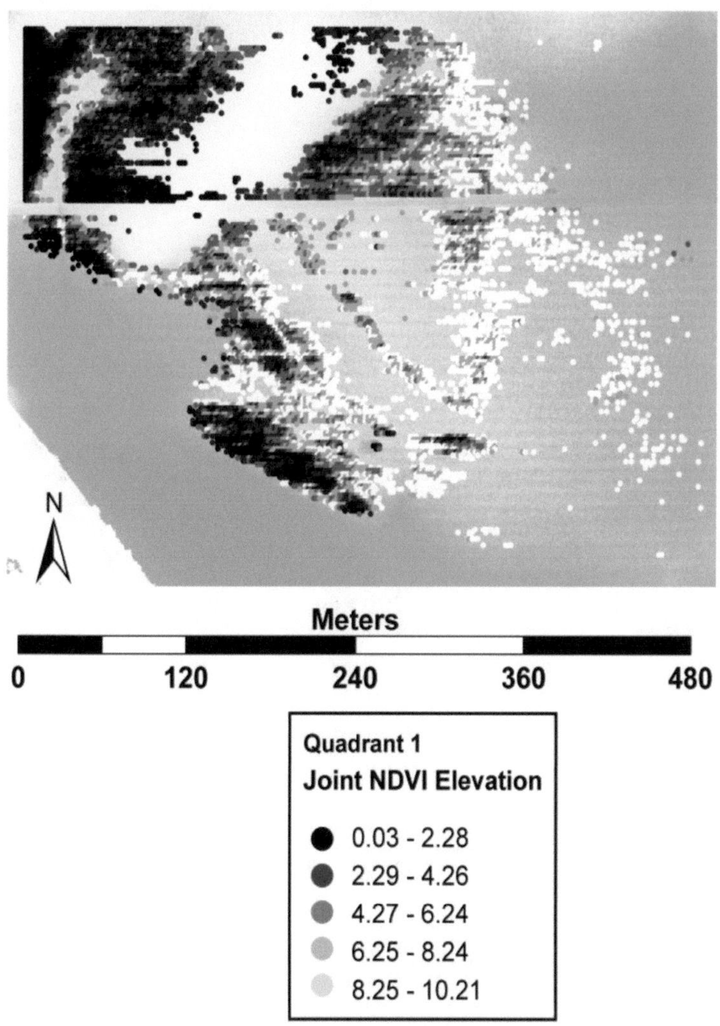

Fig. 11. Joint improper probability map for Quadrant 1.

While working with values from laser scanning elevation and other proximal and remote sensing data streams with equation (5), it was learned that the attributes for the two raster input layers are not required to be of the same units (for example normalized difference vegetation index is unit-less, elevation is in meters (HAE or MSL), and EC_a data is in mS/m). An interesting fact found while using (5) was that many improper probabilities predicted on the left-hand side were of similar magnitudes, whose frequency histogram was concave in shape, often symmetrical, and exhibited higher frequencies to the left and right of a central minimum frequency. Since (5) is the Mahalanobis distance (McLachlan, 1999), and by its form, involves the squaring of positive and negative distances from the centroid, the distance differences of the predicted value of any point pair does not indicate direction with respect to the centroid mean.

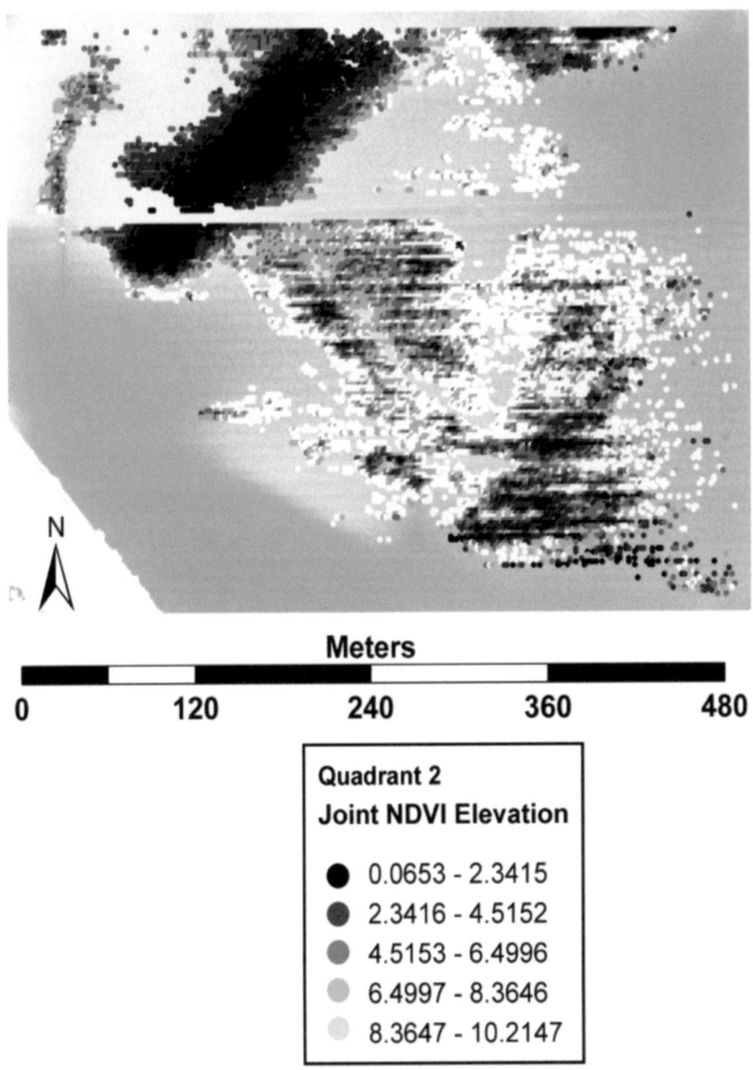

Fig. 12. Joint improper probability map for Quadrant 2.

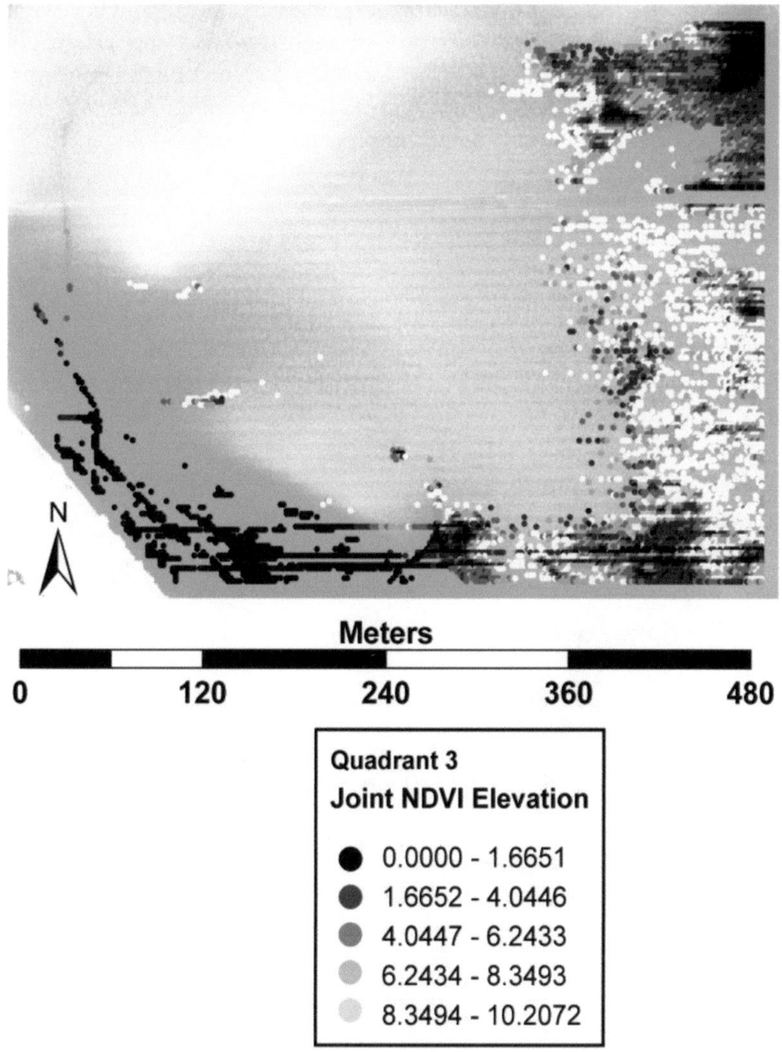

Fig. 13. Joint improper probability map for Quadrant 3.

The second step required was to then determine the Cartesian quadrant for each output pixel with respect to the centroid origin of the data space comprised of the normalized difference vegetation index and elevation. This Cartesian quadrant attribute referenced each output pixel using the traditional labeling (I, II, III, and IV) of a Cartesian coordinate system (Pignani & Haggard, 1970) and defined a new attribute named QUADRANT for the left-hand side predictions (these labels refer to the nominal partitioning of the input raster's data space).

Using these codes the predicted values on the left-hand side of (5) could be displayed in the geographical data space (the UTM coordinate grid), as shown in Figures 11-14. It was remarkably insightful to see that these predicted values, when geographically sorted by their nominal quadrant labels, depicted an irregular but spatially distinctive pattern of dispersion. Such results indicate the advantage gained by the researcher who employs a laser scanning DEM and works concurrently in both the data and geographical spaces. Therefore, it does pay to examine the older literature to learn useful concepts which can refine applications of a newer technology such as laser scanning. After all, often ideas are explored in theory long before technology can produce the methodologies to verify, use, or disprove the ideas.

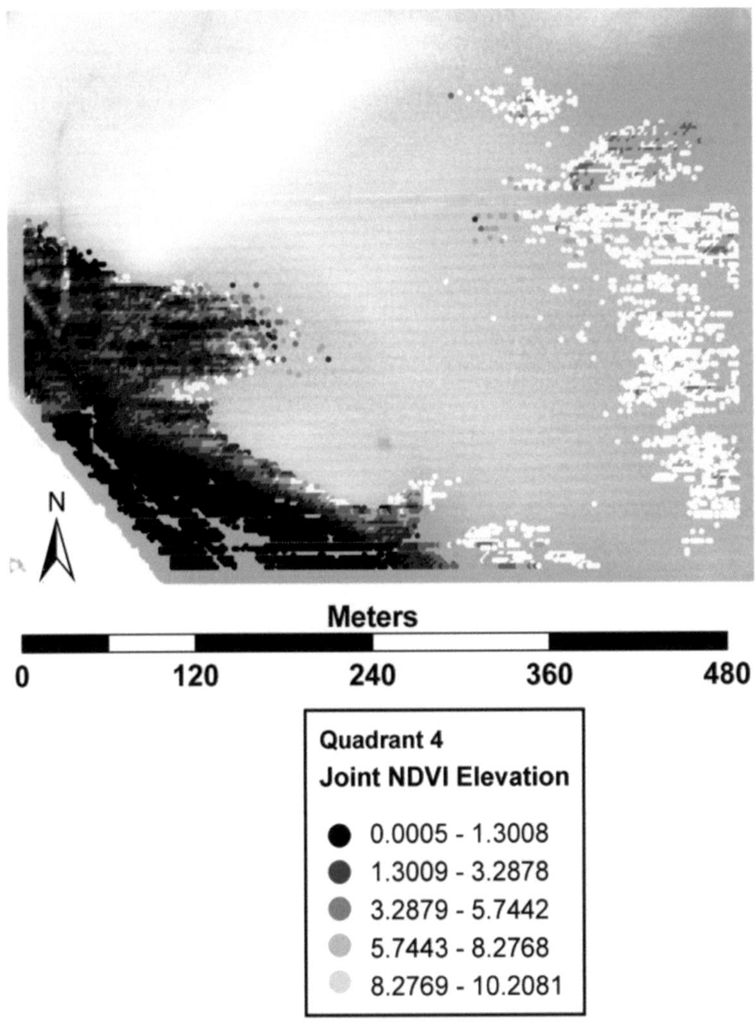

Fig. 14. Joint improper probability map for Quadrant 4.

2.5 Application - nitrogen and corn yields in a Nebraska field

Another area of possible use for laser scanning is for nitrogen (N) management in corn production. Nitrogen management to optimize crop production is a complex process involving such factors as applied N, soil nitrogen supply, crop nitrogen demand, and the economics of profit maximization, all of which can vary spatially and temporally. Because of the complexity of addressing these challenges, current nitrogen management practices generally result in low nitrogen use efficiency (NUE), estimated to be as low as 30-40% for cereal crops, such as corn (Cassman et al., 2002; Raun & Johnson, 1999). Unused nitrogen can eventually contaminate surface and groundwater, creating environmental and health concerns in addition to economic losses for agricultural producers. Low nitrogen use efficiency can be attributed to such things as poor synchronization between soil nitrogen supply and crop demand, uniform application rates of nitrogen fertilizer to spatially variable landscapes, and failure to account for temporally variable influences on crop nitrogen need (Shanahan et al., 2008).

To address the issues of low nitrogen use efficiency, research projects are evaluating the use of active crop canopy sensors to assess in-season plant nitrogen status and apply in real-time spatially-variable nitrogen applications; thereby increasing nitrogen use efficiency (Raun et al., 2002; Solari et al., 2010). Active canopy sensors generate their own source of modulated light and measure canopy reflectance in the visible (400-700 nm) and near-infrared (NIR) (700-1000 nm) parts of the electromagnetic spectrum. Solari et al. (2010) developed an algorithm on small plots in central Nebraska to direct in-season nitrogen applications in corn. Using a sufficiency index (SI), site-specific nitrogen was applied according to the equation:

$$N_{app} = 317 \times \sqrt{0.97 - SI_{sensor}}$$

where SI_{sensor} was the ratio of reflectance measurements from N-stressed to N-sufficient areas. However, they indicated the need to evaluate this algorithm across a broader range of soil and climatic conditions.

Research to address low nitrogen use efficiency has also involved the development of management zones, defined as dividing a field into sub-regions with homogeneous yield-limiting factors or regions of similar production potential (Doerge, 1999). A variety of crop or soil data layers have been used to develop management zones within fields; however, these efforts have produced mixed results, characterizing homogeneous production areas well in some years, but not in others. Schepers et al. (2004), as well as Shanahan et al. (2008), suggested a responsive in-season nitrogen application approach combining management zones and crop-based remote sensing as a possible strategy to increase nitrogen use efficiency.

In 2008, a study was conducted on an irrigated cornfield in central Nebraska to evaluate the algorithm proposed by Solari et al. (2010) against a conventional uniform nitrogen management approach, and, also, to explore the usefulness of an integrated management zone and active sensor approach for improved nitrogen management. The study location consisted of Hastings silt loam and Hastings silty clay loam soils ranging from 0 to 11% slope. The field had substantial change in elevation (~8-10 m), resulting in multiple landscape classifications within the study. Multiple spatial data layers were collected prior to planting to characterize spatial patterns of soil properties within the field. These layers included soil optical reflectance, apparent soil electrical conductivity (EC_a), laser scanning elevation, and slope. The laser scanning for this study was mapped during leaf-off conditions, at a 2-m spatial resolution. The field was also grid soil sampled (Oliver, 2010) to characterize field variation in soil chemical properties.

Hybrid selection, planting date, seeding rate, and field operations were at the producer's discretion. The sensor algorithm proposed by Solari et al. (2010) was evaluated using five different nitrogen application treatments as follows:

1. 45 kg N ha^{-1} at planting (45 At Planting)
2. University of Nebraska-Lincoln soil-based algorithm at planting + split application (per University of Nebraska-Lincoln recommendations)
3. 45 kg N ha^{-1} at planting + sensor algorithm delivered N (45 At Planting + variable-rate)
4. 90 kg N ha^{-1} at planting + sensor algorithm delivered N (90 At Planting + variable-rate)
5. High N (280 kg N ha^{-1}) reference at planting (N-Reference)

Treatments 1 and 5 were included to provide limiting and non-limiting nitrogen conditions to evaluate nitrogen response across the landscape as well as to provide the nitrogen reference (N-Ref) for calibration of the sensor algorithm. Treatment 2 served as a comparison to sensor algorithm treatments 3 and 4, with the nitrogen application rate determined via the University of Nebraska soil-based nitrogen recommendation algorithm. The sensor algorithm treatments 3 and 4 consisted of a combination of at-planting nitrogen (either 45 or 90 kg ha^{-1}) and in-season (~V13-V14 growth stage) nitrogen, with in-season nitrogen rates determined by the sensor algorithm (Solari et al., 2010). A uniform base amount of nitrogen was applied at-planting because previous work (Varvel et al., 1997) has shown that, in high yielding conditions, nitrogen stress prior to the V8 growth stage causes yield losses that cannot be corrected with additional in-season nitrogen application. The purpose of including the two at-planting nitrogen rates (45 and 90 kg nitrogen ha^{-1}) was to determine the appropriate amount of at-planting nitrogen required to avoid an early season nitrogen stress before delivery of in-season nitrogen using the sensor algorithm. Treatment 5 (N-Reference) received 280 kg ha^{-1} at-planting to provide an adequate reference for in-season nitrogen application.

The experimental design consisted of field-length strips (12 cornrows per strip) of each treatment replicated 3 times across the variable landscape. For treatments 1, 2 and 5, nitrogen was applied around planting time at spatially uniform rates. All treatments were applied at the appropriate times and rates using a high-clearance applicator, with the sensor algorithm treatments (3 and 4) being applied at approximately the V13/V14 growth stage at all fields. To determine the in-season nitrogen application rates for the two sensor algorithm treatments, active canopy reflectance sensor readings were first mapped for the N-Ref strips in each replication. Sensor reflectance in visible (VIS_{590}) and near infrared (NIR_{880}) was used to calculate chlorophyll index (CI_{590}) values according to Gitelson et al. (2003, 2005) using the equation:

$$CI_{590} = \frac{NIR_{880}}{VIS_{590}}$$

To acquire sensor readings, four sensors were mounted on the front of a high-clearance vehicle approximately 0.8 to 1.5 m above the crop canopy. The output from each sensor included pseudo-reflectance values for the two parts of the spectrum needed for CI_{590} calculation.

In-season variable nitrogen rates for 45AP + variable-rate and 90AP + variable-rate treatments were determined based on the algorithm described by Solari et al. (2010). This was done by calculating average CI_{590} for each N-Ref treatment. Next, 45AP + variable-rate and 90AP + variable-rate treatments were mapped and additional nitrogen need was determined on-the-go using a sufficiency index (SI) calculated by:

$$SI_{590} = \frac{CI_{target}}{CI_{N\ Ref}}$$

where CI_{target} is the CI_{590} value of a nitrogen target area and $CI_{N\ Ref}$ is the CI_{590} value of a non-nitrogen limiting area. At physiological maturity, the field was harvested by the producer using a commercial combine equipped with a yield monitor and differential global positioning system.

2.5.1 Ordinal categorical partitioning of the data space

In Section 2.3, a brief elaboration of the concepts of the geographical space and the data space was presented. At this time, an additional partitioning of the data space will be introduced — the ordinal categorical partition, which is only made possible through a high resolution, laser scanning digital elevation model.

To establish an ordinal categorical partition for the Cartesian coordinate data space of interest, the origin of reference is that formed by the centroid of the attribute means of any two topographical characteristics. The attribute values of one are plotted on the abscissa while the attribute values of the other (where both are co-located in the geographical space) are plotted on the ordinate axis. In the present case (Figure 15), the mean (x, y) pair, (4.22, 4.76) defines the centroid origin, where x is the natural logarithm of the range transformed (Lillesand et al., 2008, p. 504) apparent soil electrical conductivity ($EC_{a)}$ readings and y is the natural logarithm of the range transformed laser scanning elevation values (feet mean sea level).

Once plotted for ecological investigations, it is useful to recode the elevation and apparent soil electrical conductivity (EC_a) data space into an ordinal, categorical data partition (Figure 15) as opposed to the nominal categorical partition discussed in Section 2.4. To establish this ordinal partition, one examines the sign pairs of the Cartesian coordinate systems data space with respect to the centroid mean. For agriculture, it is reasonable to ordinally recode (as described in Willers et al. 2012) these quadrants in the following order: (a) associate the sign pair (+,+) to topography quadrant Q-IV, (b) the sign pair (+,-) to topography quadrant Q-III, (c) the sign pair (-,+) to topography quadrant Q-II, and (d) the last sign pair (-,-) with topography quadrant Q-I. Consequently, with respect to the statistical analysis domain, these ordinal topography quadrants represent 'topography blocks' within the design structure of the site-specific experiment (Mead, 1988; Milliken and Johnson, 2009).

The abscissa is defined by the natural log of range transformed attributes for the apparent soil electrical conductivity EC_a readings and the ordinate axis is defined by the natural log of range transformed attributes for elevation (feet mean seal level); thus, the origin of this Cartesian system is the centroid means of these two attributes. Each individual point pair in the scatter plot shows the corn yield value according to 15, natural breaks, color ramped classes (see legend inset at left of figure).

Data from Hunnicutt08 was analyzed previously (Roberts et al., 2012) using different classification techniques than those outlined in this chapter. In their work, Roberts et al. (2012) evaluated the relationship between crop response variables (CI_{590} and Yield) and apparant soil electrical conductivity (EC_a), soil optical reflectance, and landscape topography. In the Hunnicutt08, apparent soil electrical conductivity was significantly related to CI_{590} and yield, and was subsequently used to delineate management zones using the software Management Zone Analyst (University of Missouri, USDA-ARS, Columbia, MO). Management Zone Analyst delineated 2 zones within the field, with spatial patterns closely aligned with topography quadrants 1 & 3 and 2 & 4 of Figure 16. Higher positions in the landscape for this field ($Zone_{MZA}$ 1 and topography quadrants 2 & 4) corresponded to higher organic matter and more productive soils, while lower areas in the landscape corresponded to eroded drainage ways ($Zone_{MZA}$ 2 and topography quadrants 1 & 3).

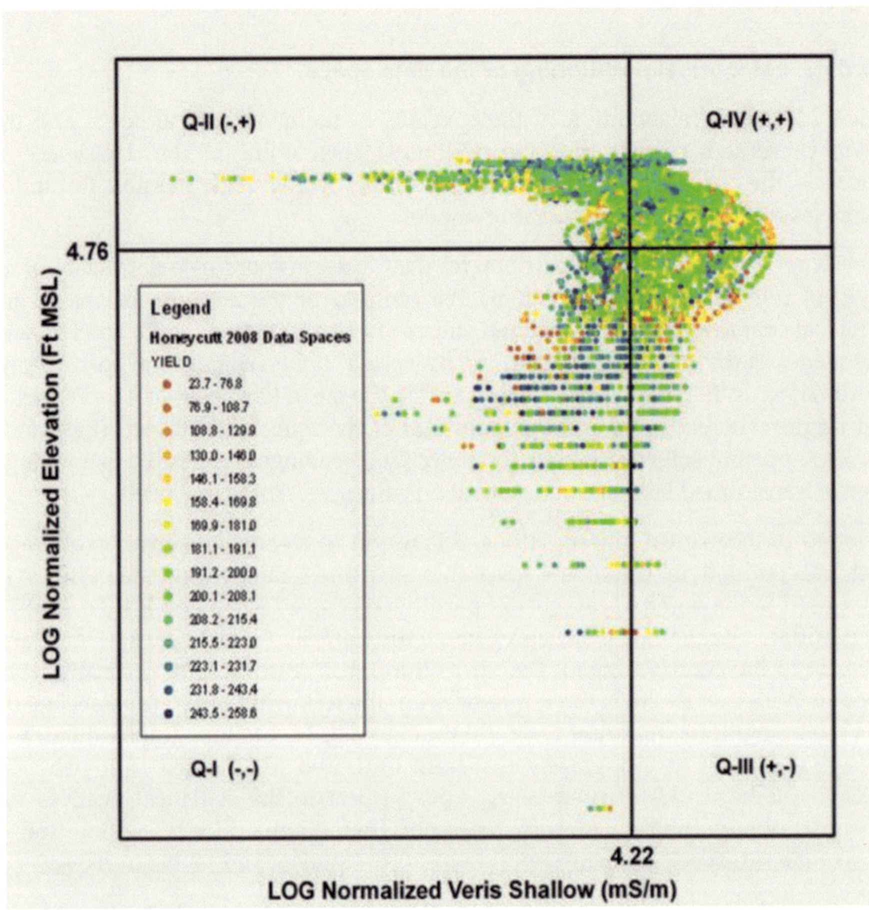

Fig. 15. The Hunnicutt 2008 nitrogen experiment structured in the data space according to (1) topography quadrants.

Roberts et al. (2012) concluded that the sensor-based algorithm used in their study may need to be adjusted according to management zones to account for differences in crop nitrogen response. In addition to the proximal ground-based sensors used to delineate zones by Roberts et al. (2012), spatial patterns identified in Figure 16 suggest that laser scanning digital elevation models would also be useful to identify spatial patterns of soil variability and crop response to nitrogen.

The floating plots in Figure 16 would require a variable-rate ground sprayer that can apportion its application swath into polygons that are 9.2 m wide by 18 m long, to apply the alternate management practice in each specific topography zone (that is, the four zones indicated by the red, yellow, green and blue colors).

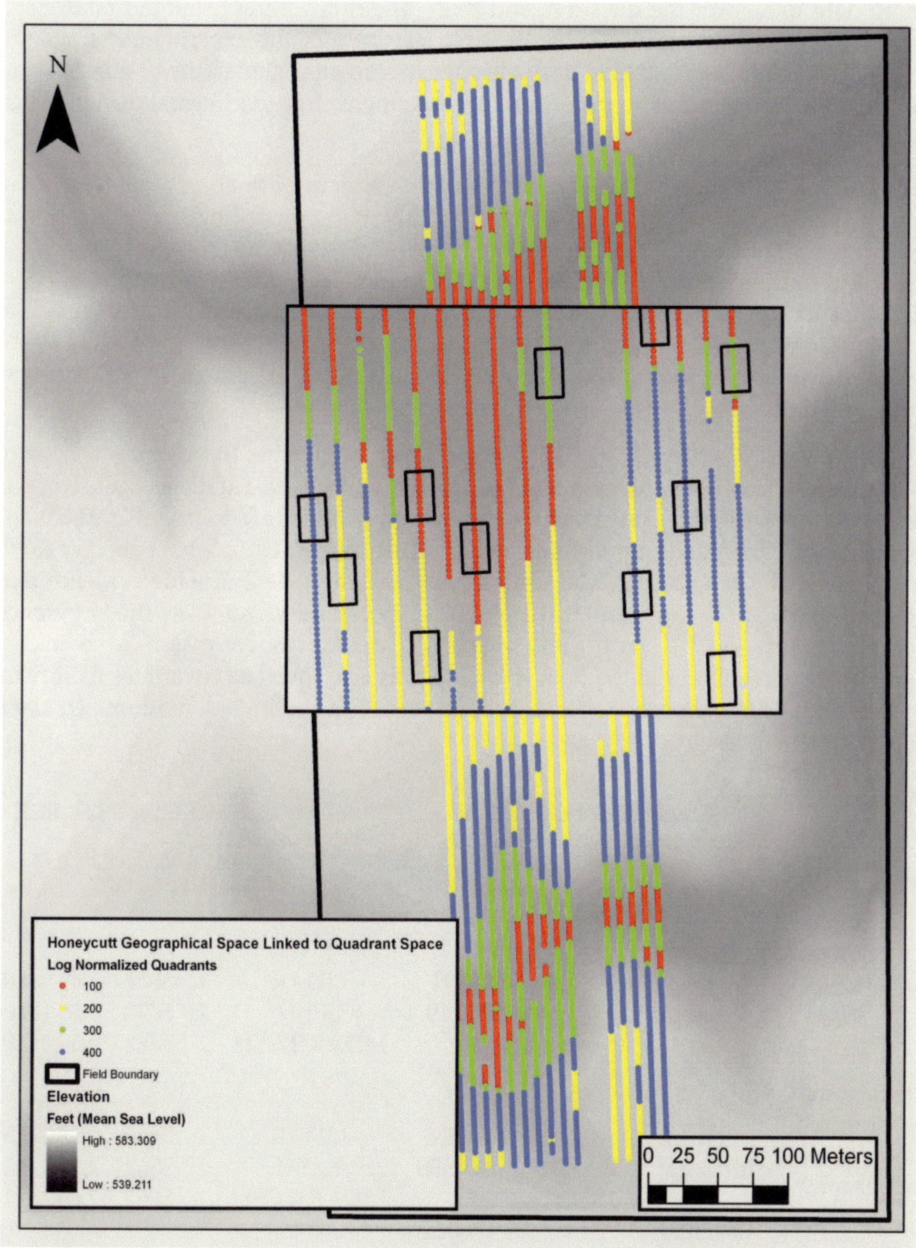

Fig. 16. The strip-plot plan of the Hunnicutt 2008 corn nitrogen experiment showing the topography blocks (see text) and examples of imbedded floating plots within the harvest paths of the combine.

2.5.2 Value of topographical partitions for site-specific experimental designs

The chief aim of a good experimental design is to (1) define the question to be tested, (2) define the experimental units and apportion these experimental units into homogenous populations, (3) define and describe the appropriate treatments or treatment combinations to obtain data to answer the question, and then (4) employ an appropriate randomization scheme to assign the treatments to the sensible structure of the experimental units (Mead, 1988; Milliken & Johnson, 2009). In agricultural experiments, availability of a laser scanning digital elevation model leads to significant improvements in experimental design. Evidence of this capability is presented in this section.

In Section 1.2, information addressing the issue of a system of floating plots was discussed (see also Milliken et al., 2010). The methodology involving an ordinal, categorical partition of two topographical attributes represents the first description of how to establish the geographical location of these floating plots in commercial fields. This method of choosing floating plot locations exploited the data and geographical spaces of information obtained by a laser scanning system and a second type of sensor system. More research is necessary to define the minimum size of these floating plots for optimal efficiency in a site-specific experimental design.

This same procedure generates another process which establishes the geographical extent of an asymmetrical, irregularly shaped set of topography blocks as a statistical construct useful for inclusion within the design structure component (Milliken & Johnson, 2002, 2009) of a site-specific experimental design (Milliken et al., 2010; Oliver, 2010; Schabenberger & Pierce, 2002; Willers et al., 2004, 2008b). The results presented in Table 2 provide evidence that the topography zones (as geographical 'blocks') successfully remove the influence of topography effects on the crop yield response variable as compared to where these topography layer attributes in the data space are only employed as covariates (compare sets of P-values at the far right column) and if the analysis is a traditional, randomized complete block experimental design.

Experimental Design Type	Covariance Parameters			Tests of Fixed Effects (Type 3)				
	Cov Parm	Subject	Estimate	Effect	Num DF	Den DF	F Value	Pr > F
Traditional (Randomized Complete Block)	Intercept	BLOCK	1.8393	LIDAR	1	5871	5.43	0.0198
	Residual		1094.19	ECSH	1	5871	9.59	0.0020
				LIDAR*ECSH	1	5871	9.38	0.0022
Site-Specific (Randomized Complete & Topography Blocks)	Intercept	T_BLOCK	2.8635	LIDAR	1	5862	0.28	0.5974
	Intercept	BLOCK* T_BLOCK	23.6590	ECSH	1	5862	2.15	0.1423
				LIDAR*ECSH	1	5862	2.08	0.1493
	Residual		1080.23					

Table 2. Summary statistics for two situations, where the topography covariates are employed for a traditional randomized complete block experimental design or are employed as covariates for a site-specific topography block experimental design.

3. Conclusion – agricultural laser scanning's '*trains and troops of works*'

This chapter is the logical supplement of two previously published works (Willers et al., 2009a; Willers & Riggins, 2010). A common theme of the collection is the illustration of how geo-spatial information of appropriate spatial, temporal and spectral resolution is a valuable resource for agro-ecological investigations. This work concludes with two major points.

The first point is that without the development of a suite of adequate tools and procedures to manage the copious flows of information for the experimenter, consultant, farm technician, farm supplier, or producer, the *fruit* that laser scanning provides for insight into the structure and function of agro-ecological systems will never be harvested. At the present time, the application of laser scanning and other remote sensing tools resides in the domain of specialists and not in the domain of the agriculturalist. The answer(s) needed to achieve a shift in the domain of usage and audience is not an easily resolved problem.

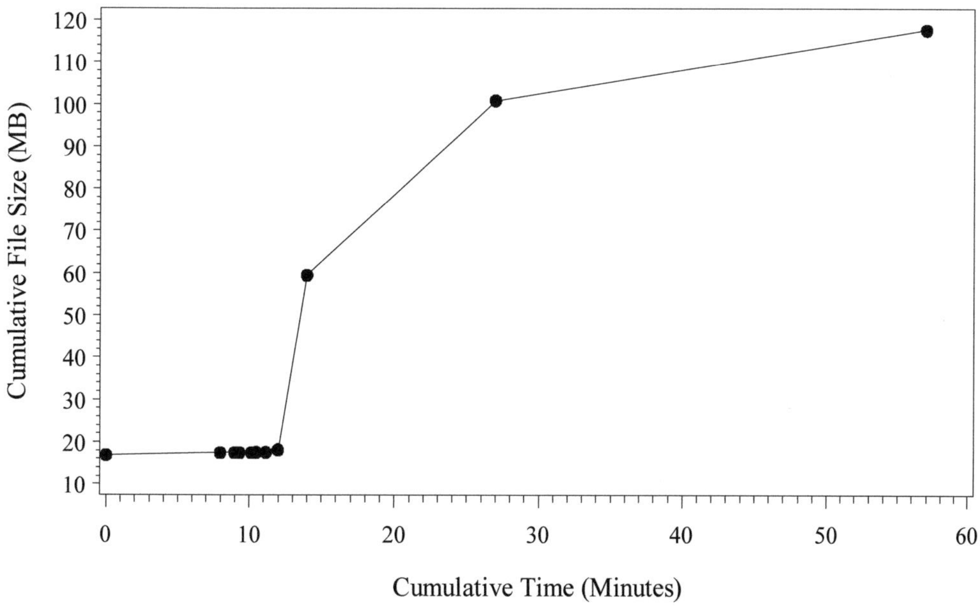

Fig. 17. Illustration of the relationship between cumulative time and cumulative file size for a task combining laser scanning elevation model with a raster layer of arctan normalized difference vegetation index values for one field on a commercial farm.

One simple example provides some indication of the kind of focus required among cross disciplinary skills and task(s) to make such a shift in usage. Employment of methods like those of Milliken et al. (2010) or Oliver (2010) indicates that usage of these mathematical approaches is limited because a comprehensive data processing and computing infrastructure for on-farm, in the field agricultural activities, does not exist (Schuster et al., 2011). We provide on example of the problem. Figure 17 shows the cumulative time required for a specialist to complete a task involving the combining of a digital elevation model layer with a normalized difference vegetation index (NDVI) layer to produce an output layer similar to Figures 11-14. The figure shows that the total time required can approach 60 minutes, while the cumulative file sizes involved increase to just a few megabytes. In real-farm applications, gigabytes of data are collected. Interestingly, the step involving some automation (the sharp inflection point near 12 min) is a step completed rather promptly.

Unless a formal infrastructure for precision agriculture is developed that significantly reduces processing time and establishes interoperability, all of the theories, ideas, techniques, data, and mathematical models developed through years of government and university research and industry investment will be underutilized or fall into disuse. Consequently, a "Henry Ford" type of construct is needed to reduce the time required to process and produce meaningful analyses for clients and reduce the amount of labor needed. For such a complex, multifaceted problem, it will take multiple consortiums of investigators to discover ways to make laser scanning information and other remote sensing data streams affordable and easily available to agricultural systems. Aside from establishing the capability of gathering data from using sensors in the field and on farm machinery, there is the overriding need to promptly use the huge amounts of data for rapid decision-making. At its core, the fundamental limitation on data-intensive agriculture is the lack of interoperability for data in different formats and the time constraints between data collection and results being available for the end user.

The second concluding point is the opportunity and need for additional confirmatory experimentation, built on exploratory experimentation procedures introduced in Sections 2.2 - 2.5. If a probability plot examination of the features of an agricultural landscape indicate the presence of more than one data distribution (D'Agostino & Stephens, 1986), then concurrent processing in the geographical space is required. Creation of a Cartesian coordinate system whose origin is a centroid formed by the arithmetic means of the data space obtained from two sensor systems, where at least one is elevation mapped by laser scanning, should reveal different autocorrelations among groups established by the categorical data partition of such a centroid. If differences in spatial autocorrelation among categorically derived groups are evident, then such evidence dictates that different management zones exist in the field and each requires different rules for their site-specific management. Without access to laser scanning information, investigators could model the spatial autocorrelation of their data attributes with an isotropic semivariogram and consequently not recognize the reality that more than one spatial random field (Oliver, 2010; Schabenberger & Pierce, 2002) determines the properties of the first (the mean) and second (the variance) moments of the data space comprised of the measured variables of interest. Thus, the modifiable areal unit problem, when examined in the '*light*' provided by laser scanning digital elevation models, is actually an indication of opportunity (not problems) with respect to the goals and philosophy of precision agriculture (Barnes et al., 1996; Moran et al., 1997; Oliver, 2010; Plant et al., 2001; Willers & Riggins, 2010).

4. Acknowledgment and disclaimer

Appreciation is expressed to Ronald E. Britton for assistance with preparation of the chapter and for providing helpful suggestions to improve the manuscript.

Mention of trade names or commercial products in this publication is solely for the purpose of providing specific information and does not imply recommendations or endorsement by the US Department of Agriculture.

5. References

Aguilar, F.J., Mills, J.P., Delgado, J., Aguilar, M.A., Negreiros, J.G., & Pérez, J.L. (2010). Modelling vertical error in LiDAR-derived digital elevation models. *ISPRS Journal of Photgrammetry and Remote Sensing*, 65, pp. 103-110

Anonymous. (2000). *Spectral Visions' 2000 Year End Report to NASA's Geospace Applications and Development Directorate*, Stennis, MS, USA, December 11, 2000

ASPRS. (2004). *ASPRS Guidelines: Vertical Accuracy Reporting for Lidar Data.* Version 1.0. American Society for Photogrammetry and Remote Sensing (ASPRS) Lidar Committee (PAD), May 15, 2004

Baltsavias, E. P. (1999). Airborne laser scanning: basic relations and formulas. *ISPRS Journal of Photogrammetry and Remote Sensing,* 54, pp. 199–214

Barnes, E. M., Moran, M., Pinter, Jr., P., & Clarke, T. (1996). Multispectral remote sensing and site-specific agriculture: Examples of current technology and future possibilities, *Proceedings of the International Conference on Precision Agriculture*, p. 845-854, Minneapolis, MN, June 1996

Berry, J. (1998). Topic 7: Linking Data Space and Geographic Space, In: *Beyond Mapping III*, Berry, J.K., Retrieved from www.innovativegis.com/basis/

Birrell, S.J., Sudduth, K. A., & Borgelt, S. C. (1996). Comparisons of sensors and techniques for crop yield mapping. *Computers and Electronics in Agriculture*, 14, pp. 215-233

Bugayevskiy, L., & Snyder, J. (1995). *Map Projections. A Reference Manual*, Taylor and Francis, Philadelphia, PA

Burris, E., Burns, D., McCarter, K., Overstreet, C., Wolcott, M., & Clawson, E. (2010). Evaluation of the effects of Telone II (fumigation) on nitrogen management and yield in Louisiania delta cotton. *Precision Agriculture,* 11, pp. 239-257

Campanella, R. (2000). Testing components toward a remote-sensing-based decision support system for cotton production. *Photogrammetric Engineering and Remote Sensing*, 66, 10, pp. 1219-1228

Cassman, K.G., Dobermann, A. & Walters, D.T. 2002. Agroecosystems, nitrogen-use efficiency, and nitrogen management. *Ambio,* 31, pp. 132–140

Crombaghs, M.J.E., Brügelmann, R., & de Min, E.J. (2000). On the adjustment of overlapping strips of laser altimeter height data. *International Archives of Photogrammetry and Remote Sensing,* 33 (B3/1), pp. 230–237

D'Agostino, R. B., & Stephens, M. A. (1986). *Goodness-of-Fit Techniques*, Marcel Dekker, ISBN 0-8247-7487-6, New York

de Smith, M., Goodchild, M., & Longley, P. (2007). *Geospatial Analysis. A Comprehensive Guide to Principles, Techniques and Software Tools,* Matador, ISBN 1-905886-60-8, Leicester, UK

Doerge, T. (1999). Management zone concepts. SSMG-2. In: *Site Specific Management Guidelines,* Clay et al., International Plant Nutrition Institute, Norcross, GA, Retrieved from http:// www.ipni.net/ppiweb/ppibase.nsf/$webindex/article=375FAC448525695A00559405CF15E6B8

Donner, M., Eble, K., & Helbling, R. (1968). Francis Bacon. The New Organon. In: *The Intellectual Tradition of the West. Vol. 2*, Donner, M., Eble, K. E., & Helbling, R. E., pp. 106-120, Scott, Foresman and Company, Glenview, IL

Dupont, J., Campenella, R., Seal, M., Willers, J., & Hood, K. (2000). Spatially variable insecticide applications through remote sensing, *Proceedings of the Beltwide Cotton Conference*, Vol. 1, pp. 426- 429, San Antonio, TX

Elvidge, C., & Chen, Z. (1995). Comparison of broad-band and narrow-band red and near-infrared vegetation indices. *Remote Sensing of the Environment*, 54, pp. 38-48

FEMA. (2003). Appendix A: Guidance for Aerial Mapping and Surveying. In: *Guidelines and Specifications for Flood Hazard Mapping Partners*, Federal Emergency Management Agency (FEMA), April, 2003

FGDC. (1998). *Geospatial Positioning Accuracy Standards. Part 3: National Standard for Spatial Data Accuracy*. Subcommittee for Base Cartographic Data, Federal Geographic Data Committee (FGDC), c/o U.S. Geological Survey, Reston, VA

Fleischer, S. J., Blom, P. E., & Weisz, R. (1999). Sampling in precision IPM: when the objective is a map. *Phytopathology* 89, pp. 1112-1118

Frigden, J., Seal, M., Lewis, M., Willers, J., & Hood, K. (2002). Farm level spatially-variable insecticide applications based on remotely sensed imagery. *Proceedings of the Beltwide Cotton Conference*, Atlanta, GA (Unpaginated Proceedings on CD)

Fritsch, D., & Kilian, J. (1994). Filtering and calibration of laser scanner measurements. *International Archives of Photogrammetry and Remote Sensing*, 30, 3, pp. 227–234

Gitelson, A.A., Gritz, Y., & Merzlyak, M. N. (2003). Relationships between leaf chlorophyll content and spectral reflectance and algorithms for non-destructive chlorophyll assessment in higher plant leaves. *Journal Plant Physiology* 160, pp. 271-28

Gitelson, A.A., Vina, A., Ciganda, V., Rundquist, D. C., & Arkebauer, T. J. (2005). Remote estimation of canopy chlorophyll content in crops. *Geophysical Research Letters* 32(L08403), pp. 1.-4

Gotway, C. A., Bullock, D. G., Pierce, F. J., Stroup, W. W., Hergert, G. W., & Eskridge, K. M. (1997). Experimental design issues and statistical evaluation techniques for site-specific management. Pierce, F. J., and Sadler, E. J., (Eds.) *The State of Site Specific Management for Agriculture*, ASA-CSSA-SSSA, Madison, WI

Gotway, C. A., & Hartford, A. H. (1996). Geostatistical methods for incorporating auxiliary information in the prediction of spatial variables. *Journal of Agricultural, Biological, and Environmental Statistics*, 1, pp. 17-39

Gotway, C. A., & Stroup, W. W. (1997). A generalized linear model approach to spatial data analysis and prediction. *Journal of Agricultural, Biological and Environmental Statistics*, 2, pp. 157-178

Gotway, C. A., & Young, L. J. (2002). Combining incompatible spatial data. *Journal of the American Statistical Association*, 97, pp. 632-648

Han, S.H., Heo, J., Sohn, H.G, & Yu, K. (2009). Parallel processing method for airborne laser scanning data using a PC cluster and a virtual grid. *Sensors*, 9, pp. 2555-2573

Hargrove, W., & Hoffman, F. (1999). Using multivariate clustering to characterize ecoregion borders. *Computing in Science and Engineering*, (July/August 1999), pp. 18-25

Hogben, L. (1968). *Mathematics for the Million* (4th edition), W. W. Norton & Company, New York, NY

Huising, E. J., & Pereira, L. M. (1998). Errors and accuracy estimates of laser data acquired by various laser scanning systems for topographic applications. *ISPRS Journal of Photogrammetry and Remote Sensing*, 53, 5, pp. 245–261

Jensen, J. R. (2000). *Remote Sensing of the Environment: An Earth Resource Perspective,* Prentice-Hall, ISBN-0-13-489733-1, Upper Saddle River, NJ

Kennedy, M. (1996). *The Global Positioning System: An Introduction,* Ann Arbor Press, Chelsea, MI

King, J. R. (1980). *Frugal Sampling Schemes, Technical and Engineering Aids for Management,* Tamworth, NH

Lillesand, T., Kiefer, R., & Chipman, J. (2008). *Remote Sensing and Image Interpretation,* John Wiley & Sons, ISBN 978-0-470-05245-7, Hoboken, NJ

Littell, R. C., Milliken, G. A., Stroup, W. W., Wolfinger, R. D., & Schabenberger, O. (2006). *SAS® for Mixed Models,* SAS Institute, ISBN 978-1-59047-500-3, Cary, NC

Long, J. (1997). *Regression Models for Categorical and Limited Dependent Variables,* Sage Publications, ISBN 978-0-8039-7374-9, Thousand Oaks, CA

Luethy, L., & Ingensand, H. (2001). How to evaluate the quality of airborne laser scanning data. *International Archives of Photogrammetry, Remote Sensing Spatial Information Science, 36,* 8/W2, pp. 313–317

Maas, H. (2002). Methods for measuring height and planimetry discrepancies in airborne laserscanner data, *Photogrammetric Engineering & Remote Sensing,* 68, pp. 933-940

McKinion, J. M., Jenkins, J. N., Willers, J. L., & Zusmanis, A. (2009). Spatially variable insecticide applications for early season control of cotton insect pests. *Computers and Electronics in Agriculture,* 67, pp. 71-79

McKinion, J. M., Willers, J. L., & Jenkins, J. N. (2010a). Spatial analyses to evaluate multi-crop yield stability for a field. *Computers and Electronics in Agriculture,* 70, pp. 187-198

McKinion, J. M., Willers, J. L., & Jenkins, J. N. (2010b). Comparing high density LIDAR and medium resolution GPS generated elevation data for predicting yield stability. *Computers and Electronics in Agriculture,* 74, pp. 244-249

McLachlan, G. J. (1999). Mahalanobis distance. *Resonance,* June, pp. 20-26

Mead, R. (1988). *The Design of Experiments: Statistical Principles for Practical Application,* Cambridge University, ISBN 0-521-28762-6, Cambridge, UK

Milliken, G. A., & Johnson, D. E. (2002). *Analysis of Messy Data, Vol. 3. Analysis of Covariance,* Chapman and Hall/CRC, ISBN 1-58488-083-X, New York

Milliken G.A., & Johnson, D.E. (2009). *Analysis of Messy Data, vol. 1, 2nd edn. Designed Experiments,* Chapman and Hall/CRC, ISBN 978-1-58488-334-0, New York

Milliken, G., Willers, J., McCarter, K., & Jenkins, J. (2010). Designing experiments to evaluate the effectiveness of precision agricultural practices on research fields: Part 1. Concepts for their formulation. *Operational Research International Journal,* 10, 3, pp. 329-348

Moran, M. S., Inoue, Y., & Barnes, E.M. (1997). Opportunities and limitations for image-based remote sensing in precision crop management. *Remote Sensing in the Environment,* 61, pp. 319-346

NDEP (2004). *Guidelines for Digital Elevation Data. Version 1.0,* National Digital Elevation Program (NDEP), May 10, 2004

NOAA. (2008). *Lidar 101: An Introduction to Lidar Technology, Data, and Applications,* National Oceanic and Atmospheric Administration (NOAA) Coastal Services Center, Charleston, SC

Oliver, M. (Ed.) (2010). *Geostatistical Applications for Precision Agriculture,* Springer, ISBN 978-90-481-9132-1, Dordrecht, The Netherlands

Piepho, H-P., Richter, C., Spilke, J., Hartung, K., Kunick, A., & Thöle, H. (2011). Statistical aspects of on-farm experimentation. Crop and Pasture Science, 62, pp. 721-735.

Pignani, T. J., & Haggard, P. W. (1970). *Modern Analytic Geometry,* D. C. Heath and Company, Lexington, MA

Plant, R. E., Munk, D. S., Roberts, B. R., Vargas, R. N., Travis, R. L., Rains, D. W., & Hutmacher, R. B. (2001). Application of remote sensing to strategic questions in cotton management and research. *Journal of Cotton Science,* 5, pp. 30-41

Pouncey, R., Swanson, K., & Hart, K. (Eds.) (1999). *ERDAS Field Guide* (5th edition), ERDAS, Atlanta, GA

Raun, W.R., & Johnson, G.V. (1999). Improving nitrogen use efficiency for cereal production. *Agronomy Journal,* 91, pp. 357–363.

Raun, W.R., Solie, J.B., Johnson, G.V., Stone, M.L., Mullen, R.W., Freeman, K.W., Thomason, W.E., & Lukina, E.V. (2002). Improving nitrogen use efficiency in cereal grain production with optical sensing and variable rate application. *Agronomy Journal* 94, pp. 815-820

Richards, J. A., & Jia, X. (1999). *Remote Sensing Digital Image Analysis. An Introduction* (3rd edition), Springer-Verlag, ISBN 3-540-64860-7, Berlin

Roberts, D. F., Ferguson, R. B., Kitchen, N. R., Adamchuk, V. I., & Shanahan, J. F. (2012). Relationships between soil-based management zones and canopy sensing for corn nitrogen management. *Agronomy Journal* (In Press)

Rouse, Jr., J. W., Haas, R. H., Deering, D. W., Schell, J. A., & Harlan, J. C. (1974). *Monitoring the Vernal Advancement and Retrogradation (Greenwave Effect) of Natural Vegetation,* NASA/GSFC Type III Final Report, Greenbelt, MD

Sadler, E. J., Busscher, W. J., Bauer, P. J., & Karlen, D. L. (1998). Spatial scale requirements for precision farming: a case study in the southeastern USA. *Agronomy Journal,* 90, pp. 191-197

Sadler, E. J,. & Russell, G. (1997). Modeling crop yield for site-specific management. In: *The State of Site-Specific Management for Agriculture,* Pierce, F. J., and Sadler, E. J. (Eds.), ASA, pp. 69-79, Madison, WI

Schabenberger, O., & Pierce, F. J. (2002). *Contemporary Statistical Models for the Plant and Soil Sciences,* CRC Press, ISBN 1-58488-111-9, Baca Raton, FL

Schepers, A.R., Shanahan, J.F., Liebig, M. A., Schepers, J. S., Johnson, S. H., & Luchiari, Jr, A. (2004). Appropriateness of management zones for characterizing spatial variability of soil properties and irrigated corn yields across years. *Agronomy Journal,* 96, pp. 195-203

Schuster, E.W., Lee, H-G., Ehsani, R., Allen, S.J., & Rogers, J.S. (2011). Machine-to-machine communication for agricultural systems: an XML-based auxiliary language to enhance semantic interoperability. *Computers and Electronics in Agriculture,* 78, pp. 150-161

Shanahan, J. F., Kitchen, N. R., Raun, W. R., & Schepers, J. S. (2008). Responsive in-season nitrogen management for cereals. *Computers and Electronics in Agriculture,* 61, pp. 51-62

Skaloud, J., & Lichti, D. (2006). Rigorous approach to bore-sight self-calibration in airborne laser scanning. *ISPRS Journal of Photogrammetry and Remote Sensing,* 61, pp. 47–59

Solari, F., Shanahan, J.F., Ferguson, R.B., & Adamchuk, V.I. (2010). An active sensor algorithm for corn nitrogen recommendations based on a chlorophyll meter algorithm. *Agronomy Journal,* 102, pp. 1090–1098

Stewart, I. (2008). *Taming the Infinite. The Story of Mathematics,* Quercus, ISBN 978-1-84724-181-8, London

Strahler, A. H. (1980). The use of prior probabilities in maximum likelihood classification of remote sensing data. *Remote Sensing of Environment,* 10, pp. 135-163

Theobald, D. M. (2003). *GIS Concepts and ArcGIS® Methods,* Conservation Planning Technologies, Fort Collins, CO

USGS. (2010). *Version 13, USGS National Geospatial Program Lidar Guidelines and Base Specification,* U.S. Geological Survey (USGS) National Geospatial Program (NGP), February 22, 2010

Varvel, G.E., Schepers, J.S., & Francis, D. D. (1997). Ability for in-season correction of nitrogen deficiency in corn using chlorophyll meters. *Soil Science Society America Journal* 61, pp. 1233-1239

Vosselman, G. (2002). On the estimation of planimetric offsets in laser altimetry data. *International Archives of Photogrammetry and Remote Sensing,* 34, 3A, pp. 375–380

Vu, T.T., Yamazaki, F., & Matsuoka, M. (2009). Multi-scale solution for building extraction from LiDAR and image data. *International Journal of Applied Earth Observation and Geoinformation,* 11, pp. 281-289

Wang, Y., Weinacker, H., & Kock, B. (2008). A Lidar point cloud based procedure for vertical canopy structure analysis and 3D tree modelling in Forest. *Sensors,* 8, pp. 3938-3951

Willers, J.L., Boykin, D.L., Hardin, J.M., Wagner, T.L., Olson, R.L., & Williams, M. R. (1990). A simulation study on the relationship between the abundance and spatial distribution of insects and selected sampling schemes, *Proceedings of the Conference on Applied Statistics in Agriculture,* pp. 33-45, Kansas State University, Manhattan, KS

Willers, J. L., Jallas, E., McKinion, J. M., Seal, M. R. & Turner, S. (2009a). Precision farming, myth or reality: selected case studies from Mississippi cotton fields, In: *Advances in Modeling Agricultural Systems,* Papajorgji, P. J., & Pardalos, P. M. (Eds.), pp. 243-272, Springer

Willers, J. L., Jenkins, J. N., Ladner, W. L., Gerard, P., Boykin, D. L., Hood, K. B., McKibben, P. L., Samson, S. A., & Bethel, M. M. (2005). Site-specific approaches to cotton insect control. Sampling and remote sensing techniques. *Precision Agriculture,* 6, pp. 431-452

Willers, J. L., Jenkins, J. N., McKinion, J. M., Gerard, P., Hood, K. B., Bassie, J. R., & Cauthen, M. D. (2009b). Methods of analysis for georeferenced sample counts of tarnished plant bugs in cotton. *Precision Agriculture,* 10, pp. 189-212

Willers, J. L., Jin, M., Eksioglu, B., Zusmanis, A., O'Hara, C. G., & Jenkins, J. N. (2008a). A post-processing step error correction algorithm for overlapping LiDAR strips from agriculture landscapes. *Computers and Electronics in Agriculture,* 64, pp. 183-193

Willers, J. L., Ladner, W. L., McKinion, J. M., & Cooke, W. H. (2000). Application of computer intensive methods to evaluate the performance of a sampling design for use in cotton insect pest management, *Proceedings of the Conference on Applied Statistics in Agriculture,* pp. 119-133, Kansas State University, Manhattan, KS

Willers, J. L., Milliken, G. A., Jenkins, J. N., O'Hara, C. G., Gerard, P. D., Reynolds, D. B., Boykin, D. L., Good, P. V., & Hood, K. B. (2008b). Defining the experimental unit for the design and analysis of site-specific experiments in commercial cotton fields. *Agricultural Systems,* 96, pp. 237-249

Willers, J. L., Milliken, G. A., O'Hara, C. G. & Jenkins, J. N. (2004). Information technologies and the design and analysis of site-specific experiments within commercial fields, *Proceedings of the 16th Conference on Applied Statistics in Agriculture,* pp. 41-73 Manhattan, KS, April, 2004

Willers, J. & Riggins, J. (2010). Geographical approaches for integrated pest management of arthropods in forestry and row crops,. In: *Precision Crop Protection - the Challenge and Use of Heterogeneity*, Oerke, E., Gerhards, R., Menz, G. & Sikora, R. A. (Eds.), pp. 183-202, Springer, ISBN 978-90-481-9276-2, Dordrecht, The Netherlands

Willers, J. L., Seal, M.R. & Luttrell, R. G. (1999). Remote sensing, line-intercept sampling for tarnished plant bugs (Heteroptera: Miridae) in mid-South cotton. *Journal of Cotton Science,* 3, pp. 160-170

Willers, J. L., Wu, J., O'Hara, C. & Jenkins, J. N. (2012) A categorical, improper probability method for combining NDVI and LiDAR elevation information for potential cotton precision agricultural applications. *Computers and Electronics in Agriculture, Computers and Electronics in Agriculture*, 82, pp. 15-22

Willers, J.L., Yatham, S. R., Williams, M. R. & Akins, D. C. (1992). Utilization of the line-intercept method to estimate the coverage, density, and average length of row skips in cotton and other crops. *Proceedings of the Conference on Applied Statistics in Agriculture*, pp. 48-59. Kansas State University, Manhattan, KS